Pinot Noir

JOHN SAKER

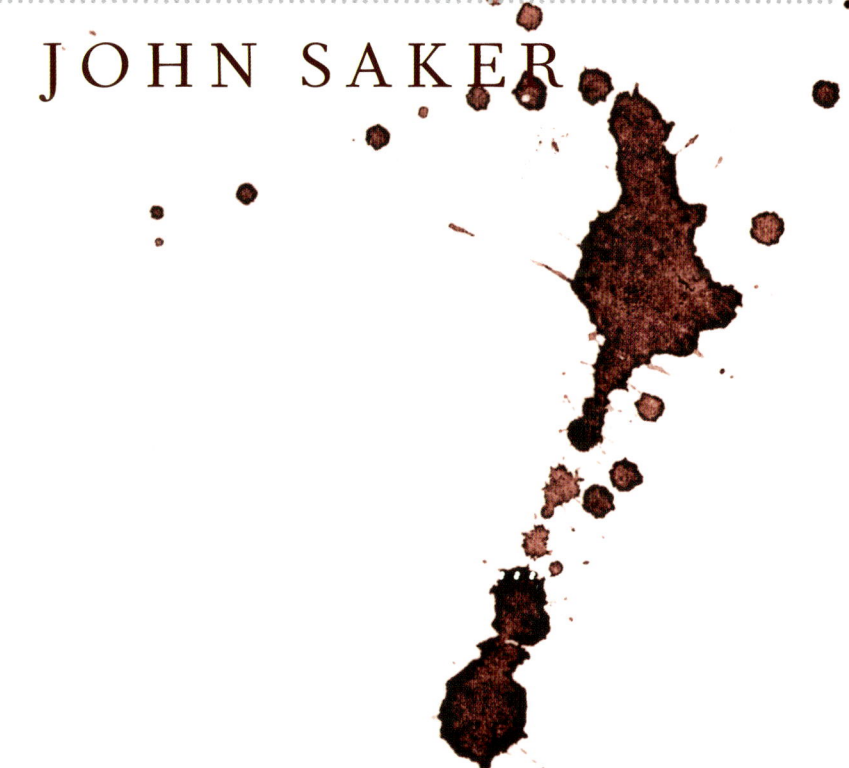

RANDOM HOUSE
NEW ZEALAND

TO THREE WONDERFUL CHILDREN
— PAUL, ZOE AND POPPY.

ACKNOWLEDGEMENTS

Many people have helped me with this project in various ways and have my gratitude. In particular, I would like to thank the following:

My sister Nicola Saker, whose early discovery of the joys of New Zealand Pinot sparked my own interest; John Comerford, Dr Warren Moran and Larry McKenna for historical insight and reading parts of the manuscript; Wairarapa archivist Gareth Winter; Brother Gerard Hogg and Ken Scadden of Marist Archives, Wellington; Blair Walter, Danny Schuster and Michael Cooper for the loan (or gifting) of books; John Hawkesby, Nick Nobilo, Stephen Nobilo, Laurie Bryant and Humphrey Myers for help with arranging bottle shots; Clive Paton, Phyll Pattie, Helen Masters and the rest of the Ata Rangi team for all manner of things, including putting up with me for part of the 2008 vintage; the boards of the Pinot Noir 2001, 2004, 2007 and 2010 conferences for valued opportunities to add to my knowledge; Aaron McLean for his terrific photographs; Random House New Zealand deputy publishing director, Jenny Hellen, for her patience and encouragement; the many Pinot producers who have been generous with tastings and information; and, finally, my wife Rolla, for her editing skill, unflagging support and love.

A RANDOM HOUSE BOOK
published by Random House New Zealand

18 Poland Road,
Glenfield,
Auckland,
New Zealand

For more information about our titles go to www.randomhouse.co.nz

A catalogue record for this book is available from the National Library of New Zealand

Random House New Zealand is part of the Random House Group

New York   London   Sydney
Auckland   Delhi   Johannesburg

First published 2010

© 2010 text John Saker, photography Aaron McLean

The moral rights of the author have been asserted

ISBN 978 1 86979 279 4

This book is copyright. Except for the purposes of fair reviewing no part of this publication may be reproduced or transmitted in any form or by any means, electronic or mechanical, including photocopying, recording or any information storage and retrieval system, without permission in writing from the publisher.

Printed in China by Everbest Printing Company Ltd

Design: Athena Sommerfeld

CONTENTS

1. Pinot mystique 16
2. A plant, a grape, a drink 30
3. Early promise 42
4. Rebirth 54
5. Viticulture & vinification 68
6. Terroir Aotearoa 88

7. **Wairarapa 104**
   - Ata Rangi 110
   - Craggy Range 118
   - Dry River 120
   - Escarpment Vineyard 127
   - Martinborough Vineyard 130
   - Palliser Estate 134
   - Voss Estate 138

8. **Nelson 140**
   - Greenough 144
   - Neudorf 152

9. **Marlborough 158**
   - Churton 164
   - Cloudy Bay 168
   - Dog Point 170
   - Fromm Winery 174
   - Mahi 180
   - Pernod Ricard 182
   - Seresin Estate 188
   - TerraVin 196
   - Villa Maria 198

10 CANTERBURY 202
    Bell Hill 206
    Kaituna Valley 214
    Mountford 216
    Muddy Water 218
    Pegasus Bay 220
    Pyramid Valley Vineyards 226

11 CENTRAL OTAGO 228
    Carrick 240
    Chard Farm 242
    Felton Road 251
    Gibbston Valley Wines 258
    Mount Edward 266
    Mt Difficulty 268
    Peregrine 270
    Quartz Reef 276
    Rippon Vineyard 279
    Rockburn 284
    Valli 287

    Index 290
    Bibliography 295

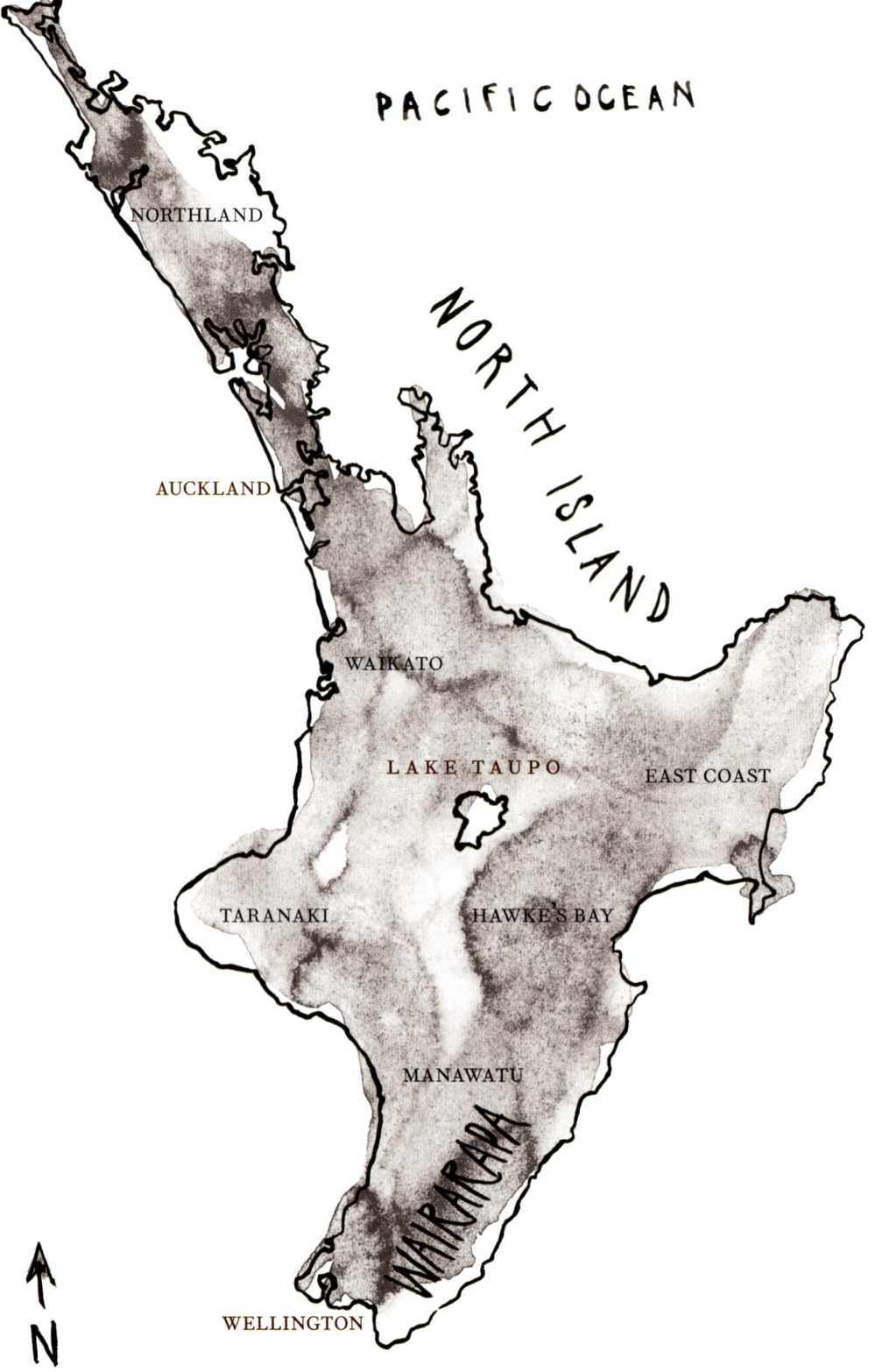

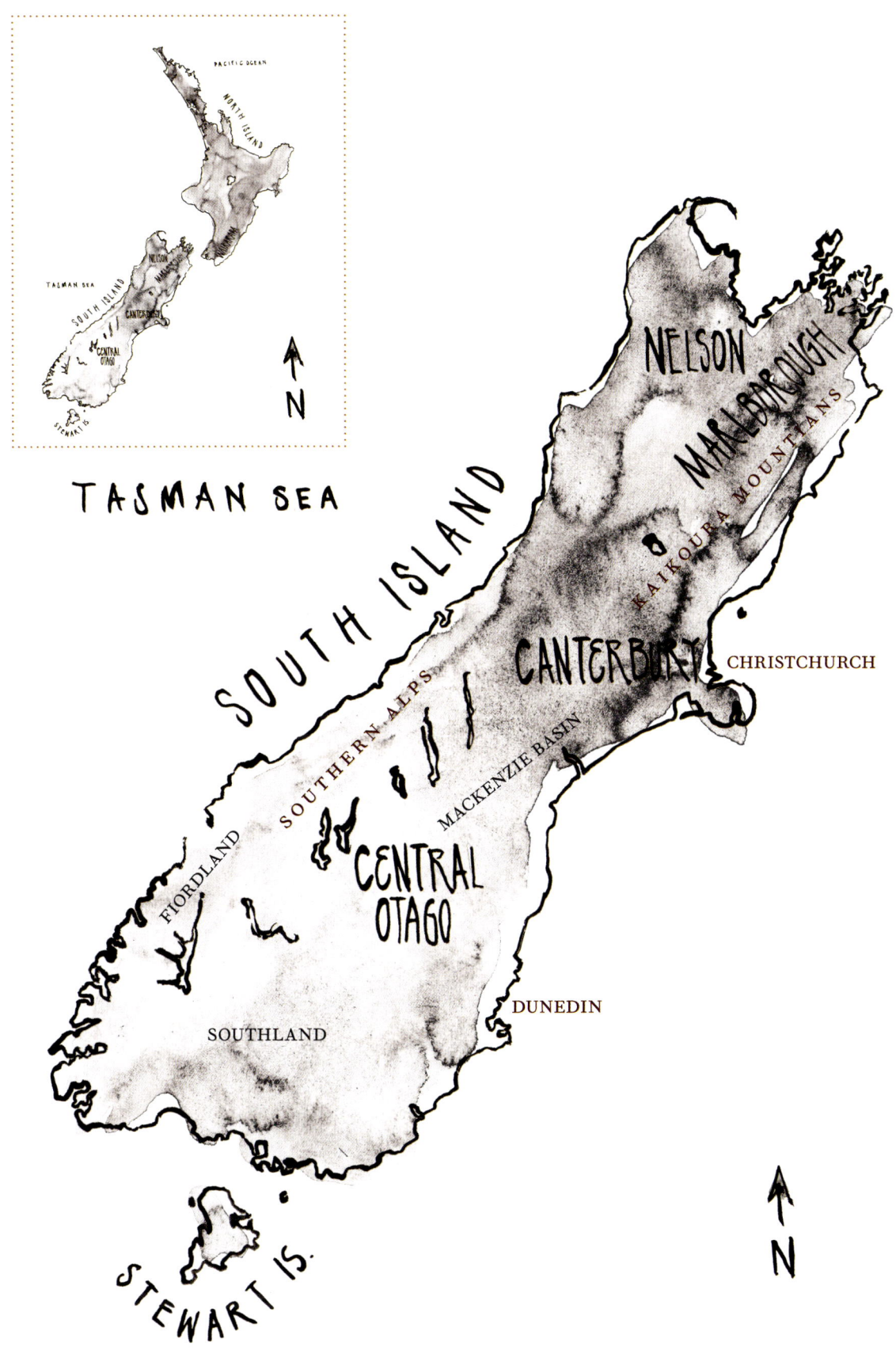

INTRODUCTION

Writing this book has been both a privilege and a pleasure.

The deeper I got into the project, the more convinced I became that this was a wine book that needed to be written.

And from start to finish, the capacity of the New Zealand Pinot tribe to surprise me has never wavered.

The book was nearing completion when I was given the Waiwera Estate Pinot Noir 2000. The twinkle in winemaker Dave Heraud's eye as he handed me the bottle hinted at his pride in the wine. Tasting it back in Wellington a few weeks later, I was amazed. The fruit was intact and gorgeous, and the years had added their layers of quiet charm. Here was a very fine 10-year-old New Zealand Pinot, produced by a tiny winery in Golden Bay of which I had heard almost nothing.

That kind of delight in discovery has been a constant in the New Zealand Pinot Noir story. It was expressed by Governor Lord Ranfurly when he tasted William Beetham's Pinot-dominant red blend more than a hundred years ago and is a regular presence at New Zealand Pinot tastings around the world.

This book follows a standard path: some background and an overview, followed by a journey through the regions. The selection of the key producers in each region is a personal one and would have been larger if space had permitted. When choosing which wineries to include, track record was a key factor. I kept in mind the exhortation of JD Salinger's fictional character Seymour Glass to his brother Buddy, an aspiring writer. In critiquing Buddy's work, Seymour says he knows Buddy can write a 'rattling good story', but he wants more than that. 'I want your *loot*,' he tells his brother. There is many a rattling good Pinot out there; the vignerons who interest me most are digging deeper into themselves and their land for the real loot.

It is my hope that this book does justice to them and their craft.

John Saker
Wellington, 2010

ONE. PINOT MYSTIQUE

Wine is sometimes described as a living substance because of its capacity to endure, in many cases improve, over time. I see it more as an afterlife. The grapes are alive and kicking when plucked from the vine, but from that moment their days as breathing earthlings are all but over. Once fermented, their juice becomes a sterile liquid, technically as dead as the glass in which it is entombed. And yet — in a mysterious, removed, ghostly way — it continues to reflect the character of that juicy, bursting fruit from which it was made. From bottle to bottle, or glass to glass, this phantom presence never makes two visitations that are quite the same, which tantalisingly suggests something vital still flickers.

Hardly surprising, then, that we sometimes describe wines as 'ethereal' or talk of being 'haunted' by them. Such language is applied to some wines more than others. Pinot Noir, for me, leads that category.

A global cult has developed around Pinot Noir over the past two decades or so, based upon a thirst both for the drink and for an understanding of what lies behind it. Cult membership rose steeply in the wake of the hit movie *Sideways*, released in 2004. The obsession of the film's hopelessly Pinot-struck leading male, Miles, rubbed off on cinema audiences everywhere, inducing an upward swing in demand that came to be known as the '*Sideways* effect'.

The clip of Miles' now-famous paean to the grape has played at Pinot conferences around the world, and many thousands of times on YouTube. It shows Miles placing Pinot squarely among the world's wonders: '… only the most patient and nurturing of growers can do it, really. Unless someone really takes the time to understand Pinot's potential and coax it into its fullest expression … then … oh, its flavours are just the most haunting and brilliant, thrilling and subtle — and ancient — on the planet.'

Heaven knows exactly how many thousands of rushed trips to bottle stores were launched by Miles' impassioned murmurings. ACNielsen reported that in the US alone, sales of Pinot Noir rose 45% during the year following the film's release.

The Pinot cult — like the wine itself — has many layers. It has history, too. The discoverers of the greatness of Pinot Noir were the Cistercian monks of Burgundy in the Middle Ages, themselves a cult.

Commerce was not the Cistercians' principal imperative when they set off on their systematic and thrilling cross-examination of what the grape and Burgundy's Côte d'Or (the limestone escarpment on which Burgundy's great vineyards are planted) could achieve together. The Cistercians had the time and the work ethic to bore into viticulture and winemaking as these subjects had never been bored into before. They are credited with being the first to note the way different vineyards, or blocks of vines, produced wines with distinct, consistently recognisable qualities. They saw these differences as a divine gift, something to celebrate rather than lose through blending. According to one legend, the good monks tasted the earth itself in the hope of fully understanding their different vineyards. The Cistercians set the tone for centuries to come, establishing the idea that Pinot was a trove of secrets, only accessible through the labours of a dedicated few.

The artisanal wine culture that grew up around landlocked Burgundy contrasted starkly with that of its great rival to the southwest. Bordeaux's vineyards occupy a large, sprawling area that is four times the size of Burgundy. The large Bordeaux estates surround an important ocean port and have always been outward-looking and cosmopolitan. Aquitaine, of course, was under English rule for 300 years in the Middle Ages. The compulsion to trade and make money is in Bordeaux's DNA, which partly explains why its famous reds are blends of several grape varieties. A difficult year for, say, Cabernet Sauvignon, can be offset by better results for Merlot and/or the rest of the pack. A sound hedging policy, by any accountant's reckoning. Burgundy, by contrast, has largely been about tiny parcels of vineyard and the brilliant solo voice of Pinot Noir, a far riskier proposition. I say 'largely', because other varieties have played roles at different times. Pinot's pale mutations, Pinot Gris and Pinot Blanc, were present in varying proportions (quite sizeable ones) several centuries ago. More alien additions have come in the form of 'blending material' trucked in from outside the region. This practice was only made illegal in the 1930s.

At the epicentre of Pinot's mystique is, of course, the wine itself. Beyond the flavours and textures that can be qualified with comparative descriptors lies its ability to trip off an emotional response. This is where the adjectives like 'ethereal' are called upon. And where, in the words of British wine writer Oz Clarke, Pinot becomes 'above rational discourse'.

For Clive Paton of Martinborough's Ata Rangi winery, the drink's mystique owes much to the fact that its personality can be so hard to nail down. No obvious party line is being peddled. For every claim, one senses, there is a counterclaim. Is it heavy or is it light? Sweet or savoury? Refined or earthy? Masculine or feminine? In this sense, it can be like a person, or a great piece of literature — a complex web of shades of grey. You can digest a few truths and be entertained along the way, but at the end you're still left wondering if Hamlet did or didn't get what he deserved.

This duplicity is summed up by the well-worn comparison of Pinot to 'an iron fist in a velvet glove'. It's equally evident in the diversity of views expressed by wine scribes.

Declares Burgundy expert Clive Coates: 'Here is a wine which can sing like a nightingale, shine forth like a sapphire, intrigue like the most complex of chess problems, and seduce like the first kiss of someone you are about to fall in love with.' Meanwhile, Anthony Hanson, no less an expert, approaches the subject from a different angle: 'Great Burgundy should taste of shit.'

Rudi Bauer, owner and winemaker of Central Otago winery Quartz Reef, once told me he felt the intense response Pinot elicited from people was due in part to a kinship between its scents and tastes and those of humans — more so than with any other variety. The old Burgundian practice of *pigeage*, where people strip off and wade waist-deep through fermenting vats of Pinot to help extract colour and tannins from the grape skins, can only help in that regard.

The role of place as an influence on the finished wine is another strong strand in the romance of Pinot. According to Aubert de Villaine, the head of Burgundy's Domaine de la Romanée-Conti, arguably Pinot's most sacred patch: 'Pinot Noir doesn't really have a taste. On its own, it has no interest. It is only interesting if married to a terroir.'

Terroir is a subject for later discussion. But what we can take from de Villaine's statement is this: Pinot Noir is a chameleon. The conditions in which it grows are readily and openly translated to the glass. Of all grape varieties, it is arguably the most site-centric.

The added twist here is that not just any place will do. Pinot's spread beyond Europe has always lagged behind that of most other classic French grapes. The conventional wisdom, which the Burgundians loftily put about for a very long time, was that the rest of the world could forget Pinot Noir if the goal was to make quality still red wine. Pinot was a precious, uncertain child of brilliance who threw tantrums if asked to perform on any stage outside the Côte d'Or.

Even when its provenance is Burgundy, a bottle of Pinot is very capable of delivering a lousy performance. I was introduced to Burgundy when I lived in France in the mid-1970s and found it a disappointing experience. Burgundy, I noticed, was a pricier drop than Bordeaux. When I enquired as to why, I was fobbed off with 'it's Burgundy and there's not much of it'. Yet the few times I treated myself to a bottle of red Burgundy, I felt gypped. They were pallid, acidic, ungenerous wines. I learned to avoid them, yet continued to hear stories of the spellbindingly attractive wines that Burgundy produced.

A lot of poor Pinot Noir is produced in Burgundy, much of it labelled Bourgogne Rouge. You have to venture right up the price scale, know who's doing good work and who isn't, to start having seriously good experiences with Burgundy. Even there you can find wine that leaves you feeling short-changed. The same can be true with Pinot from other parts of the world.

Pinot also has the maddening trait of deciding on any given day not to turn up. It often goes into a funk after being moved from barrel to bottle, more so than most other wines. And even once supposedly settled in the bottle, its moods can swing and it won't taste like the same wine you might have had a month ago. Its distaste for travel is famous. Opening a bottle of Pinot that has just come off a long journey is never a good idea.

Elusive quality belongs to the Pinot mystique. An experience with a great bottle is something to remember and cherish, because who knows when the planets will be thus aligned again?

All of this was enough for the early practitioners in the New World to steer clear of Pinot Noir, other than to use it as a sparkling wine component. Early pacesetters in the New World surge, California's Napa Valley and Australia's Barossa Valley, were anyway warm-climate regions that didn't suit Pinot. By the early 1970s, Cabernet Sauvignon had emerged as the star red of the Californian wine industry, while Shiraz (aka Syrah) had top billing in Australia. New Zealand at this time had been loosening itself from its long embrace with American hybrids such as Albany Surprise, only to become enamoured with Müller-Thurgau, the high-yielding white crossing described by British wine writer Jancis Robinson as 'the bane of German wine production'.

Serious reds were in short supply. Many tipped Cabernet Sauvignon to be the great red hope, given the encouraging results that Tom McDonald had achieved with the variety at McWilliams in the 1960s.

But before the decade was out, producers in Oregon's Willamette Valley had caused serious cracks to appear in the conventional wisdom. Pinots made by Willamette pioneer David Lett outperformed a number of big-name Burgundies at an international wine Olympiades in Paris in 1979. There was also promising activity in New Zealand around the same time, starting with Nick Nobilo's 1976 Pinot. This was backed up in the early

1980s by interesting efforts from Canterbury, Martinborough and parts of California. Pinot Noir had been issued with a passport.

Rather than diminish the veneration accorded Pinot and Burgundy, these successes away from home did the opposite. The cult simply got more airplay and scope. Burgundy buffs around the world didn't see the new colonies as puncturing Burgundy's aura of exclusivity. Great Burgundy — like great Champagne — would always be inimitable. The New World expressions were fascinating add-ons; unmistakably Pinot Noir, and increasingly serious Pinot at that. Following their progress became a rewarding and affordable way to indulge a Pinot craving. For many consumers, the versions from Oregon, California and New Zealand provided an entry into 'planet Pinot'. The wine, intrigue and slightly unhinged devotion they discovered made many of them firm followers, eager to know and taste more.

Of course, it's not as if Pinot Noir's spread has ever assumed tidal-wave proportions. Though parts of Victoria and Tasmania in Australia have also begun making very good Pinot Noir over recent years, the club remains select and each of Pinot's new outposts is small. The set of conditions it requires are far from commonplace.

In *Sideways*, Miles alluded to Pinot Noir being a demanding cuss to grow. Its aromatic brilliance can only be realised in cooler climates on the margins of grape-growing lands. And even in these zones there are vintages when the thin-skinned, delicate grapes struggle to ripen.

The challenge it presents to growers, together with its inherent loveliness and veil of mystery, has led Pinot to be dubbed the Holy Grail. The use of that term is apt on several levels, one being that the Grail is, in many tellings, a phantom presence. Another is in regard to the human side of the equation. As put by Wikipedia: 'In most versions of the [Grail] legend, the hero must prove himself worthy to be in its presence.'

'No mean-spirited bastard ever made a decent Pinot Noir,' thundered Oz Clarke at the 2003 Pinot Noir Conference in Wellington, igniting delighted applause. Such bombast is rife at these occasions (along with tribal T-shirt declarations like 'I'd rather have a case of the clap than a case of Cabernet'). Behind Oz's hyperbole is the truth that a certain type of person is attracted to the Pinot club.

'Unhinged devotion' is a phrase I've already used. Most quality Pinot producers,

like their Cistercian forerunners, aren't drawn to it by the money. In New Zealand, the national average return on capital for Sauvignon Blanc is 10%; for Pinot Noir, the figure is 2%. Dreamers, free spirits, sensualists, poets, hippies, obsessives — committed Pinot producers have been called all of these. Theirs is a winemaking subculture notable for its hunger to find out more, to share ideas, to taste and compare terroirs. Ultimately, their goal is not so much to make the best wines possible as to make the most truthful wines possible.

I don't think it's coincidental that every certified organic wine producer in New Zealand (at the time of writing) makes Pinot Noir. Nor is it by chance that those who embrace the equally, green, though more off-centre, biodynamic approach are not only all Pinot producers, but are also among our most prestigious and committed. The biodynamics movement has its critics, but you can't doubt the sincerity of its advocates' desire to do the best by the land they are cultivating and let it have its say in its own, undistorted voice. Burgundy and Oregon similarly have strong followings in this area.

It's fascinating to compare the people element in the Burgundy-Oregon-New Zealand Pinot triangle. Burgundy is the least pretentious of the great French regions, known for its close-to-the-land farming mentality, the relatively casual demeanour of its vignerons, and their spirit of generosity. The same could be said about Oregon and New Zealand.

In a relatively short time, the three places have developed close ties, closer than anyone would have thought possible only 20 years ago. The rise of Oregon and New Zealand Pinot has served to highlight the underlying magnanimity of Burgundy. Any preconceptions of an arrogant, introverted Old World region have been trumped by the interest and respect a new generation of Burgundians has shown towards the Pinot diaspora. The intern winemaking traffic between the regions is intense. Most young New Zealand winemakers with an interest in Pinot aspire to work a vintage or several in both Burgundy and Oregon. The same goes for their counterparts in the two northern-hemisphere regions.

Consumers have more contact with winemakers today than possibly at any time in history. This is mostly due to marketing, the thinking being that for a maker of a product of the land, especially one as ancient and universally revered as wine, commerce demands you get out there and 'tell your story'.

As all the above indicates, Pinot people have great stories to tell and tend to tell them with passion. The winemakers have become the Pinot Noir cult's priests and priestesses, the official nurturers and guardians of this gift of nature.

The commitment and zeal they have for what they do is palpable. There is a humility there, too. You will often hear them declare that the more they know, the more they realise they have yet to learn. Pinot Noir's capacity to continue astonishing those who work with it every day, and their telling of it, are important Pinot mystique nutrients. It's not lost on anyone that mystique sells. But that shouldn't diminish the validity of the mystique itself.

Following pages: pp24-5 The generous sweep of the vineyard planted by Danny Schuster in the Omihi Hills, Waipara; p26 A post at the end of a row of Pinot vines identifies the clone/rootstock combination that has been planted; p27 The pure limestone near Waikari on which the Bell Hill Vineyard is planted; p28 French oak barrels remain integral to the production of fine Pinot; p29 Seresin Estate shows off some biodynamically nurtured compost.

TWO. A PLANT, A GRAPE, A DRINK

Six centuries before the angst-ridden Miles of *Sideways* presented cinema audiences with his mean case of Pinot fever, there was Philip the Bold. The first of the Valois dukes of Burgundy could be labelled the world's first pinotphile. There is a theory he even dreamed up the name Pinot himself. Whoever did coin the term was apparently struck by the grape bunch/pine cone visual echo, which is the word's most commonly cited etymology.

Its presence in a number of Philip's official documents that have survived from the late fourteenth century provides the first references we have to the word and give it a venerability few other vine variety names share. The most common spelling in those medieval texts was 'pineau', which is still in use in some parts of France. It was only in the late nineteenth century that Burgundy officially adopted the 'ot' ending that we use today.

Philip's attachment to the variety was made famous in a document dated 1395 in which he expressed his distaste for Gamay (today's Beaujolais grape), describing it as 'vile and noxious'. He went on to order the wholesale deracination of Gamay vines in Burgundy so that more Pinot could be planted.

At that time there were other names in use for the variety we now know as Pinot

Noir, among them Noirien and Morillon. The latter is still used in parts of France today. It has been suggested Philip's new name was specifically created to identify a better form, or sub-variety, of this vine. Whether that is correct or not, his strong preference for Pineau over Gamay was extraordinary in a number of ways.

One is the earliness, historically, of such deliberate varietal selection. No written reference was made to individual grape varieties in Bordeaux until the eighteenth century. Philip was well ahead of his time in this line of viticultural decision-making.

He also showed precocity in recognising the wider advantages of growing a local wine of distinction. The Valois dukes of Burgundy were a line of wily politicians and power-brokers whose self-serving instincts were legendary. It was the Burgundians, remember, who all but lit the match under Joan of Arc by handing her over to the English.

Under their century-long stewardship, Burgundy the wine began to achieve the fame it has more or less enjoyed ever since. A number of historians have suggested this was a deliberate political ploy. Philip understood Pinot would make better wine, and that quality Burgundian wine would make Burgundy — and his own power base — stronger. Today they'd call it brand-building.

It was probably around the time of the Valois dukes that Pinot Noir began to spread to other parts of France, and then further afield to Germany, Switzerland and Italy. The most famous of these outposts became, of course, Champagne, well-known for another wine of enigmatic qualities.

The writings of Columella, one of the first scribes to make the agricultural beat his own, suggest the grape that Philip called Pineau was well settled in Burgundy 2000 years ago. In *De Re Rustica* ('On Country Matters', published in AD65) Columella wrote extensively on viticulture and wines around the Roman Empire. He describes a grape grown in what is now Burgundy that had small berries, stood up well to cold conditions and produced wine with reasonable longevity. Many a wine writer/historian has joined the dots between Columella's grape and Pinot Noir.

As we go further back in the search for the grape's origins, speculation largely takes over from fact. It has long been held that the wine-growing grapevine *Vitis vinifera* can call Transcaucasia home. In this barren land to the west of the Caspian Sea (roughly today's Georgia and Armenia) ancient grape pips have been found that point to vine cultivation — and, it is assumed, the making of wine — about 8000 to 10,000 years ago. From this Central Asian birthplace, the domesticated, hardy grapevine spread in all directions. The carriers were mostly colonists, traders and empire-builders, the Greeks and Romans being the most commonly cited during the last BC millennium. The vine was thus a gift given to Europe by Central Asia. In Jancis Robinson's 1986 book *Vines, Grapes and Wines* there is a map showing sweeping arrows — including one representing the path taken by Pinot Noir — extending from the eastern side of the Mediterranean across to France and western Europe.

Another view has gained currency. It doesn't overrule the general idea of east–west vine traffic flow, which certainly occurred to some extent. What it does suggest is that there were grapevines in western Europe well before that known period of movement around the Mediterranean, and that Pinot Noir was an indigenous Gallic variety of grapevine. In support of this theory are varying strands of evidence which together form a solid case.

First, there is no shortage of archaeological evidence suggesting that the ancient Gauls enjoyed wine as early as 500BC — well before the arrival of the Romans.

In his book *North American Pinot Noir*, John Winthrop Haeger also points to the fact that wild *Vitis vinifera* vines grew throughout the forests of western Europe until being savaged by phylloxera in the late nineteenth century. It is unlikely these untamed and widely distributed grapevines were the nature-scattered progeny of cultivated vines introduced several centuries before the birth of Christ.

We can say with certainty that Pinot Noir is a very old grape. This has been confirmed by the pioneering research on grapevine DNA carried out at the University of California, Davis (UCD) over the last decade. Among a raft of fascinating genetic discoveries, researchers at UCD have identified a host of grape varieties parented by Pinot Noir. These include Chardonnay, Gamay and Auxerrois. However, the DNA sleuthing does not expose any parents for Pinot Noir itself. This suggests Pinot is perhaps only one generation removed from wild vines. And why not Burgundian wild vines? Professor Carole Meredith, the now-retired geneticist who led the research at UCD, expresses the view that 'Pinot Noir most likely came from northeastern France or southwestern Germany'.

Pinot Noir's closeness to feral vines is underscored by the grape's inherent genetic instability. It mutates with spectacular ease in the vineyard. Pinot Gris, Pinot Meunier and Pinot Blanc, although commonly regarded as grape varieties in their own right, are in fact mutations of Pinot Noir. It is not unusual for a Pinot Noir vine to throw out a shoot at random bearing Pinot Gris or Pinot Blanc fruit.

This unsettled genetic make-up — or readiness to adapt, which is another way of looking at it — has also led to the wide array of Pinot Noir clones in existence. A clone, in the wine sense, is simply a cutting (the word clone derives from the Greek word for twig) from a vine that has been chosen for its superior performance, perhaps showing particular, sought-after strengths. It follows that all cuttings from a mother vine will be genetically identical to one another. Truly speaking, these are all clones, although usage has seen the word clone become a collective noun for all the progeny of a single mother plant. So when we refer to this clone or that clone, we are talking about a genetically distinct sub-variety.

The identification and propagation of clones is quite a recent development in wine production. It began in France in the 1950s, a time when disease was wreaking havoc in a number of regions, most particularly in Burgundy. The fittest vines were singled out and reproduced by the taking of cuttings.

Today there are hundreds of different clones of Pinot Noir, each offering something different in flavour, structure, ripening curve, suitability for different sites and so on. In New Zealand, there are fewer than three dozen Pinot clones in current circulation. It has become standard viticultural practice for serious Pinot producers to plant a selection of clones, and the decision about which ones to use is given careful consideration. Consumers with a keen interest in Pinot have had to familiarise themselves with the common clones and their different properties, because winemakers so often invoke their influence on a finished wine. The eyes of anyone outside the Pinot circle tend to glaze over very quickly when the talk turns to the different properties of clones, which have names like '10/5' and '777'.

In 2007, our knowledge of the genetic make-up of Pinot Noir increased dramatically. A team of Italian and French researchers mapped the genome of the grape, making it only the second food crop (rice was the first) to have its complete genetic code compiled. The variation they found was extraordinary. The grape has close to 30,000 genes

(compared to the human total of 20,000 to 25,000). Among these were an unusually high number of genes whose function was to create flavour. Science confirmed what human palates been registering for a long time: Pinot Noir is a complex drink.

There is nothing suggestive of greatness in the physical appearance of a cluster of ripe Pinot Noir grapes. The bunch itself is generally small, conical and compact. The dark blue berries are also small, far less imposing than the larger, fit-to-burst orbs you see with ripe Sauvignon Blanc, for example. But that undemonstrative presence does hint at Pinot Noir's famously sensitive nature and the effort it demands of growers. Not without reason is it referred to in North America as the 'heartbreak grape'.

It is intolerant of any climate that is not cool and marginal — unlike Chardonnay, with which it shares the vineyards of Burgundy. Too much sun and heat drains Pinot of its aromatic charm and finesse, resulting in clumsy, 'stewed' wines. It ripens early, but doesn't like to be rushed in the process.

Pinot Noir also buds early (leaving it vulnerable to spring frosts), and is prone to viruses, mildews and grey rot — the latter particularly if conditions become humid as harvest approaches.

A number of the challenges it presents are due to its thin skin. The skin of a red grape is where tannins and colour reside, as do significant levels of flavour compounds. Pinot's skin holds fewer than half the colour pigments (anthocyanins) and tannins of most other red grapes, a key reason for its relative lightness. The thin membrane, easily broken by rain near harvest, also contributes to the grape's susceptibility to some of the ailments mentioned above.

On the plus side, that skin is also host to a number of phenolic components known to be good for us, among them the antioxidant resveratrol. Present in all grape skins, resveratrol has been linked to lower blood pressure, a lower risk of Alzheimer's, lower levels of bad cholesterol (LDL) and it is thought to have anti-carcinogenic properties. Pinot Noir is particularly rich in resveratrol. It has several times more than other red wines such as Syrah or Merlot, varieties which in turn contain 10 times more resveratrol than white grapes. The reason, apparently, is the stress that Pinot has to endure because of its fragility and marginal habitats. Resveratrol is a natural protector, part of the grape's inbuilt immune system, and Pinot needs all it can get to defend itself.

More than any other variety, Pinot Noir does not abide high yields. If cropped excessively it becomes a dilute dry red, its unique fragrances and characters lost. This is largely why it is more expensive to buy a bottle of Pinot than most other varietals, and why contract growers in New Zealand mostly eschew it in favour of Sauvignon Blanc, a far less bothersome and more bountiful proposition. A typical yield for New Zealand Sauvignon Blanc is 12 tonnes per hectare; for still Pinot Noir it is 5 tonnes per hectare.

Describing what we perceive in a glass of Pinot is a difficult, imperfect business; the most carefully chosen words only take us so far. But of course that doesn't stop us — wine writers especially — having a crack, expending battalions of adjectives and metaphors in the attempt. There is no 'typical' Pinot Noir — different terroirs and the grape's transparency see to that. What follows is a personal, very generalised attempt at describing the liquid issue of a Pinot Noir vine from a New Zealand perspective.

First — and many would add foremost — are the smells a glass of ripe, well-made Pinot releases. Your nostrils are likely to receive an effusive, aromatic wave that has you thinking instantly of roses, violets and sweet summer fruits, perhaps also a hint of coffee. There may also be deeper, more savoury scents reminiscent of mushrooms, truffles and meats. Smokiness and vanilla, also sweet spice, can come from the use of oak, but should never override the fruit aromatics.

Its appearance in the glass should be less 'noir' than that of most other international red varieties such as Syrah, Merlot and Malbec. It is generally agreed that Pinot should never be opaque, offering instead a gleaming, ruby-like translucence.

In the mouth, most noticeable are berryfruit flavours — red or dark or both. Plums can also enter the spectrum. Sweetness is a hallmark, but it is best tempered with savoury, earthy perhaps fungal tastes. Herbal flavours are not uncommon. Greener herbal notes point to a lack of ripeness; a dried herb flavour can be indicative of certain clones and sites. Texturally, Pinot tends towards light and clean, although plusher, plumper styles are popular in New Zealand. It should leave you feeling refreshed. In the words of Jancis Robinson: 'Pinot dances across the palate.'

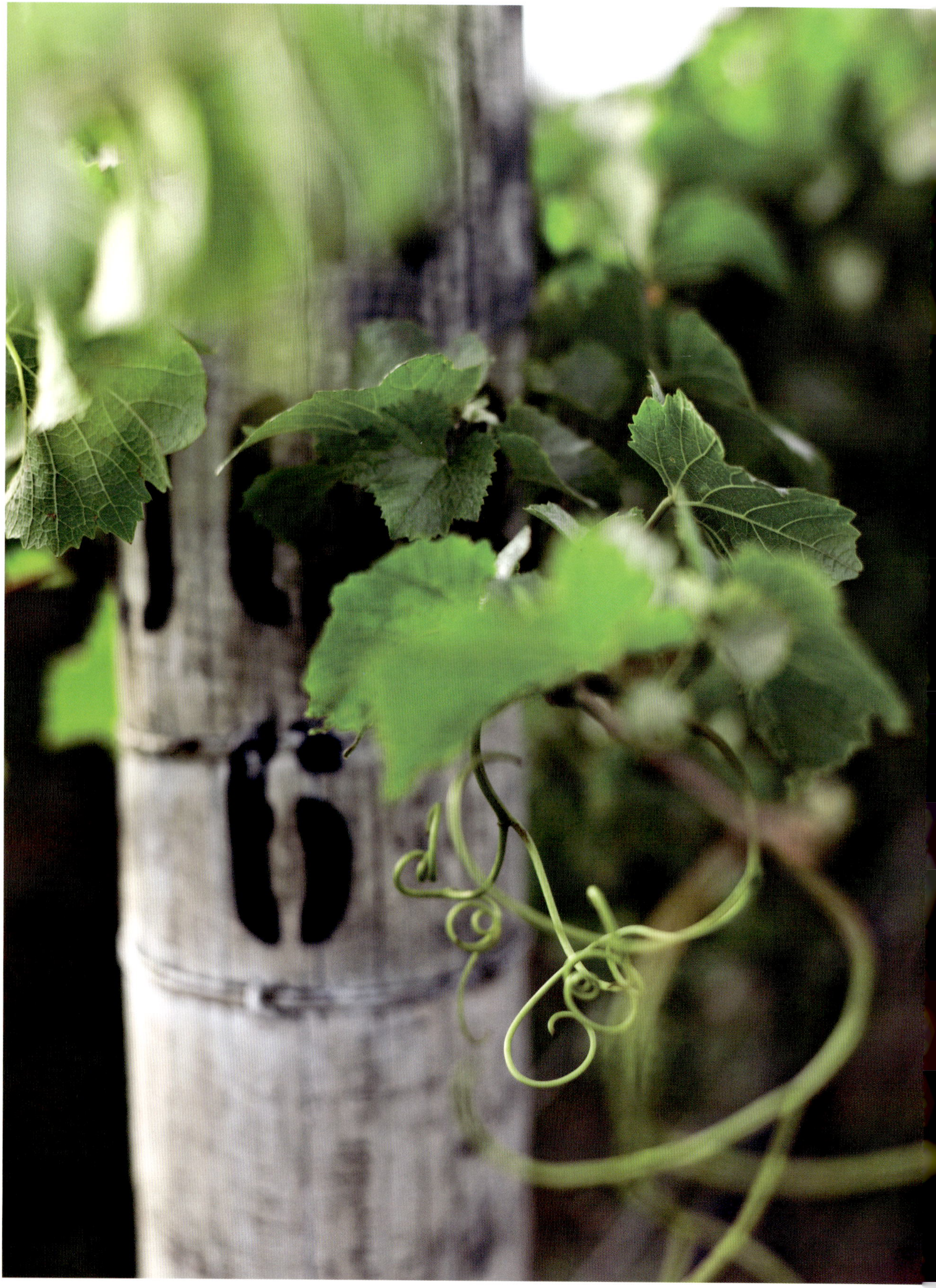

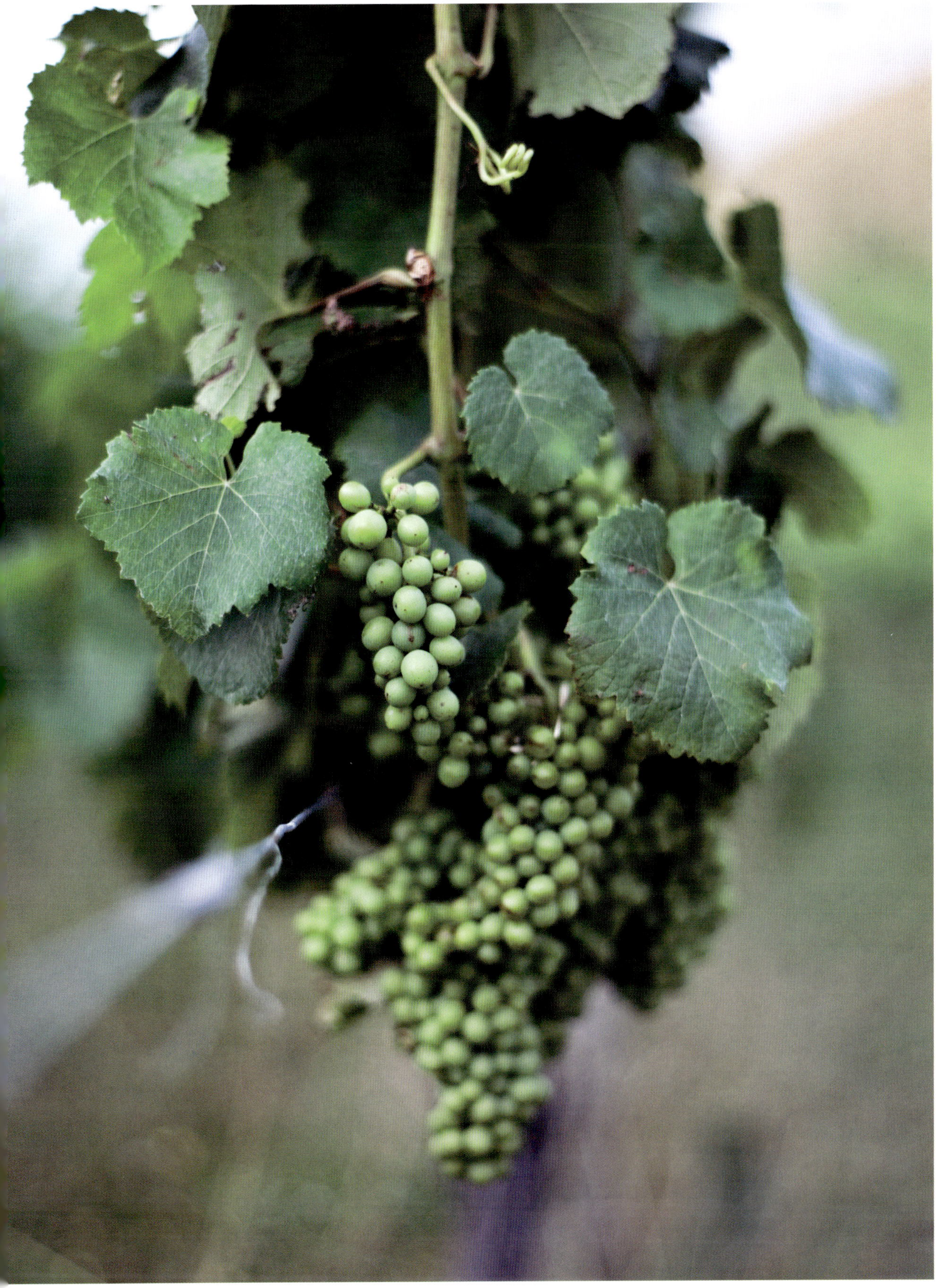

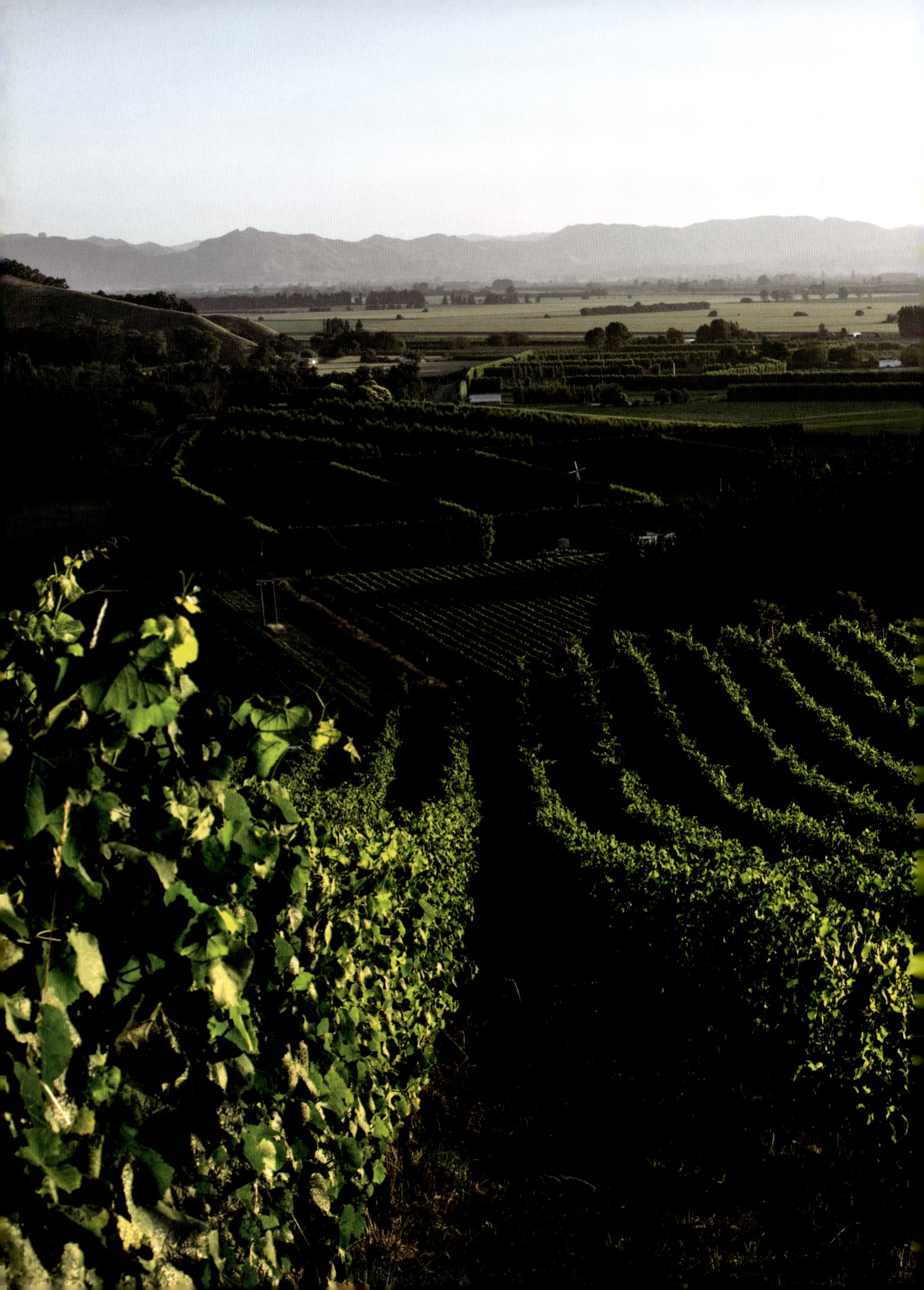

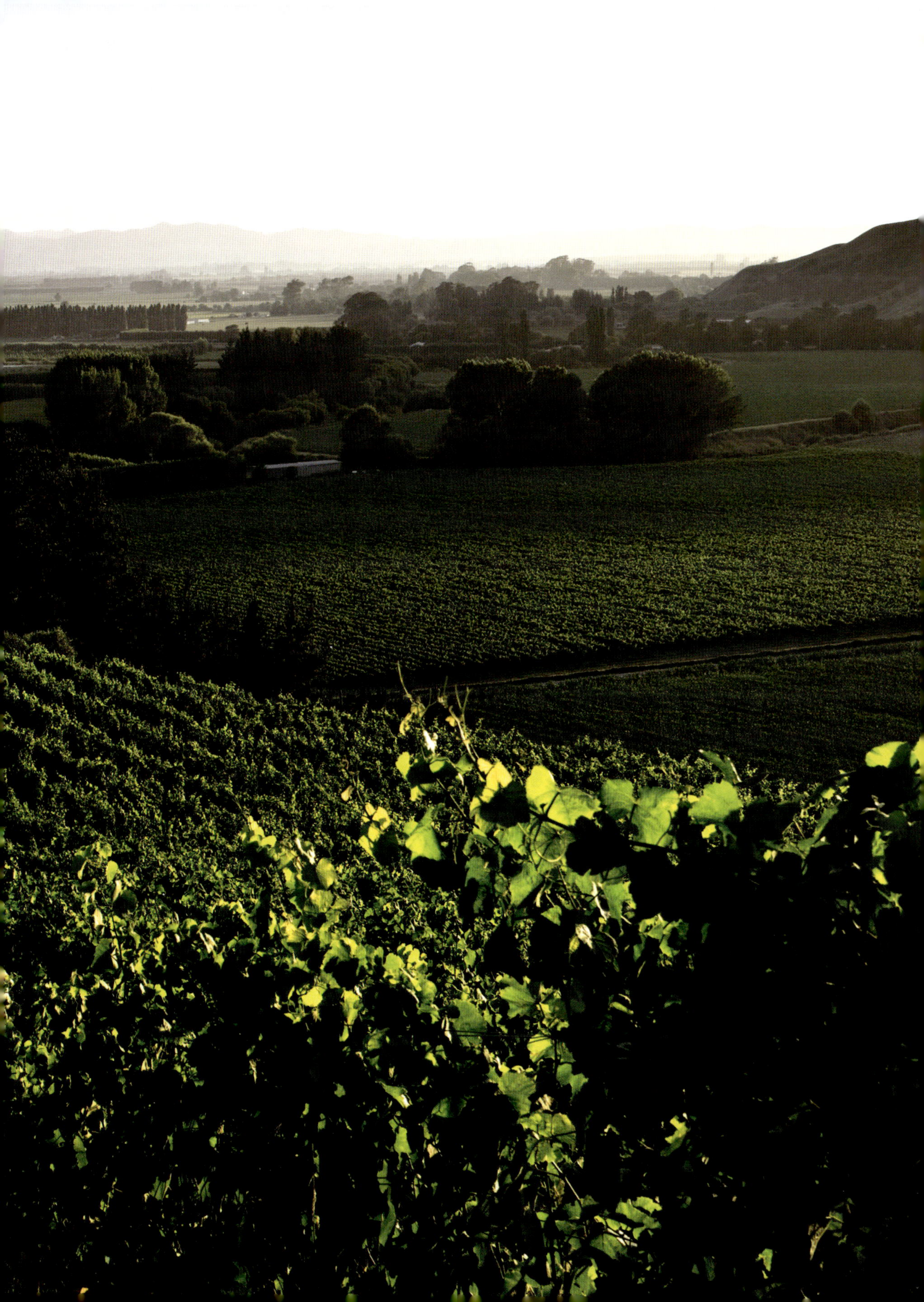

THREE. EARLY PROMISE

WE CAN'T BE SURE WHEN, WHERE OR BY WHOM THE FIRST PINOT NOIR VINE WAS PLANTED IN NEW ZEALAND. WHAT WE DO KNOW IS THAT BY THE LATE NINETEENTH CENTURY IT WAS THRIVING IN A NUMBER OF VINEYARDS, MAKING IT A RELATIVELY EARLY AND SIGNIFICANT PLAYER IN OUR VITICULTURAL HISTORY.

Frustrating the quest for knowledge in this area is the scant interest many of our viticultural pioneers had in the varieties they were planting. This was not at all unusual for the time — the variety-centric wine culture we find ourselves in today is quite a recent development.

The chances are slim that Pinot Noir was among the first collection of *Vitis vinifera* cuttings to be put ashore in New Zealand. Their importer, Samuel Marsden, appeared to have had little inclination to use their fruit for anything but table grapes, whereas the planting of Pinot Noir generally points to serious winemaking intent. The 'whipping parson', as the Anglican missionary was known at his base in New South Wales, recorded in 1819 that he 'planted about a hundred grape-vines of different kinds' at Kerikeri in Northland. Black Hamburg is a variety commonly linked to Marsden.

The nation's first proven winemaker, Edinburgh-born James Busby, co-author of the Treaty of Waitangi, looked to have been caught up in the Victorian obsession with horticulture and plant identification, with a particular interest in the grapevine. Yet no record remains of the varieties that grew in the vineyard Busby planted in 1836 on the open, gentle slope before his dwelling, the famous Treaty House at Waitangi. All we are left with today is the well-tended lawn that would rival Eden Park for the title of 'most hallowed turf in the nation'. There is no clue anywhere that it is the site of New Zealand's first wine-producing vineyard.

Busby's Waitangi vines were drawn from an extensive collection, mainly gathered during a four-month tour of France and Spain. As a result of these travels, which Busby used to acquire wine-producing knowledge as well as vines, he was able to transport an astonishing 362 varieties by convict ship to Sydney. Only some of these then crossed the Tasman after Busby's appointment to the Bay of Islands as British Resident. That Pinot Noir was among his original collection there can be no doubt. Busby spent several respectful, excited days in Burgundy during his continental wine tour. These included a visit to the famous Clos de Vougeot vineyard, where he noted the white grape 'Chaudenay' was losing ground to the black grape 'Pineau', as wine made from the latter had double the earning power. The Burgundy chapter of Busby's account of his travels finishes thus: 'After having received … a small bundle of each of the kinds of vines cultivated in the Clos de Vougeot … I took my leave.'

Alas, by 1845 Busby's vineyard had been destroyed, purportedly by troops camped at Waitangi. The little we know about the wine it briefly produced has come to us via the diary of French explorer Dumont d'Urville, who was given a taste of it in 1840. For penning this first-ever review of a New Zealand wine, the French matelot deserves to be regarded as the father of Kiwi wine writers. He makes clear his enjoyment of the wine, which he describes enigmatically as being *plein de feu*.* He also states that it was a white wine.

Despite the occasional unsubstantiated claim to the contrary, there is no indication that Pinot was put in the ground by the French settlers at Akaroa, who arrived with vines in the Treaty year. The same goes for the rest of the country's mid-nineteenth-century *vinifera* plantings. Most of these were short-lived projects, an exception being

---

* This phrase in d'Urville's tasting notes has spawned a much-published, erroneous translation that renders the words in English as 'very sparkling'. 'Sparkling' has a specific meaning in relation to wine, and plein de feu *(literally, 'full of fire') can hardly be a reference to the presence of bubbles. When I have quizzed French people in the wine industry about the phrase, most have thought it would have meant 'vibrant' or 'brightly acidic', or perhaps referred to the wine's alcoholic strength.*

the efforts of the first Catholic missionaries, Bishop Pompallier and his accompanying troop of Brothers from the Society of Mary (the Marists). For these French men of God, wine was both a professional and a personal necessity, and they began planting vines from the time of their arrival in the Far North in 1838. Many a mention was made of vineyard progress in their letters back to Marist HQ in Lyon (enticingly close to Burgundy), but in true French fashion, variety names didn't seem to be part of their vocabulary. The closest we get is when a Brother Elie-Regis at the mission station at Whangaroa wrote of his intention to make 'a Côte Rotie wine'. If the description was inspired by grape variety, it points to Syrah and Viognier being the chosen vines.

The Marist Brothers' subsequent southward migration saw them establish the wine tradition in Hawke's Bay that continues today as Mission Estate, still under the Society of Mary's ownership. It is here the Pinot trail gets warmer. After a couple of decades of steady but uninspired winemaking, featuring unremarkable varieties such as Chasselas and Sweetwater, the Marists' Hawke's Bay winery stepped up a gear with the arrival of Brother Cyprian Huchet in 1871.

The son of winegrowers in the Loire, the cheerful, robustly built Brother Cyprian brought a new energy and sense of *métier* to the Marist vineyard at Meeanee. He set about slowly extending the area in vines and building a new press house, and under his stewardship wine began to be sold commercially for the first time. Helping the Brothers was a useful product endorsement: doctors in the area began recommending their patients drink Mission wines for health reasons.

It is in 1889 that we finally find a record of 'Pinot varieties' being planted. We know that by the late 1890s Pinot Noir, Pinot Meunier, Pinot Blanc and what the Brothers called Pinot Chardonnay (Chardonnay to us) were all established contributors to the Mission portfolio. Some of the credit for the arrival of these varieties is given to Dean Nicolas Binsfeld, a *Luxembourgeois* who instigated the wave of vineyard expansion that occurred in the early 1890s. However, there is no evidence that he actually made calls on what was being planted. Brother Cyprian was still cellarmaster at this time.

It would be tempting to bestow on Brother Cyprian Huchet the mantle of New Zealand's original Pinot pioneer, were it not for activities to the south.

Jean Désiré Feraud was the son of a doctor from Vence, close to the Côte d'Azur. Soon after arriving in Central Otago in 1863, he struck gold on the west bank of the Clutha, just south of Alexandra (an area now known as Frenchman's Point). After 1865 Feraud hunted no more for gold; he decided instead to plant vines and fruit trees on a block he bought near Clyde. Offering him encouragement were two other Frenchmen, the brothers Bladier from Toulouse, who had viticultural experience and had bought a neighbouring property. This small Gallic confederacy obtained vine cuttings from Australia and planted Central Otago's first vineyards.

Feraud called his property Monte Christo. Despite the slight difference in spelling, perhaps he felt a connection between himself and Dumas' fortune-finding, redemption-seeking hero in *The Count of Monte Cristo*. The popular novel first appeared just 20 years before Feraud's arrival in Otago. Feraud was certainly seen as a man of vision, resourcefulness and action in his adopted community, becoming Clyde's first mayor in 1866.

In 1865, the *Otago Witness* reported that a bunch of fruit had appeared on a vine planted by Feraud and the Bladiers. The grapes were identified as Chasselas, the only direct reference we have to the varieties they planted. Feraud then built a winery from local stone. Still intact, it is New Zealand's oldest remaining winery building.

By 1870, for unknown reasons, the Bladiers had dropped out of the picture. Feraud had 1200 vines in the ground that were producing fruit and he had begun selling beverages with names like 'Aniseed Liquor' and 'Constantia Wine'. Of interest to us is the one he called 'Burgundy Wine', for which he won a prize in an Australian show. If its maker had been Anglo-Saxon, we might more easily dismiss this 'Burgundy Wine' as probably being red and made from any variety at all. But would a Frenchman apply the name of a great French region so loosely? We're left guessing. If it was Pinot Noir that Feraud had in the ground in Central Otago in the 1870s, he was not only our first Pinot producer, but also a vigneron well ahead of his time.

The nineteenth-century Pinot Noir trail also leads to the Wairarapa, another region now established as a Pinot stronghold. Beetham is one of a number of names synonymous with the opening up of the Wairarapa to colonial farming. William Beetham arrived in the colony with his parents and siblings as a teenager in 1855. Twenty years on, a

Pineau Noir, his first favourite, still surpasses all others.

partnership that included Beetham and his brothers had acquired and cleared tens of thousands of Wairarapa acres for grazing land. Brancepeth, one of New Zealand's great rural properties, was quick to flourish.

In middle age Beetham was a man of means and turned his attention to a range of things. He founded the Masterton Club (of which he was president for 29 years), he liberated trout in the Waipoua River, and he travelled to France, returning in 1882 with a French wife.

Little is known about Marie Zélie Hermanze Frère, known as Hermanze, and there are contradictory accounts of what part of France she was from. It seems likely that Mareuil-sur-Ourcq, in Picardy, was her hometown. We do know that as well as returning to New Zealand with Hermanze on his arm, Beetham had vines in his luggage. It may have been the wish of his new bride to plant vines and make wine in her adopted land; a neighbour later referred to her as being 'the one behind the whole thing'.

Ripening the French vines proved difficult in the Wairarapa. Beetham, though, was sufficiently encouraged to expand his vinous stake. In 1890, *The Wairarapa Daily* reported that 'this spring [Beetham] planted a second patch of a similar size with cuttings from Napier, of, we believe, the Pineaux Noirs variety, a grape which produces in Hawkes Bay a first class wine'. It seems very likely that Brother Cyprian and his Marist cohorts were the source of Beetham's Pinot cuttings.

Pinot Noir became Beetham's darling. There are several references over the next few years to his success with, and fondness for, the variety. 'Mr Beetham finds that pineau noir is the ideal grape for this climate,' reported *The Wairarapa Daily* in 1901. 'It contains a high percentage of grape sugar, is usually dead ripe about the first of April and produces a splendid Burgundy wine. He has tried other varieties … but the "pineau noir", his first favourite, still surpasses all others.' At the turn of the century Beetham's Lansdowne vineyard (a short distance to the northeast of the centre of Masterton) was producing about 12,000 bottles of wine a year. The country's aristocratic Governor of the day, Lord Ranfurly, was a wine devotee — he arrived at his Wellington posting accompanied by a private stash of 7200 bottles. In his opinion Beetham's wine equalled, if not surpassed, the best Australia could produce at that time.

There are bottles of William Beetham's wine still lying in the cellar of the Brancepeth homestead. In 1985 New Zealand wine writer Geoff Kelly was part of a small group that tasted a bottle of Lansdowne Claret 1903 at Brancepeth. Kelly felt the use of the name 'claret' referred to the intended wine style, rather than the variety make-up, which he was sure owed nothing to Bordeaux. And he was agreeably surprised by its quality. 'The flavour is clearly old burgundy in style, with the oak standing firm,' Kelly wrote, 'yet amazing fruit, body and freshness for the age. The finish is superb, long and lingering. The wine is still satisfying, though very, very dry. It must have had excellent extract and balance and flavour when young, to have lasted so long. There seemed little doubt it was made from the Pinot Noir and Hermitage recorded for the vineyard; it was Burgundian.'

The Beethams' success influenced others. On the southern side of Masterton, Tararua Vineyard was planted in 1898 by a Captain AC Turner. At 3.2 hectares it was double the size of Beetham's Lansdowne vineyard and planted solely in Pinot Noir. Several years later the vineyard was sold to William Lamb, who left his job as a bank clerk to throw himself into the wine business, expanding the vineyard and building a winery and cellar. The Premier, 'King' Dick Seddon, came calling to congratulate Lamb on his success. He sampled Lamb's claret (which we can presume was made from Pinot grapes) and declared confidently that it reminded him of a wine he had tasted in Naples.

After visiting the Beethams and tasting their wine, Henry Tiffen, 71 years old, raced back to Hawke's Bay and planted a large vineyard at Greenmeadows. Before long his wine operation was held up as a model for others to follow. A few years earlier another Hawke's Bay landowner, Bernard Chambers, had become the progenitor of today's Te Mata Estate near Havelock North.

Pinot Noir had a presence in both these vineyards. We are fortunate that a journalist's personal encounter with one of Tiffen's wines, a Burgundy made from Pinot Noir and Pinot Meunier, was recorded in the *New Zealand Farmer* in 1897. '[I] came across a bottle of Burgundy in private life and it bore on the bottle the Greenmeadows label,' wrote the correspondent. 'The excellence of the wine had drawn my curious attention to the label, and I was both surprised and pleased to find wine so matured and of such high-class quality produced, so to speak, at one's elbow. For good sound, light wine we have really no occasion to go outside the colony.'

Even further north, there is evidence that William Heathcote's vineyard beside the Kaipara Harbour was host to 'Pinots', as was Joseph Soler's in Wanganui.

This promising *fin de siècle* ripple of Pinot production found a ready cheerleader in Romeo Bragato, the moustachioed, lazy-eyed son of the Austro-Hungarian Empire (he was born in what is today Croatia) who arrived in 1895 at the behest of the New Zealand Government. Trained in viticulture and oenology in Italy, Bragato travelled the length of the country visiting, tasting, studying, discussing and appraising, a tour which resulted in the publication of a famously prescient report.

In the course of his raptures over New Zealand's vine-friendly disposition, Bragato gave the thumbs up to almost all today's leading wine regions and recommended most of the classic varieties around which the New Zealand industry is now based. Yet he wasn't the perfect seer; Marlborough and Sauvignon Blanc were notable omissions.

In his *Report on the Prospects of Viticulture in New Zealand* Bragato makes mention of 'Pinot' or 'Pinots' (due to the fact that Pinot Noir and Pinot Meunier were usually planted, and often blended, in tandem at that time) more than any other variety. He was impressed with their performance in both Beetham's and Tiffen's vineyards. Of the latter, he wrote: 'In this vineyard the Pinot, the proper grape for champagne, yielded a magnificent crop of the finest grapes seen by me.'

The praise grew richer towards the end of his visit when he addressed a packed house at the Chamber of Commerce in Dunedin. Reported the *Otago Daily Times*: 'After having travelled throughout Europe, and having never been satisfied with the growth of grapes of the Burgundy type, [Bragato] was astonished when he came to this colony to see how well they grew; in fact, there was no country on the face of the earth which produced better Burgundy grapes than were produced in Central Otago and in portions of the North Island.'

That strongly suggests Pinot Noir was in the ground in Central Otago at the time of Bragato's visit, which in turn lends a strand of support to the authenticity of Feraud's Burgundy. Of all the regions Bragato visited, he seems to have been most taken with Central Otago, describing its wine potential as having 'enormous wealth-producing capabilities'.

Twin tempests, phylloxera and prohibition, broke over the New Zealand wine industry during the early part of the twentieth century. The prohibition movement hit the wine industry in 1908 when Masterton and Eden, two districts where wine was grown, voted to go 'dry'. Among the victims was Lamb's Tararua Vineyard. Lamb was prosecuted for selling wine direct from his vineyard and his business collapsed as a result. A horse and chain tore out his 7000 Pinot vines and he sold his equipment for next to nothing. Other producers survived by creating outlets for their products just outside the limits of their dry districts — sometimes just a few yards down the road.

The hair's breadth result in the national poll of 1919, when only the votes of returning soldiers kept the whole of New Zealand from going dry, was the prohibition movement's high-water mark. Its influence dwindled, although New Zealanders continued to vote on the issue at every general election until 1990.

Phylloxera is a disease spread by a winged aphid that attacks a vine's roots. When it was discovered in a vineyard in France's southern Rhône region in 1865, it had just embarked on a destructive world tour. By the time the first infection was found on Burgundy's Côte d'Or a decade later, the entire French wine industry was staring at collapse. (As an aside, a couple of plots of Pinot Noir in Champagne, owned by Bollinger, are among a select few pockets of European vines that have to this day mysteriously survived the depredations of phylloxera.)

Romeo Bragato, during his 1895 tour of duty, confirmed the phylloxera scourge had made landfall in New Zealand. A stumbling response followed, during which most of the country's vines became infected. When Bragato returned to the country in 1902 as our first Government viticulturist, combating phylloxera governed the agenda. Applying the remedy that had restored the vineyards of Europe, Bragato grafted phylloxera-resistant rootstock on to classic European varieties and began distributing them. The take-up was good. Pinot Noir, however, somehow suffered in this shuffle. Bragato's belief in Pinot's suitability for New Zealand didn't waver right up until his final departure in 1912, but there is little evidence of the grape being much planted at this time. This was in stark contrast to its close cousin, Pinot Meunier. The hardy Meunier thrived, gaining popularity particularly among the Dalmatians, many of whom had just turned from digging kauri gum to grape growing. Joe Babich remembers his father's first vineyard, planted in Henderson in 1919, actually being called the 'Pinot

Vineyard', due to the dominant presence of Pinot Meunier. In 1917, Pinot Meunier accounted for two-thirds of all the grapes used to make wine in this country.

But classic varieties were not all that was being planted. A fondness began to be shown for American hybrids. Derived from indigenous American *Vitis labrusca* varieties, these vines are high-yielding and disease-resistant. Unfortunately, they are also resistant to producing table wine of any quality, a shortcoming that went unchallenged by the unsophisticated local market. New Zealand wine sank into decades of mediocrity during which the staple products were sherries and fortified wines.

By 1965, two American hybrids, Baco 22A and Albany Surprise, were New Zealand's most-planted varieties. Pinot Meunier plantings had receded dramatically to less than 3% of the national vineyard. Pinot Noir does not figure anywhere on the list, not even among the minor plantings (under 10 acres each).

And yet, it had not disappeared completely. Two rows of Pinot Noir vines, together with 12 of Pinot Meunier, stood for most of the twentieth century at the Mission Vineyard at Greenmeadows, on the slope where the annual Mission concerts are now held.

Remembered as being 'pretty gnarled and low cropping' by Brother John Cuttance, the vines are thought to have been planted by Henry Tiffen, from whom the Marist Brothers bought the property and vineyard in 1897. Through the 1950s and 1960s, the fruit of the vines contributed to the Mission Reserve Pinot, a table wine popular among wine cognoscenti.

Brother Joe Lamb, another who worked with the vines, recalls them being pulled out a year or two after he left the winery in 1979. It seems criminal that such an important link with New Zealand nineteenth-century Pinot Noir was wiped out so recently. The old Mission Pinot vines survived just long enough to see in the rebirth of the variety in New Zealand.

....

Following pages: pp52-3 Et in Arcadia Pinot — Millton's Naboth's Vineyard, midsummer.

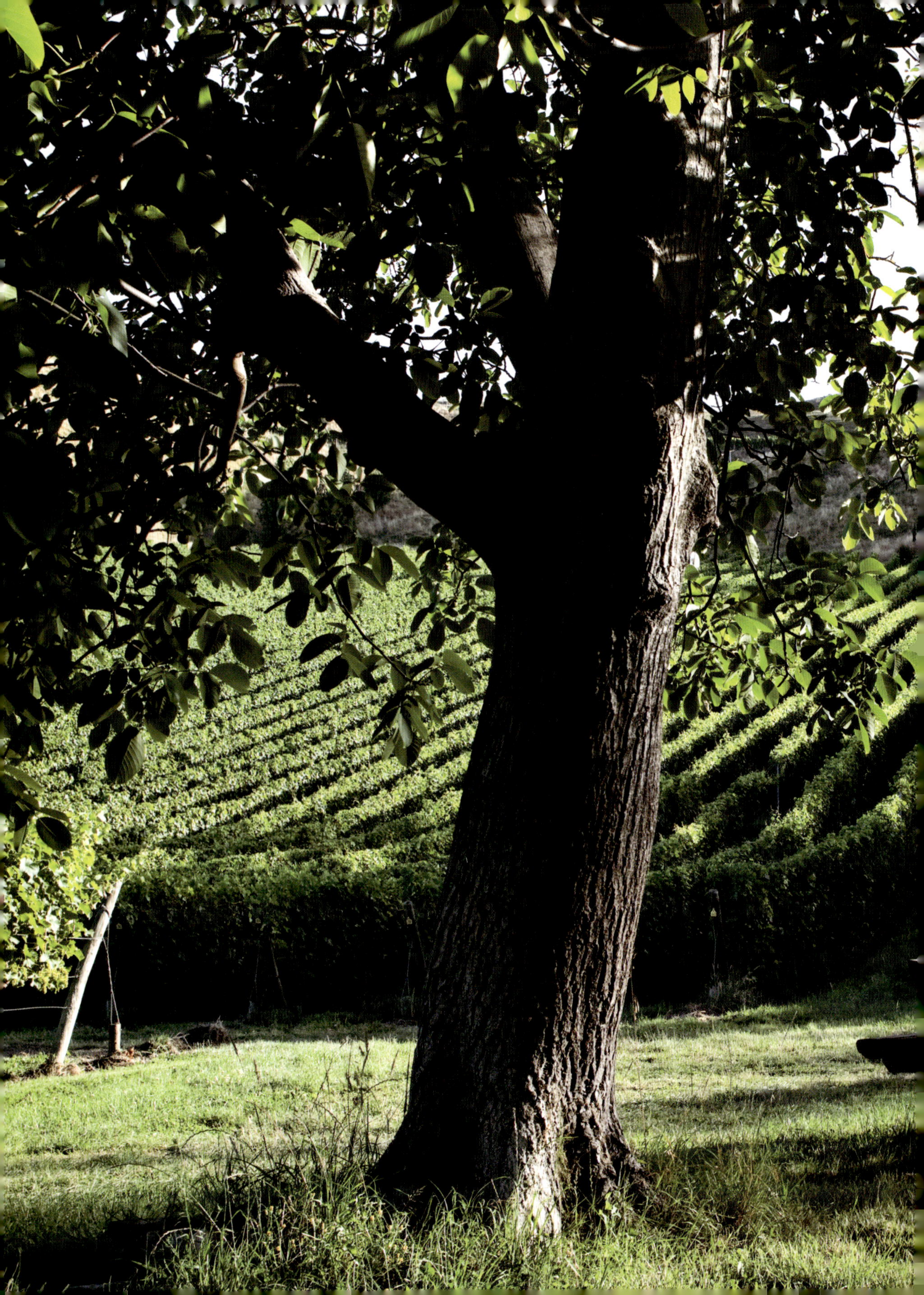

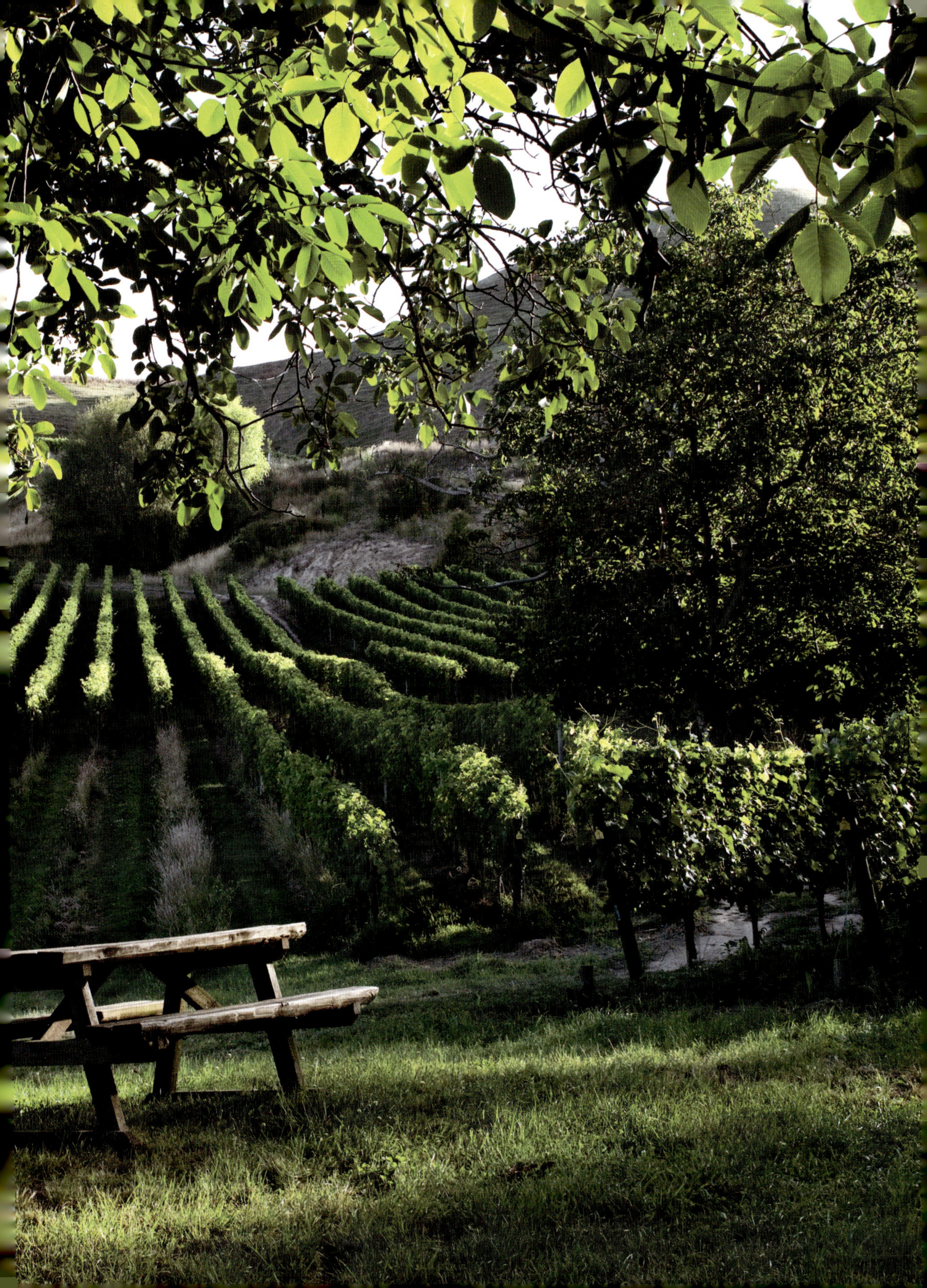

FOUR. REBIRTH

Youthful zeal and fresh ideas found fertile growing conditions in the 1960s. Nick Nobilo Junior came into the decade with plenty of both. The son of Nikola and Zuva Nobilo, immigrants from the winegrowing town of Lumbarda on the Croatian island of Korcula, Nick went straight from school in 1960 to help his father on the family vineyard at Huapai, West Auckland.

Nikola had begun planting vines in 1942. By the early 1960s his vineyard was, like its West Auckland neighbours, producing principally fortified wines and sherries, along with a small quantity of table wine. Nobilo's Dry Red, besides being the table wine of choice chez Nobilo, had a loyal following among New Zealanders with experience of Europe, a group that included returned soldiers who had served in Italy during World War II. Other dry table wines from the likes of San Marino (today's Kumeu River) and McWilliams were similarly making small impressions. McWilliams Bakano was launched in the mid-1950s. Made from hybrid grapes and a token iota of Cabernet Sauvignon, Bakano went on to become the first red table wine brand to achieve something close to household-name status in New Zealand. Young Nick Nobilo sensed a rising interest in more serious wine styles, and it was a future he wanted to help shape.

The path led by necessity to a return to classic varieties. Successive régimes at the

Government's small and under-resourced Te Kauwhata Viticultural Station had been trying to loosen the industry's embrace of the likes of Albany Surprise, Baco 22A and the other non-*vinifera* American hybrids. Small inroads were being made. The talented Hungarian viticulturist and winemaker Denis Kasza produced a successful Chardonnay at Te Kauwhata during the 1950s, prompting several wineries to plant the variety.

Kasza's superior at Te Kauwhata was the dedicated Frank Berrysmith. In the early 1960s, as Government viticulturist, Berrysmith embarked on a busy vine importation programme. By the time Nick Nobilo came knocking, Berrysmith had a diverse range of *vinifera* varieties growing, including the French classics. Among Berrysmith's collection were three Pinot Noir clones: Bachtobel, Oberlin and AM 10/5 (from here referred to simply as 10/5). All three had been obtained from the Waedenswil viticultural research centre in Switzerland in 1962.

Encouraged by his father and his brother Mark, who had also joined the family concern as a viticulturist, Nobilo was eager to experiment. He spent some time at Te Kauwhata with Berrysmith, observing the vines and discussing their relative performances. The outcome was a trial plot at Huapai made up of a variety mosaic that included the three Pinot clones.

Nobilo was not alone in obtaining classic varieties from Berrysmith at this time. But he was the first to start taking the Pinot Noir he got from Te Kauwhata seriously. Impressed by the fruit and flavours he was getting from the three Swiss clones, and encouraged by some rough trial wines he made from them, he included them in a new 5-hectare vineyard he and his father planted at Huapai. This led to the production of New Zealand's first commercial Pinot Noir of the modern era, the Nobilo Pinot Noir 1973.

It came as no surprise that, by Nobilo's own admission, his early efforts were hardly impressive. He admired some of the delicate flavours in the wine, but was concerned about its lack of colour. Tasting good Burgundies made him realise just how far off the pace he was, but also strengthened his desire to make something that approached them in quality and style.

The breakthrough came with the 1976 vintage when Nobilo altered his winemaking methodology. He later recalled:

We hand-picked the grapes, then put them all — some whole bunches — in tall draining tanks. I left them there for 10 days and drained off some of the free-run juice. What was left was fermented en masse and we closed it, creating a carbonic maceration effect. Then I separated the berries and pressed them out ... and got this beautiful wine. I was out of sync with the rest of the winemaking community . . . I was passionate.

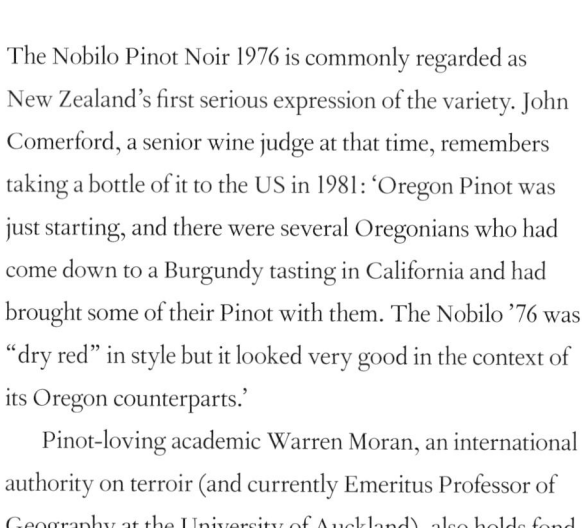

The Nobilo Pinot Noir 1976 is commonly regarded as New Zealand's first serious expression of the variety. John Comerford, a senior wine judge at that time, remembers taking a bottle of it to the US in 1981: 'Oregon Pinot was just starting, and there were several Oregonians who had come down to a Burgundy tasting in California and had brought some of their Pinot with them. The Nobilo '76 was "dry red" in style but it looked very good in the context of its Oregon counterparts.'

Pinot-loving academic Warren Moran, an international authority on terroir (and currently Emeritus Professor of Geography at the University of Auckland), also holds fond memories of the wine and vintage. 'It was a special year climatically in West Auckland … a long "Indian summer". I tasted the 1976 Nobilo Pinot over a 15-year period and was always impressed with how it maintained a tight structure and gained in elegance, bouquet and on the palate.'

In 1986, Moran was present when a close cousin of the wine — the Nobilo Claret 1976, a blend of Pinot Noir and Pinotage — appeared on a table in Burgundy. Moran and his wife were having dinner at the house of their friend Claude Chapuis, a member of an old Aloxe-Corton winegrowing family. In the late 1970s, Chapuis had worked in the Nobilo vineyards and had returned home with a few bottles of the '76 Claret. The other eight guests that night were all Aloxe-Corton wine folk and, in true Burgundy fashion, a number of wines, six in all, were served blind and discussed.

'The Nobilo Claret was served in a sock [as is the Burgundy tradition] as the second wine and the shoulders of its Bordeaux bottle were discernible,' recalls Moran.

'The vignerons were puzzled by the flavour. The comments began as "un bon vin", then quickly evolved to "même un très bon vin" and finally to "très, très bon". And this, remember, was Nobilo's second Pinot with at least the two varieties in it. Following it were an Aloxe-Corton village wine and a Grand Cru Chambertin, both about a decade old. The New Zealand wine stood tall in this company, and amongst sophisticated palates and blunt tongues.'

Nick Nobilo backed up his 1976 success with another very good vintage in 1977. By this time he was favouring the Bachtobel clone above the other two, and the Nobilo Pinot Noir was eventually composed entirely of Bachtobel fruit. But the wine's last vintage was 1983. A massive hailstorm wiped out the Huapai vineyard's fruit in 1984; the mid-1980s traumas that beset the entire industry did the rest. A glut, a price war, together with the Labour Government's punishing increase in wine tax all contributed to the crisis. Along with scores of other growers around the country, the Nobilos accepted a Government financial incentive and pulled out their Huapai vines in 1986. Some of the vineyard was sold off and, along with many others, the Nobilos shifted their emphasis to other regions, notably Marlborough.

The traffic in cuttings of the new classic varieties was beginning to flow vigorously between the wineries themselves. Nick Nobilo remembered Frank Yukich of Montana coming in 'with a bloody army to get heaps of cuttings from our vineyard'. From Corbans, Joe Babich obtained Bachtobel vines which he planted at Henderson during the mid-1970s. A commercial wine was made for the first time in 1980. A year later the Babich Pinot Noir 1981 struck gold at the Royal Easter Show — the first New Zealand Pinot to go gold at a national competition.

'It was just another variety … we didn't know much about its culture,' says Joe Babich. We made it the same as the other red wines, and not much of it either … around

800 cases. It was light in style and New Zealanders were used to Australian red wines at that time, so consumers had an adjustment to make. But after we won that gold medal we quickly sold out of it.'

As is often the way, what was happening with Pinot Noir in Auckland was happening in other regions at roughly the same time. Of particular significance was 'a bunch of hooligans getting together with the idea of growing grapes in the South Island'.

Those were the words of Danny Schuster, the Prague-born, European-trained winemaker who arrived in New Zealand in 1973. Schuster teamed up with Dr David Jackson at Lincoln College (now University) for a period of winegrowing research during the mid-1970s, work which was to lay foundations both for the wine industry in both islands and for Lincoln's degree course in viticulture and oenology.

They obtained up to 60 varieties from a number of sources within New Zealand, including the Te Kauwhata Research Station and the Ministry of Agriculture and Fisheries outpost in Hastings. These were mostly the Swiss clones, which now also included clone 2/10.

'We didn't really know what we were doing in those early years, especially when it came to ripeness; we picked the grapes according to numbers, rather than by taste,' recalled Graeme Steans, a former Lincoln senior tutor who assisted with the project. Despite that, there were encouraging signs. 'I clearly remember our excitement — I think it was 1976 — on tasting several clones of Pinot,' said the late Professor Don Beaven of the Christchurch School of Medicine, a regular member of the Lincoln tasting panel. 'The fruit was very intense, they were young, there was no wood of course. I recall the 10/5 clone being the best of them, while the Bachtobel had a strong cherry-like, Beaujolais nose. We'd never seen a New Zealand red wine that was so lively.'

Word spread and the regular tastings held at Lincoln began to attract interested wine insiders. Among them were names that would later become closely associated with the development of New Zealand Pinot Noir — Dr Neil McCallum, Tim Finn, Rolfe Mills, Ivan Donaldson, Rudi Bauer.

The signs were encouraging enough for Steans, with advice from Danny Schuster, to plant an acre of Pinot Noir in the Kaituna Valley on Banks Peninsula in 1977.

This was the first private planting of Pinot in Canterbury, and it remains the oldest commercial Pinot Noir vineyard in New Zealand. It is currently owned by Grant and Helen Whelan and contributes to their characterful Kaituna Valley Pinot Noir.

A year later a new wine operation called St Helena was established north of Christchurch on Coutts Island beside the Waimakariri River. The land was owned by Trevor Mundy, a man who had done extremely well out of growing vast quantities of potatoes during and after World War II. But potatoes were failing at his property on Coutts Island. With the encouragement and participation of his sons, and drawing inspiration, vines and Danny Schuster from the Lincoln research project, Mundy went ahead and planted grapes. This was to be Canterbury's first commercial winery.

Pinot Noir was a significant presence in the vineyard. As well as the stalwart 10/5 and 2/10 clones, a mass selection of Pinot vines that Schuster had obtained from Montpellier and started growing at Lincoln was also planted.* Schuster, with his central European background and time spent in Australia and South Africa, had not had a lot of experience with Pinot. But the early efforts from Nobilo and Babich, and his years at Lincoln, had shown him what was possible with the variety in New Zealand. Taking in fruit from Graeme Steans' Kaituna Valley vineyard to supplement that grown at St Helena, with his third vintage at the winery Schuster made a Pinot Noir that is still talked about today.

'The 1982 St Helena Pinot Noir was a wonderful wine … it seemed to me bigger and more structured than the earlier Nobilo Pinots,' said Don Beaven. 'I think the Montpellier vines helped give it flesh.'

Beaven was one of seven judges who tasted the wine at the 1983 Air New Zealand Wine Awards, a year in which there were six entrants in the Pinot Noir class. The judging panel included John Avery, scion of the old Bristol wine merchant company Averys, and a man with an excellent knowledge of Burgundy. Beaven remembered:

A mass selection is a collection of cuttings consisting of vines with different characteristics, selected with no regard to individual identity. A clonal selection, by contrast, is made up of cuttings all taken from a single mother vine.

We tasted and scored the wines one by one. The St Helena Pinot split the panel. John Avery and myself gave it gold first up, one judge gave it silver, another gave it bronze and three gave it no award. We each spoke to our scores. There was an opinion among those who gave it no award that something was awry — that the wine's colour and structure had to have been achieved through double fermentation. Avery described it as magnificent and perfectly ripe, saying it reminded him of a Musigny. After the wines were retasted, it went gold.

It was a breakthrough wine, establishing Danny Schuster's reputation and placing Pinot Noir squarely in the frame as an exciting new red wine possibility for New Zealand.

There were few more interested in the wine and its success than a small cluster of fledgling winegrowers in the all but forgotten southern Wairarapa town of Martinborough.

This group had arrived in Martinborough in the wake of a 1977 report authored by Department of Scientific and Industrial Research (DSIR) soil scientist Dr Derek Milne. The report identified Martinborough as being eminently suitable for grape growing, with a heat summation similar to that of Beaune in Burgundy, low rainfall (particularly in autumn) and free-draining soils. The first Pinot in Martinborough was planted by Martinborough Vineyard (of which Milne himself was one of six founding partners) and Ata Rangi, established by Clive Paton.

From the beginning, Martinborough Vineyard saw Burgundy as the beacon and determined to grow Pinot Noir.

To this end, cuttings of clone 10/5 were obtained from St Helena. By contrast, Clive Paton, a former local dairy farmer and tearaway wing three-quarter (he beat JJ Williams to the ball to score a try for Wairarapa Bush in a match against the 1977 British Lions), initially just wanted to make red wine. He planted Bordeaux varieties and Syrah, as well as a small amount of Pinot from a clone he had received from a winegrower friend in Auckland named Malcolm Abel.

The origins of this clone — at least, what we know of its origins — have given us one of the most engaging subplots in the saga of New Zealand Pinot Noir. It goes back to a New Zealand Customs officer's discovery of a cutting in the luggage of a man arriving at Auckland Airport in the mid-1970s. The cutting was found hidden in a gumboot, which is why it is sometimes called the 'gumboot clone'. Happily for the New Zealand wine industry, the customs officer was Malcolm Abel, who at the time was establishing his own winery in Kumeu. Abel's interest was piqued. Interrogating the man, he was told the cutting had been taken surreptitiously from a legendary corner of Burgundy, the La Tâche vineyard in Domaine Romanée Conti. Abel saw to it that the clone was confiscated but not destroyed. After seeing it through the proper quarantine channels, he planted it in Kumeu. His wine enterprise did not survive long after his own premature death in 1981, but by then he had passed the cutting on to Paton, who called it the Abel clone. The first varietal Pinot produced by Ata Rangi in 1985 was made up entirely of Abel fruit. 'It looked good … that wine was still alive after 10 years,' said Paton. From there, the Abel clone has gone on to impress almost all who have planted it — and that is now a large and expanding part of the New Zealand Pinot-growing community.

Neil McCallum, another scientist and a friend of Derek Milne, had established Dry River in Martinborough in 1979. Due to his abiding affection for Alsatian varieties, and his wariness in regard to Pinot's capriciousness, McCallum was a relatively late starter with Pinot. The first Dry River Pinot Noir was made in 1989.

The early Martinborough scene was, in effect, a band of smart, committed but highly inexperienced *garagistes*, all learning on the hoof and all making their wine at Stan Chifney's small winery, the only one in town. Things began to change in the mid-1980s when Martinborough Vineyard first built a small winery of its own and then hired a professional winemaker to go with it.

Born and trained in South Australia, Larry McKenna was working in Auckland for Delegat's in the early 1980s. One day while walking through a Cabernet block in West Auckland, he noticed a rogue vine he couldn't identify. He asked his old school chum John Hancock, also a winemaker at Delegat's, what the vine might be. Hancock told him it was Pinot Noir. 'But don't worry about it,' he added. 'It only grows in Burgundy.'

McKenna was being asked to challenge that assumption when he arrived in Martinborough. Despite the seat-of-the-pants nature of the town's nascent wine industry, he liked what he saw straight away — the benefits of being in the rain shadow on the eastern side of the island, the freedom and excitement of working for a boutique operation whose goal was simply to make the best wine possible, and the chance to focus on a single red variety.

McKenna's effect was felt immediately. The Martinborough Vineyard Pinot Noir 1986 won gold and the Pinot Noir Trophy at the Air New Zealand Wine Awards. The feat was replicated with the 1987, 1988 and 1989 renditions of the same wine. Martinborough and Pinot were in business, although McKenna today marvels at the level of his success, given the limitations of his knowledge.

'I read a few magazines and books to get the general gist of it [Pinot Noir],' he remembers. 'I was trying out techniques I'd heard about, such as fermenting whole bunches. But we were really feeling our way until I went to Burgundy in 1990.'

That year, McKenna worked a vintage for Domaine de l'Arlot, living in three different Burgundian households. It was an awakening, 'an enormous learning curve', says McKenna.

> It was significant in my understanding of how to make Pinot and it set parameters that are still in place. When I left Delegat's to make Pinot Noir in Martinborough in 1986, people didn't know anything about it. Burgundy proved to me how important Pinot Noir was. I got to know the people who are involved with wine in Burgundy and their culture and it was fantastic. I'd previously been to Bordeaux and had never been so disillusioned with a wine region in my life … you couldn't see people, you couldn't taste. Burgundy was a completely different experience.

The Martinborough Pinot producers were quick to develop a close relationship with their nearest metropolitan centre, Wellington. The idea of having a Burgundian village on their doorstep appealed to the denizens of the capital city's savvy wine scene. When cash-strapped Clive Paton hatched a plan in 1985 to raise money to purchase oak barrels, the support was there. Fifty people paid $50 each up-front, enabling Paton to buy barrels for the 1986 vintage. In return, the contributors each received two bottles of Ata Rangi Pinot Noir or Célèbre (a red blend) annually for the following five years.

By the end of the 1980s the pioneers had been joined by the likes of Palliser Estate, Nga Waka Vineyard and Voss Estate. Although Palliser upped the ante in terms of scale, to this day it remains a medium-sized New Zealand producer. The others, and most of the subsequent arrivals, were tiny and family-owned, creating at Martinborough the kind of close-knit, artisan culture in which Pinot can thrive.

We now move south, further south than Canterbury, to the mountain valleys of Central Otago. Following the budburst of belief shown in Central Otago by Jean Désiré Feraud and then Romeo Bragato in the nineteenth century, winegrowing in the region had gone nowhere.

Outside a few peripheral plantings, such as those by post-World War II Dalmatian farm workers who found the place hospitable enough to grow table grapes, Central Otago was a viticultural desert for the first seven decades of the twentieth century. Helping dampen any outbreak of enthusiasm was a string of assessments issuing from the Department of Agriculture. As late as the 1960s, the department carried out a study at Earnscleugh near Alexandra. Eighteen varieties were planted. The selection was eclectic to say the least; included were such grapes as Madeleine Royale, Queen of the Vineyard, and Alphonse Lavallée. This trial vineyard's lack of success prompted a scientific officer at the Government's viticultural research station at Te Kauwhata to declare to the country's small winegrowing community in 1967 that 'conditions in the South Island of New Zealand do not appear to favour grape growing for wine production'.

David Jackson and Danny Schuster were not the only mainlanders to ignore such official findings and conduct their own experiments. Anne Pinckney, a Karitane nurse who also had an MSc in horticulture from Lincoln, planted a trial vineyard in 1976 at

Dalefield, near Queenstown. Pinot Noir was among the varieties she planted, with cuttings taken from Lincoln. Pinckney's was the first confirmed planting of Pinot Noir in Central Otago.

She was one of a small group who, oblivious to each other's existence, had begun planting vineyards to make wine in different parts of the region around the same time. Rolfe and Lois Mills and family had vines in the ground at Wanaka; so had former television journalist Alan Brady in the Gibbston Valley. Regarded as insane by the local populace, not to mention many wine industry people across Cook Strait, this dedicated handful soon began meeting, sharing notes and helping each other, sowing the seeds of what grew into the Central Otago Winegrowers Association (COWA). At the same time, Bill Grant had begun the vinous adventure near Alexandra that would become William Hill. He was followed by Verdun and Sue Edwards at Blackridge, also in the Alexandra basin.

Their relationship with Pinot Noir developed in a way similar to that experienced by their counterparts in other regions. A passion for Pinot Noir did not drive them to plant grapes; they just wanted to see if they could make wine … any wine. But once again, Pinot Noir was quick to impress and a fascination with the variety grew from there.

'Pinot Noir put its hand up early down here,' says Alan Brady. 'And being the Francophile he was, Rolfe Mills had a handle on it early.'

By the late 1980s, as with Martinborough, there had been enough promise for the amateur pathfinders to enrich their group with professionally trained winemakers. Rippon hired Tony Bish (now with Hawke's Bay winery Sacred Hill) in 1986. A year later Central Otago's first commercial Pinot was made for Gibbston Valley by Rob Hay, who had trained in Germany. Anne Pinckney was not only a strong source of inspiration at that time, she also had the only functional winery in the region. All the grapes were crushed at her Taramea winery at Speargrass Flat, near Arrowtown.

The 1989 Rippon Pinot Noir was the first Central Otago Pinot to win a medal at a national show. Bish moved on, to be replaced by Austrian-born Rudi Bauer, who continued the success with a gold medal for the 1991 Rippon Pinot Noir and then a gold and trophy for the 1992 Pinot.

Central Otago Pinot was being noticed, and not just in New Zealand. Alan Brady remembers receiving a fax out of the blue from British wine scribe Jancis Robinson. A friend of Brady's had taken a bottle of the 1990 Gibbston Valley Pinot to dinner at the home of the influential critic, prompting the fax which described the wine as 'an absolute star'.

'That kind of spontaneous, unsolicited support meant so much at that stage,' says Brady.

New Zealand Pinot Noir had, in less than a dozen years, become a small but undeniably arresting force in the world of wine. That this was mostly the work of a handful of dreamers with little previous experience of making wine generally, let alone Pinot Noir, made it all the more spectacular. Things could only get better, and the early players were impatient for that to happen.

A lack of suitable planting material was one obstacle to improved quality. Of the Swiss clones Frank Berrysmith had imported in the 1960s, 10/5, and to a lesser extent 2/10, proved to be the great survivors. The rest, particularly Bachtobel, were soon shunted sideways towards sparkling wine production for which they were more suited. New Zealand's largest wine company, Montana, had begun planting Bachtobel extensively in Marlborough, Hawke's Bay and Gisborne to cater for an increasing local appetite for bubbles.

One positive development in this area had occurred quite early, in 1977, when Bill Irwin of Matawhero Wines in Gisborne took it upon himself to import four clones from America's seat of vinous R&D, the University of California, Davis. Irwin quietly took possession of Pinot Noir clones UCD5, UCD6, UCD13 and UCD22. Of these, UCD5 — often called the 'Pommard' clone — was the most significant addition. It is widely planted today, prized for its deep colour and tannin.

Hätsch Kalberer, now with the Fromm winery in Marlborough, remembers making the first wine with the new clones. He was 23 years old, recently arrived from Switzerland, when he and Irwin planted an acre of Pinot at Matawhero — half in UCD5, a third in UCD6, the remainder UCD13. A few years later enough was harvested to make the Matawhero Pinot Noir 1987. 'It was stunning and beautiful,' Kalberer says. 'We never went into wine shows so there were no medals. It was one of those wines that never jumped into your face and never disappointed. It was a gift of nature, because over the next few vintages we never did anything again that came close. There's often something about the first vintage with Pinot; as it gets more established it changes its character.'

The next clone 'wave' took place in the early 1990s, when the Dijon clones (113, 114,

115, 667 and 777) were brought into the country from France. The importer of most of these clones was the Wairarapa Vine Improvement Group (WVIG), in essence an association of Martinborough winemakers. The benefits for the region were manifold, says Clive Paton. 'The new clones were planted in Martinborough first, so we got ahead in that sense. Then selling them on helped fund vine improvement in Martinborough, so it was a win-win situation for us.'

Going into the 1990s, the growth curve for Pinot Noir in New Zealand began to steepen; by the end of the decade, it was close to vertical. The area planted in the variety rose from 141 hectares in 1989 to close to 1100 hectares in 2000. A large proportion of these new plantings were for sparkling wine (the Montana Lindauer phenomenon effectively began in 1990 when the inexpensive méthode traditionelle won a major award in London). Until 1997, almost all of the Pinot Noir grapes from Gisborne, Hawke's Bay and Marlborough — well over half the national Pinot crop — were used for sparkling wine.

The spread of more vines across the land was matched by remarkable progress in terms of the quality of still Pinot Noir. Apart from two difficult vintages in the early 1990s (the result of the Mt Pinatubo eruption in 1991), the wines generally became more exciting as each year went by. The most commonly cited reason for that is the annual winemaker get-together that became known as the Southern Pinot Workshop.

The workshop's genesis was a trip made by Larry McKenna to the famous Oregon Steamboat conference. Steamboat, as it's generally known (it's named after a fly-fishermen's haunt called the Steamboat Inn on Oregon's Umpqua River), grew from informal beginnings into an annual conference where winemakers tasted and discussed Pinot Noir.

McKenna attended the 1988 conference, where the idea of emulating the event in New Zealand was raised. In 1991, at Nelson Lakes, the first 'Steamboat New Zealand' took place. It was small, and among those in attendance were McKenna, Tim Finn (Neudorf), Andrew Greenhough (Greenhough), Neil McCallum (Dry River), Clive Paton (Ata Rangi) and Lois Mills (Rippon).

'We copied the exact recipe,' says McKenna. 'It's about peers getting to know each other, tasting each other's wine, explaining how they were made, and everybody critiquing them. The wines create the discussion. No media. It grew to cater for 64 people, but the basic format has been followed for 18 years.'

The openness of the discussion has become the stuff of legend. McKenna concedes that 'a few winemakers have probably felt like sitting down and crying'. Perhaps as a kind of warning to first-timers, the motto 'Leave your ego at the door' has been attached to the workshop, which has never left the South Island. Over recent years it has been held at Hanmer.

Behind the workshop and its free exchange of ideas is the notion that helping one other make better Pinot ultimately helps the entire Pinot-producing community.

It has worked better than anyone hoped. Participants from all over New Zealand credit the event with improving the quality of their wine, particularly through the spread of vitculture knowledge such as correct canopy management. New Zealand Pinot has gained more international recognition as a result.

Another of the event's outcomes has been the camaraderie it has engendered among the country's Pinot Noir producers. Considering its success, it seems remarkable that other varieties have only recently begun to follow suit with similar workshops of their own.

Out of the Southern Pinot Workshop sprang two other major events, wider in scope. The triennial Pinot Noir conference in Wellington began in 2001, and was followed by the Central Otago Pinot Noir Celebration, which is held on alternate years in conjunction with the Wellington event.

Pinot Noir's revival in Aotearoa has been remarkable. In the space of just a couple of decades, the country has become a South Pacific stronghold for the grape and its attendant culture.

FIVE. VITICULTURE & VINIFICATION

When Australian viticulturist Richard Smart told the delegates at the 2004 Wellington Pinot Noir conference that he would show them exactly how to produce quality Pinot, then proceeded to scribble a quick formula on a whiteboard, you could sense an indignant stiffening in his audience. To suggest the secrets of Pinot could reveal themselves via a brief line of jottings … heresy is what you call that.

The divergence of views on how it should be grown and made is probably greater for Pinot Noir than any other grape. Different stylistic objectives, personal beliefs and experimentation all come into play. Generally speaking, however, New Zealand's leading producers have adopted Burgundian methods, though often rearranging them to suit local conditions. What follows is a broad summary of the pathway that leads from planting to bottle.

The phrase is timeless and true: good wine is made in the vineyard. The first step towards creating an exceptional vineyard is finding an exceptional site. Perhaps New Zealand's greatest Pinot Noir vineyards are yet to be found. The best so far are still revealing their personalities, but all share a number of traits: free-draining soils, low rainfall and good heat summation, this last attribute often abetted by way of a north-facing aspect.

Seeking out 'hot spots' to ripen Pinot is particularly important to New Zealand's Pinot Noir-growing regions because of our cool maritime climate. Oregon and Burgundy are also relatively cool, while California and Australia offer a complete contrast; in those places growing Pinot Noir is all about finding cooler locations that offer refuge from the heat.

There are many different opinions on soil in the Pinot-producing community, and New Zealand soils are discussed in more depth in the following chapters. Again, time will be revealing in this regard. For now, Pinot is being grown successfully on a variety of soils in New Zealand, common threads being good drainage and a lack of fertility.

Frost and wind are also important site considerations. Spring frosts have become a big threat, especially to early-budding Pinot Noir. A bad frost burns off the flowers and young buds on the vines, which can massacre a vineyard's whole crop or, at best, lead to uneven ripeness. Autumn frosts — more a problem in Central Otago than in other regions — tend to attack the leaf canopy, forcing immediate harvest regardless of the ripeness of the fruit. That some sites can be hammered more regularly than others in close proximity is often due to naturally occurring protective funnels of cool air, which is where local knowledge comes in. Planting on slopes is a tried and true frost-avoidance method; unable to cling to hillsides, the chilled air drops away freely.

Wind is not as dramatic a menace as frost (and hence less discussed), but is a nagging constant in every New Zealand wine region. Strong winds dehydrate and devigorate the vines, and if the damage to leaves is extensive enough, fruit ripening can be affected. Sites with as little exposure to the wind as possible are always desirable, and shelter belts are a common sight. But the aerating effect of the wind does help keep disease at bay, and some viticulturists point to thicker skins as another 'wind-given' benefit.

Most Pinot vineyards in New Zealand are today planted in several clones, of which there is now a broad choice available in New Zealand, with new ones arriving all the time. Each clone brings differences in taste and structural dimension to the finished wine, while different vintage conditions suit different clones. For instance, one of New Zealand's old originals, the 10/5 clone, is a late ripener that does well in warm years. The Dijon clones ripen earlier, a quality prized both in Central Otago, with its shorter growing season, and in Burgundy where rough autumn weather can hasten the harvest.

If talk of different clones can seem nonsensical to outsiders, discussion around rootstock will double the bewilderment. Grafting vines to American rootstock has become the standard method to make vines resistant to the root-munching phylloxera aphid. As with clones, there are several rootstocks available, carrying names such as 101/14, SO4 and 3309. They vary in things such as their suitability to different soils and their effect on vigour (the viticultural term for the vine's leaf canopy). As with clones, producers experiment to find the best rootstocks for their sites, but it's not a straightforward quest.

'It's hard enough for us to get our heads around,' says Helen Masters, chief winemaker at Martinborough's Ata Rangi. 'We plant different combinations of clones and rootstock, and when you add year to year variation, it's very difficult to know exactly what is causing which effects.'

Vine density — the spacing between the vines — is a serious issue for Pinot Noir producers. The early plantings in the 1980s for all varieties in New Zealand followed a standard pattern: a wide 3-metre corridor between the rows and a generous 2-metre gap separating the vines in each row. Increased contact with Burgundy, where only a metre separates rows and vines, has seen all serious New Zealand producers move towards increased vine density. The theory is that the closer the vines are packed together, the greater the competition between the roots, forcing them to go deeper in search of nutrients much faster than they would without such near neighbours. The competition also serves to lower each vine's fruit production by natural means, thereby increasing fruit quality and intensity.

However, most New Zealand Pinot growers are loath to go as far as Burgundy's tight 1 metre x 1 metre spacing. The reasoning behind this offers an insight into the difficult balancing act that governs viticultural decision-making.

It starts with the need in New Zealand to train fruit at higher levels. Frost and disease pose a real risk to low-hanging fruit in this country, especially on flat ground, so our fruiting wires are mostly set higher than in Burgundy, where lower is seen as better as it helps fruit ripen earlier. That imperative is not as strong in New Zealand, with our more reliable autumn weather. But raising the height of the fruiting wire in turn means having a higher leaf canopy, for every vine requires a certain leaf area in order to function properly. When you combine those high canopies with narrow, metre-wide rows, you get shading. That is a serious consequence; leaf and fruit exposure to sunlight is too important to compromise. So the gap between New Zealand rows remains, for most growers, at 2 metres.

Canopy management — keeping the leaf canopy trimmed, upright and opened up to the sun and air — is critical. Leaf plucking is a midsummer task, done by hand, to ensure a key ripening agent, sunlight, has direct access to the fruit.

Nothing, though, is more critical than yield. The price of avarice with Pinot Noir is high. The grape's inherent phenolic (crystalline solid and aromatic compound) lightness becomes overstated if it is overcropped, the result being light, astringent, often unpalatable wine. Lower yields mean greater concentration and colour through higher levels of phenolics, and, importantly, good tannin ripeness and developed aromatics. Another advantage of lower cropping is earlier ripening.

In some regions, low crops often occur naturally. Where that doesn't happen, clusters are dropped later in the season in what is sometimes called a 'green harvest'. Makers of fine Pinot Noir generally aim for a yield of 4–5 tonnes of fruit per hectare. The contrast with Sauvignon Blanc is stark; typically Sauvignon growers in New Zealand harvest twice that volume.

The last of the big decisions in the vineyard is the question of when to pick. New Zealand's normally settled autumns are a precious asset here, often taking the weather out of the equation and allowing growers to harvest according to ripeness alone. Over the last decade, more New Zealand vignerons have adopted the Burgundian approach, being principally guided by taste, using the numbers gained by instruments (for example, the brix, a measure of sugar levels) as back-up only. Skins are tasted for the flavours winemakers are seeking, as are seeds.

'Pinot Noir is all about seed tannins,' explains Matt Donaldson of Pegasus Bay Wines in Waipara. 'In that regard it's the opposite of Cabernet Sauvignon, which gets 60% of its tannins from its skin, and not much from the seeds. With Pinot it's the other

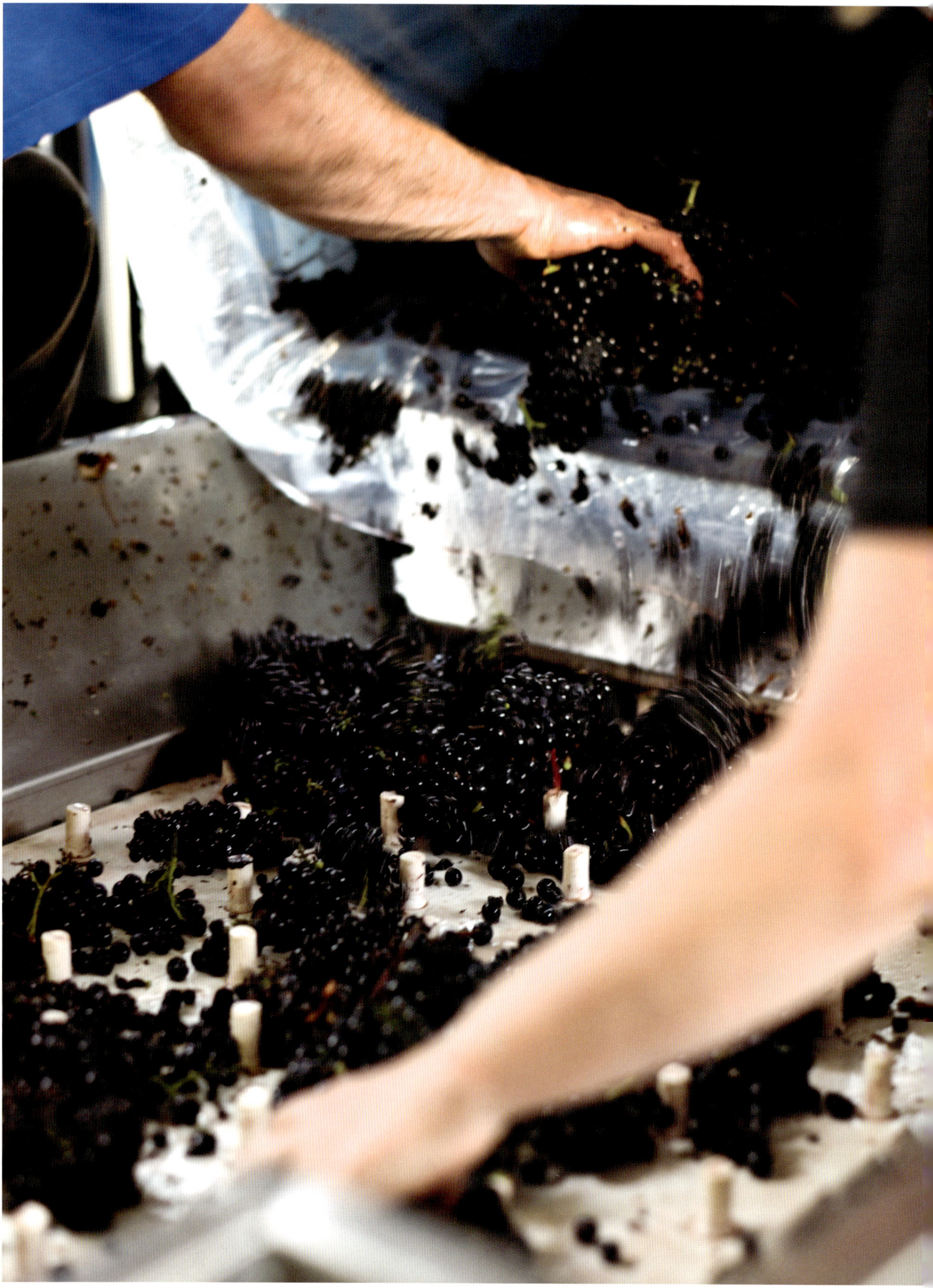

way round. It's so important to get ripe seed tannins. As the season goes on they get crunchy, nutty and smooth. You don't want them overripe — then the tannins can get too aggressive.'

High sugar levels are a given in New Zealand vineyards. By the time a winemaker is satisfied that the grapes have achieved what is often called 'physiological ripeness', sugar can be in excess of 25 brix. This leads to high alcohol levels — an ongoing conundrum for New Zealand winemakers. Because so many factors contribute to this, there is no quick fix. Pinot producers are among those trying various strategies to curb sugar creation, one being reducing the canopy size to a minimum. As the leaves are where sucrose is created before being passed on to the berries, the thinking is that fewer leaves will mean less sugar.

The vinification process, which begins once the grapes arrive in the winery, is remarkably simple.

The harvested fruit often has to run the gauntlet of a sorting table, where it is scrutinised and any unwanted grapes or bunches are passed out. Then it must be decided how much, if any, of the fruit will be left to go through fermentation as it came off the vine, the bunches intact. All wineries have destemming machines which separate stems from berries, but the option to go with 'whole bunch fermentation' is a stylistic choice. Ripe stems lend a spiced perfume to the wine as well as structural elegance through long, fine tannins. It is common practice in Burgundy, but most New Zealand Pinot producers tread more carefully, stem ripeness being the issue.

'I'll only use whole bunches on blocks where I like the tannin ripeness,' says Helen Masters of Ata Rangi. 'I chew on the stems and if there's too much greenness, we'll destem. In cooler years, I think most people in New Zealand destem.'

The next step is the 'cold soak', where the fruit sits in vats typically at 8–10°C, cold enough to hold off the fermentation process. The cold soak can last up to 10 days, with the downward pressure of the fruit, together with plunging (in small wineries this is done by hand with long tamping poles), releasing increasing amounts

Left: Harvested fruit is closely scrutinised on the sorting table. Following pages: p74 The denuded stalks' next stop is the compost heap; p75 Elbow-deep in Pinot fruit; p76-7 Hand plunging the vats requires a strong back and a good work ethic.

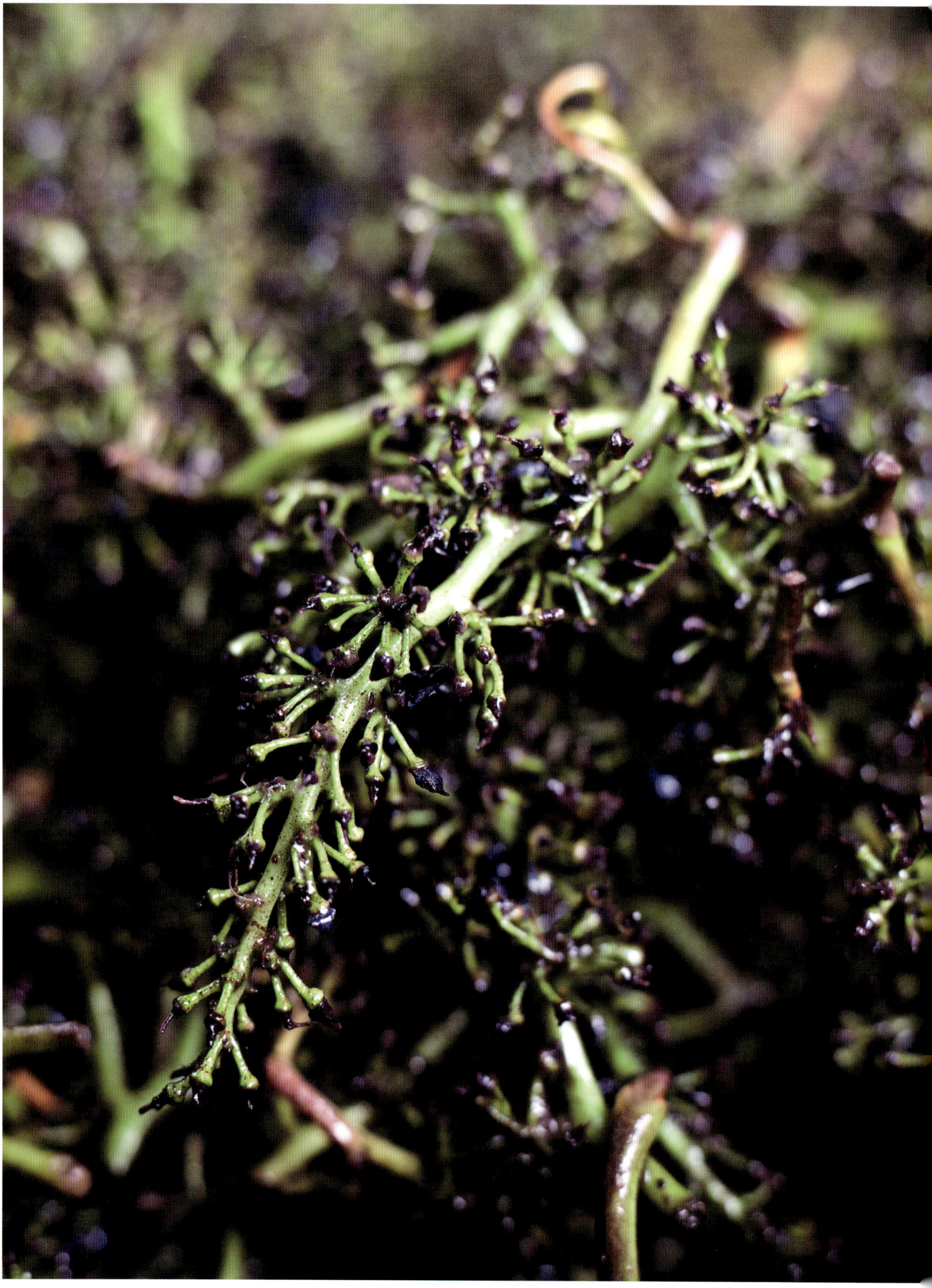

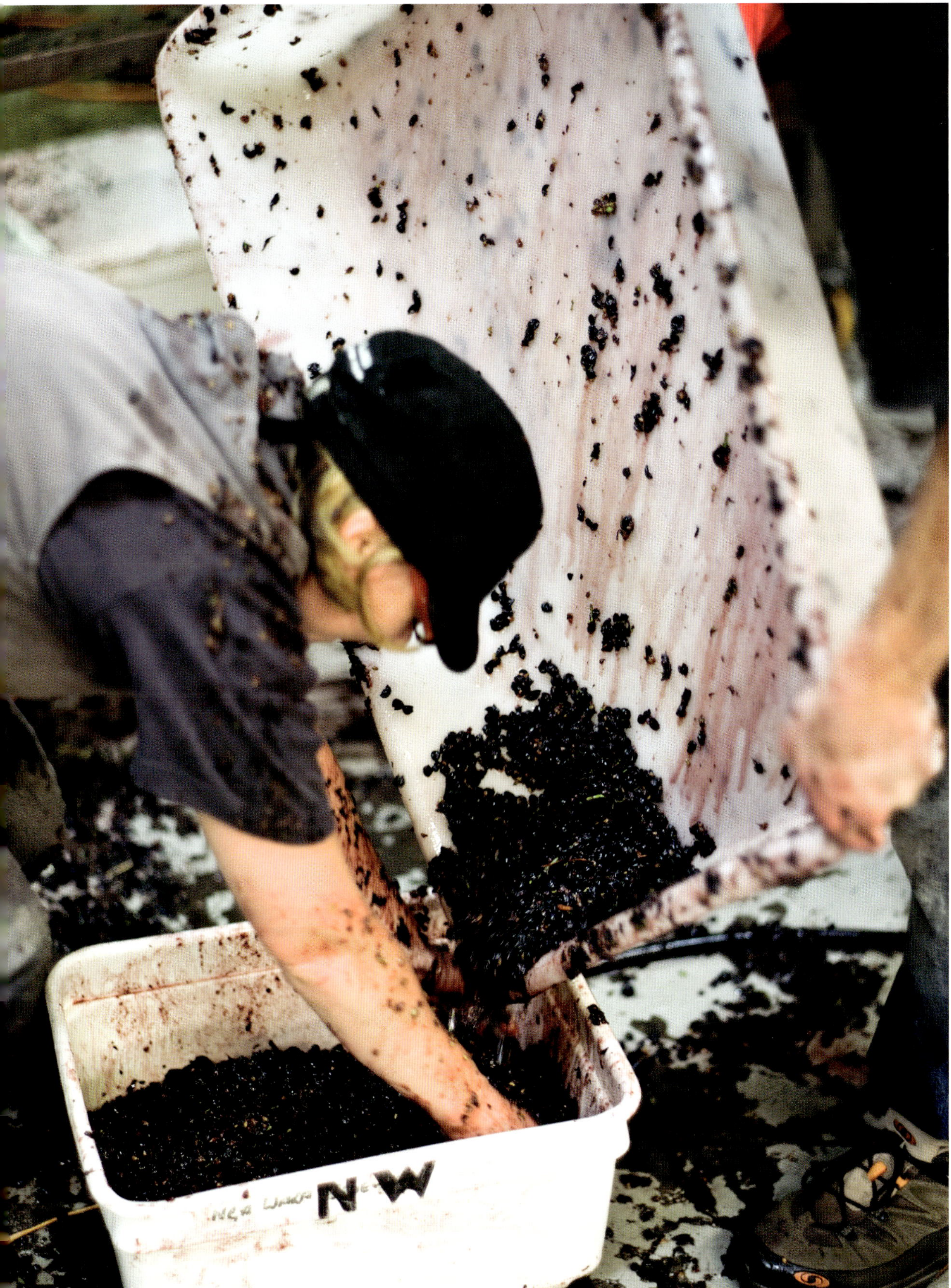
N-W

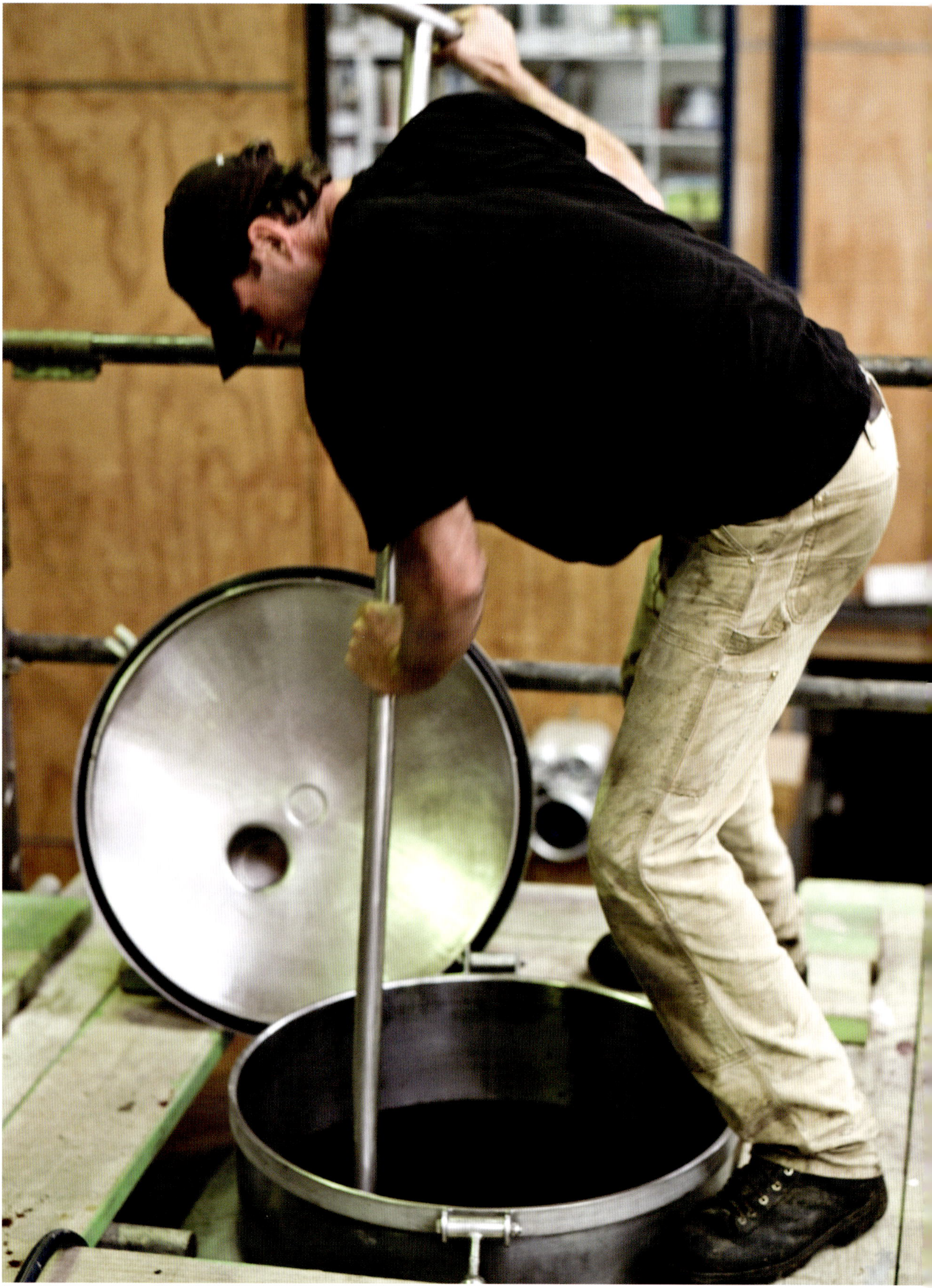

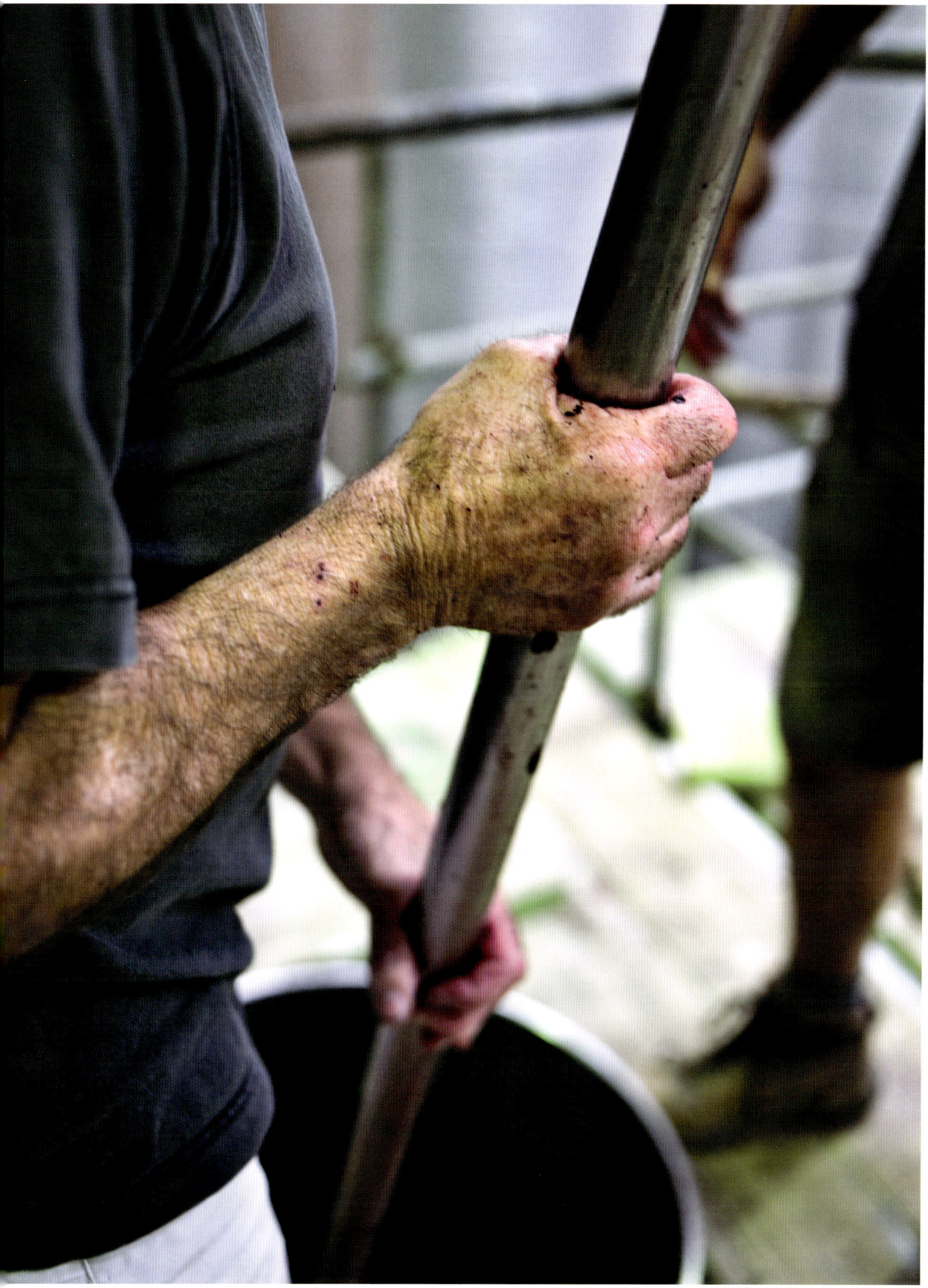

of juice. The reasons behind it are, once again, stylistic. 'The cold soak is an aqueous soak, as opposed to an alcoholic soak,' states Larry McKenna. 'It gives you rounder tannins, denser colours — especially purples — and increased fruitiness.'

The winemaker decides how to trigger fermentation. Many simply warm up the vat and allow the presence of indigenous yeasts to go to work transforming the grape sugar to alcohol. The more hands-on approach is to use cultivated yeast to inoculate the must (as the grapes and juice are called). The proponents of indigenous yeast maintain a more individual, funkier wine results, and possibly lower alcohol levels.

Fermentation temperature is also a matter of choice; some prefer cooler ferments of 16–21°C, others as hot as 35°C. It is generally thought that the hotter the ferment, the richer and bigger the wine.

The ferment usually lasts between five and 12 days. During this time the vat becomes a roiling deep-purple swamp with carbon dioxide, a by-product of the process, bubbling to the surface. The skins, seeds and possibly stems all float to the top, creating what is called the 'cap'. This is where the colour, flavour and tannins all reside, and in order to extract those elements, pushing the cap back down to make contact with the juice becomes an important, regular task during the ferment. In larger wineries, this is achieved by mechanised methods. In smaller operations, as with the cold soak, the ferment is hand-plunged. It is hard work, but is regarded as being gentler on the must. At a few New Zealand wineries, where the vats are small and open-topped, you may see cellarhands climb in and wade around, using their legs to break up the cap. This is an ancient Burgundian practice, which the French call *pigeage*. But care must be taken: the carbon dioxide fumes are, of course, toxic. Death resulting from carbon dioxide poisoning in poorly ventilated wineries is an irregular occurrence in the wine world. In New Zealand, it claimed the life of talented winemaker Mike Wolter of the Central Otago Wine Company in 1997.

In most cases, the winemaker waits until all the grape sugar has been converted to alcohol before taking the next step. For some, this will be a post-fermentation maceration — a few days during which the juice and skins simply sit, the aim being to extract more tannins and colour.

By now it will be apparent that 'extraction' plays a central role in the vinification of Pinot. It is the reason the juice will typically have four weeks' contact with the grape skins. 'Pinot's tight bunches and thin skins means we [Pinot producers] experiment with, and understand extraction more, than other red wine producers,' says Larry McKenna.

The must is then pressed and the liquid that flows from the press can now be called wine. It then leaves cold stainless steel behind for the warm caress of French oak.

Oak barrels are a major expense for any winery, a single 228-litre Burgundian *pièce* costing up to $2000. But they are necessary for their singular influence on the taste and structure of a wine. The names of the famous Burgundian coopers — Remond, François Frères, Cadus, Mercurey and Damy, to name a few — have become a familiar part of the New Zealand Pinot scene, and each winery has its favourites. Different forests, different coopers and different levels of 'toast' (the purposeful charring of the barrel's interior) all impart an imprint of their own. How much of the wine is put into new barrels, where the influences are most overt, is an important decision. In Burgundy, a lot of top Pinot Noir is given 100% new oak treatment. Their wines tend to be naturally muscular and powerful, and are thus able to handle it. New Zealand's lighter styles would be swamped by such a heavy oak régime.

The wine normally remains in barrel for 12 to 18 months. Greta Garbo's 'I want to be alone' entreaty applies to Pinot at this juncture, and the barrels are best left undisturbed. The wine has still to go through malolactic fermentation, a secondary fermentation that softens the wine by transforming tart malic acid into lactic acid. Winemakers with a hands-off philosophy allow malolactic fermentation to occur naturally in barrel, which usually happens when the warm spring weather arrives. Others prefer to get it started earlier — sometimes when the must is still in the tank — by adding cultured malolactic bacteria to trigger it.

Once the wine comes out of the barrel it is fined, often with egg whites or milk. This clarifies the wine and also eliminates hard tannins.

To filter or not to filter is the next question. Filtration brightens the fruit characters, but can lower the wine's life expectancy. A lot of the best Pinot is unfiltered, with a resulting dense, slightly cloudy appearance. The aromas are more funky, the texture more interesting on unfiltered wines.

And so to the container we know so well: the 750ml glass bottle. While the overwhelming majority of New Zealand Pinot is bottled under screwcap, a small number of very good producers (such as Dry River in Martinborough, and Prophet's Rock in Central Otago) still use cork. Others, such as Rippon in Central Otago, use cork composite. The closure debate is ongoing. There will be new twists over the years ahead, without doubt.

....

Right: Skins and seeds are the key sources of tannin. Following pages: p84 The business end of a hard plunger; p85 The ancient Burgundian practice of pigeage *is still used at some New Zealand wineries today.*

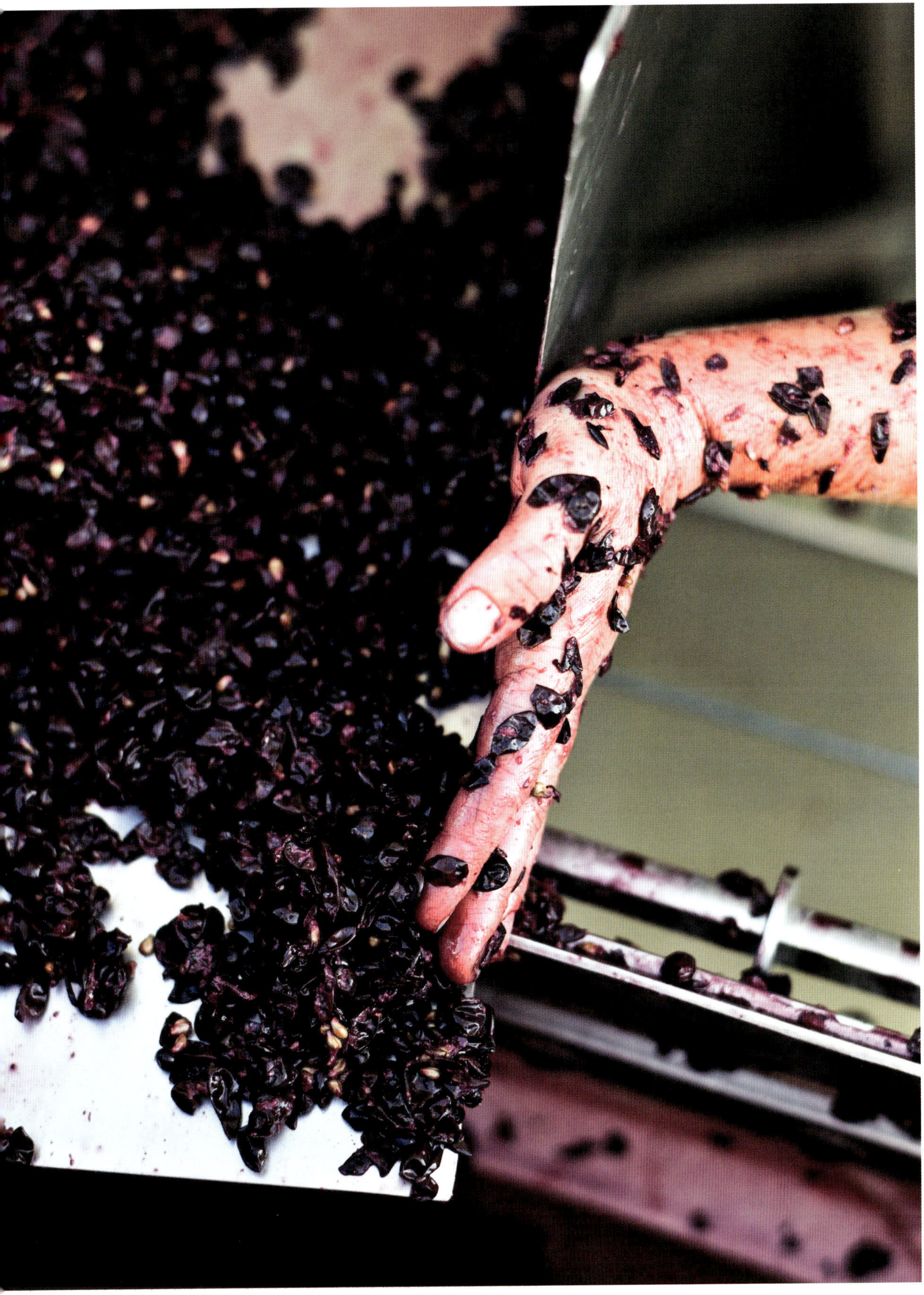

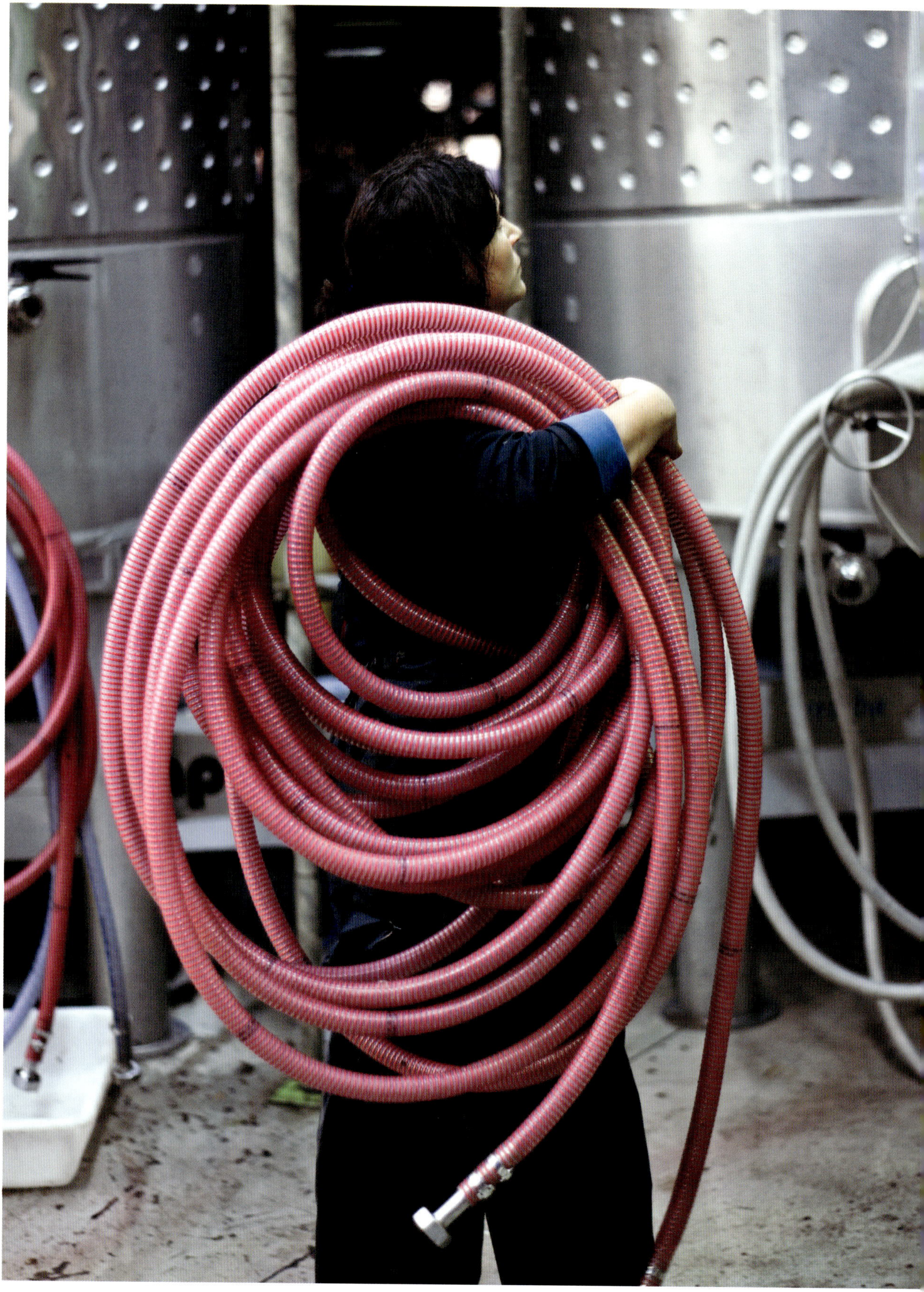

08 PN Buck 1
24/3. Seriously Nuts

Lovat + Block 7
V5
21/3 Harvested Lovat PN
" Babies
Block 7
3T
Crushed w 40ppm SO_2 + 1g/L H^+
pH=3.08 FSO_2=?
Bound=?
22/3 pH=3.12
FSO_2=8.0
Brix=23.2°
BSO_2=9.6
Brix=23.4°
23/3 pH=3.20
FSO_2=5.6
24/3 pH=3.10
Brix=23.4
25/3 pH=3.22
Brix=24
FSO_2=5.7
26/3 pH=3.23; Brix=23.8
FSO_2=1.6;

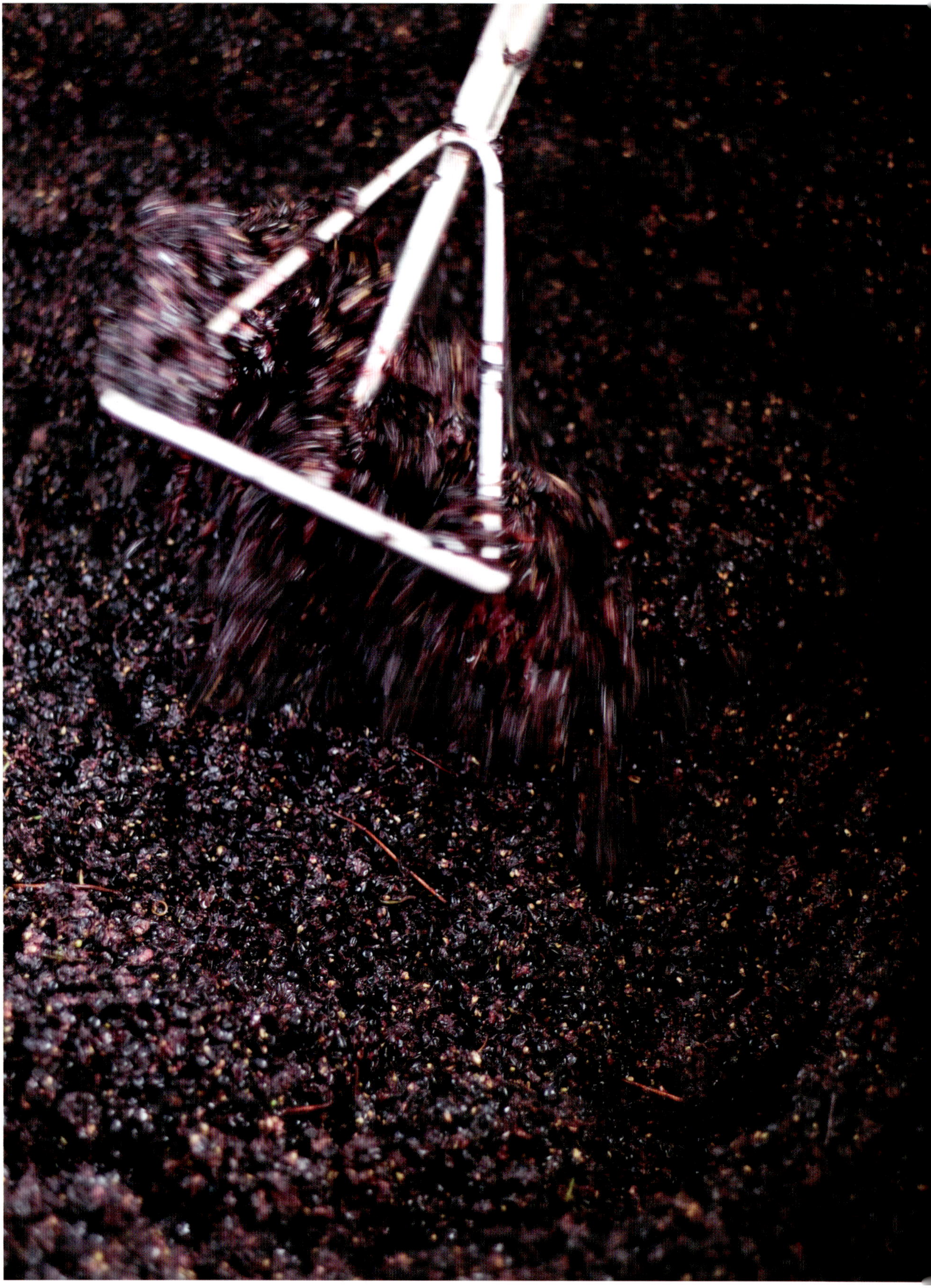

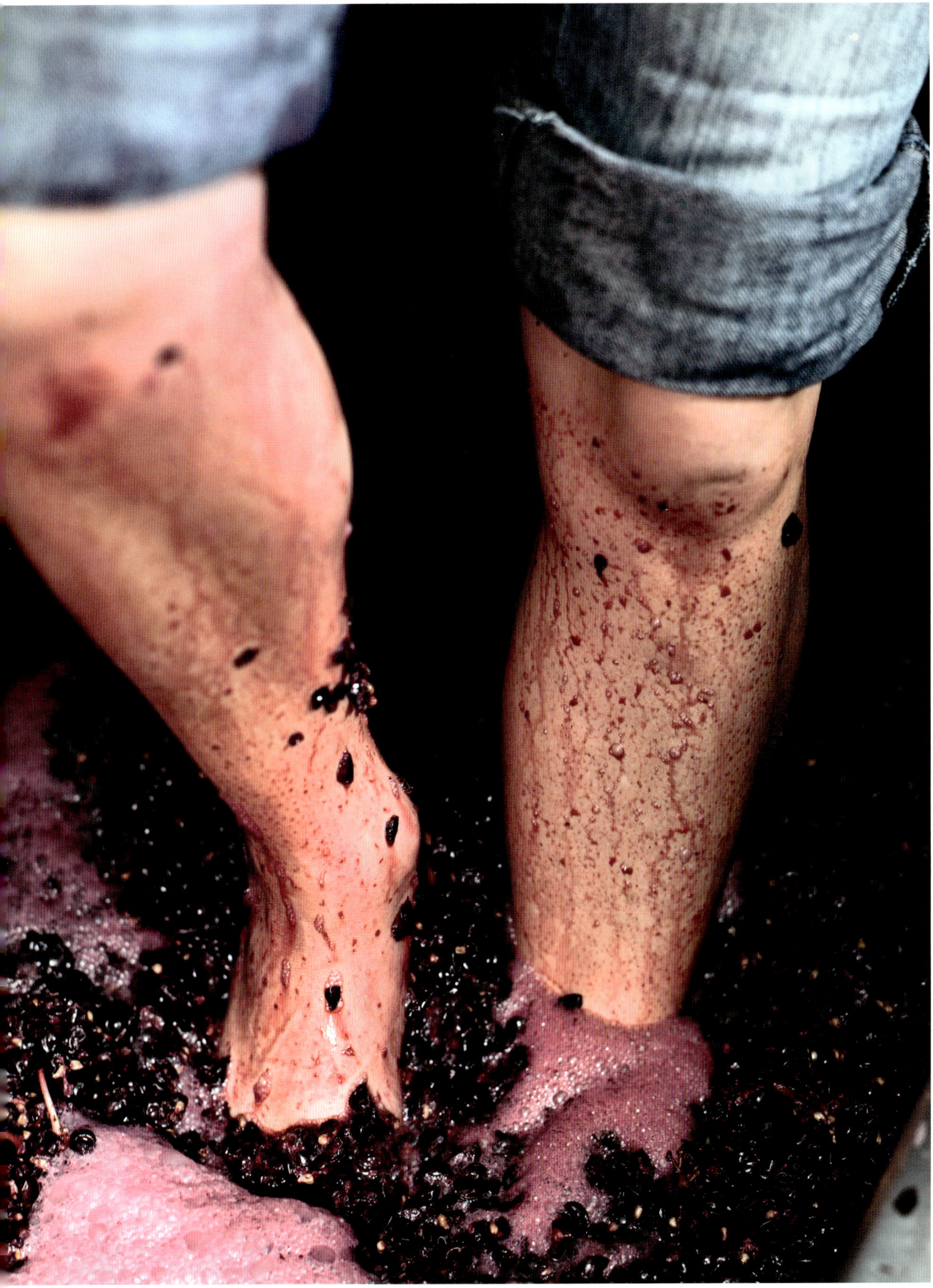

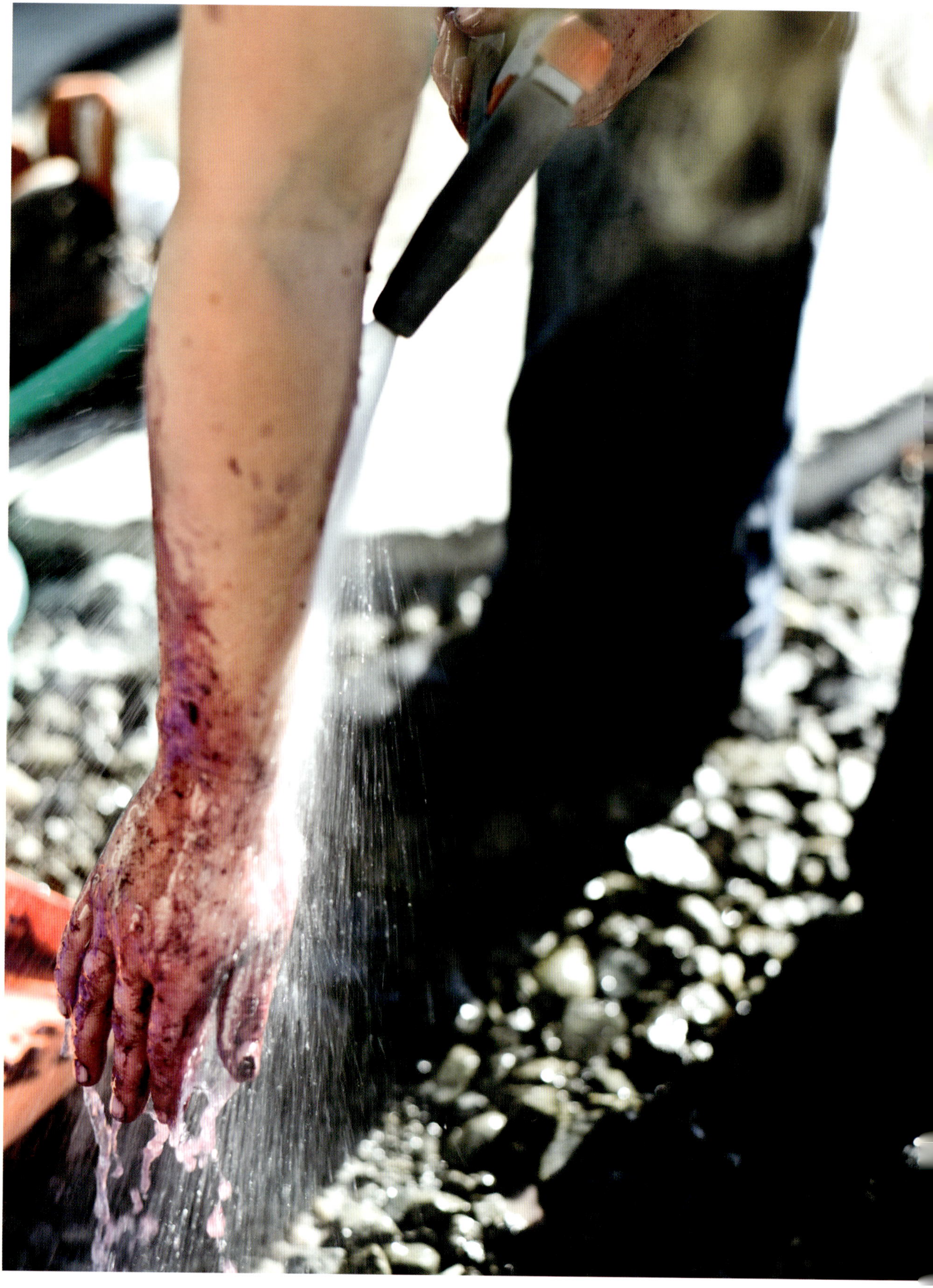

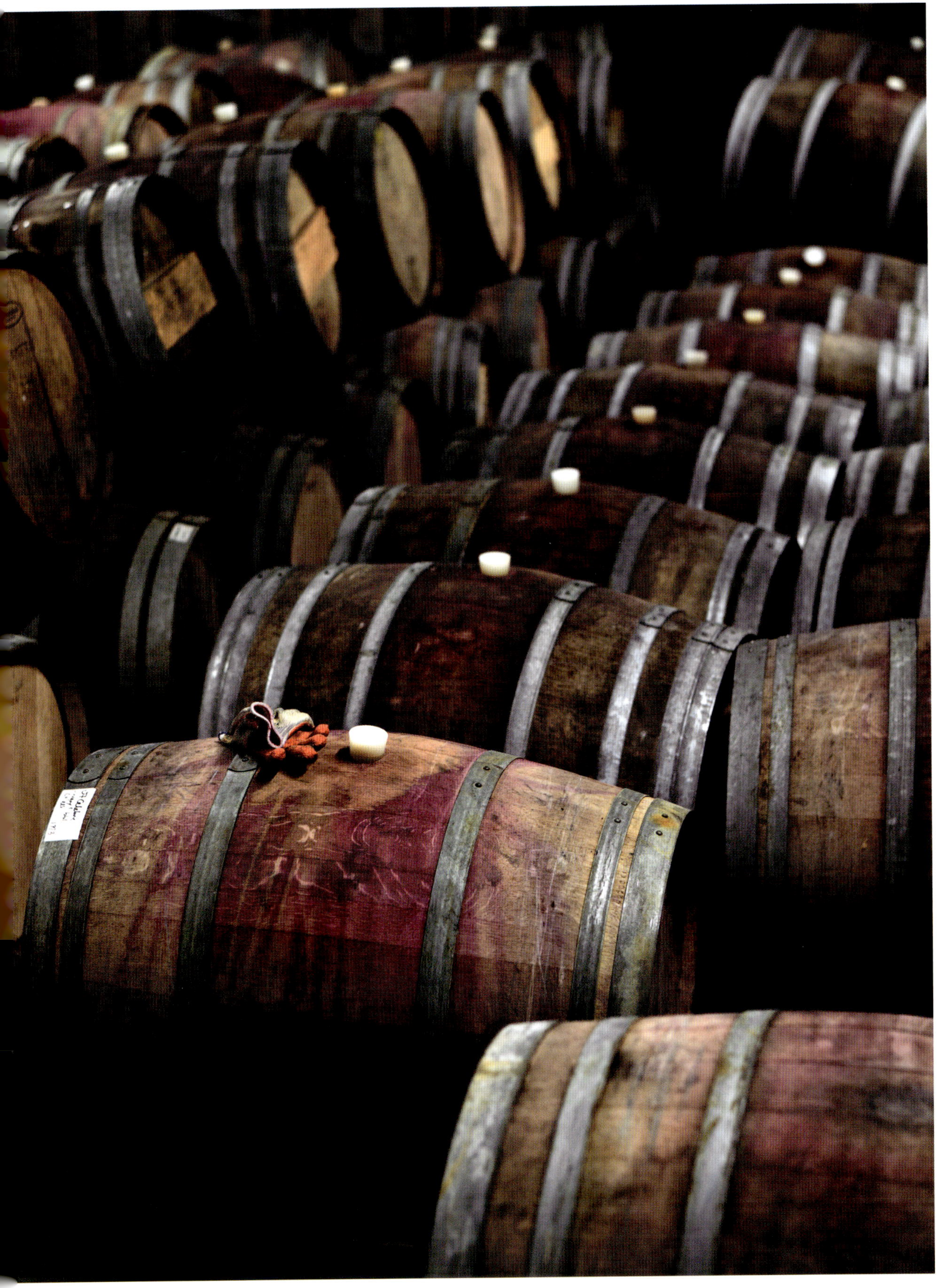

SIX. TERROIR AOTEAROA

TODAY PINOT NOIR IS NEW ZEALAND'S MOST WIDELY PLANTED RED GRAPE, BY A LONG WAY. IN 2009, THERE WERE 4777 HECTARES PLANTED IN PINOT, OF WHICH 88% WAS BEING GROWN FOR THE PURPOSE OF MAKING STILL PINOT NOIR. ACROSS ALL VARIETIES, PINOT IS SECOND ONLY TO NEW ZEALAND WINE'S CENTRAL WHITE FOUNTAIN, SAUVIGNON BLANC, WHICH COVERS 16,204 HECTARES OR 52% OF THE NATIONAL VINEYARD. PINOT'S SHARE OF THE WHOLE IS 15%.

The period of explosive growth that began in the mid-1990s continued on into the new millennium. Led by Central Otago, a 790% increase in Pinot plantings nationally took place between 1996 and 2007. Since then, the growth rate has receded to saner levels.

Of the plantings of Pinot in New Zealand, 87% are grown in an area extending between the Wairarapa in the lower North Island and Central Otago. This temperate zone falls between the latitudes of 34° South and 47° South.

Between them, two South Island regions, Marlborough and Central Otago, account for 66% of the nation's Pinot vines. They are followed by a cluster of smaller regions — principally Wairarapa, Nelson and Canterbury/Waipara. Then comes a group of disparate outposts, of varying sizes.

Of significance in this third tier is the Waitaki Valley in North Otago — a grand, mountain-flanked furrow along which the much-dammed Waitaki River runs eastward towards the Pacific Ocean. In the year 2000, the Waitaki Valley's imminent arrival as a wine region was announced, with the bold prophecy that by 2003 it would be the 'third or fourth largest grape-growing area in New Zealand'. Behind the big talk was a property-development concern spearheaded by South Island entrepreneur Howard Paterson, which had snapped up 2000 hectares on the valley's southern side, between the towns of Kurow and Duntroon. North-facing slopes, good heat summation, a long growing season and, above all, limestone soils, were all cited as vital ingredients in making the Waitaki Valley New Zealand's next Pinot Noir hotspot.

Some big names were immediately attracted to the area, among them John Forrest (Forrest Estate), Steve Smith (Craggy Range), Michelle Richardson (Richardson) and Grant Taylor (Valli), who have all produced Waitaki Valley Pinot Noirs under their own labels. The first vintage of significance was 2004. It showed undeniable promise, with a rich savoury depth evident in some of the wines. Since then, the area's marginality has bedevilled producers. So low were temperatures at flowering in December 2006 that next to no fruit was harvested in 2007. The Waitaki Valley, then, has not yet risen to the hoped-for heights, but remains an intriguing prospect.

North of the Wairarapa only a tiny number of Pinots attain a quality approaching that of their southern counterparts. One of these is from Millton Vineyard in Gisborne. In good years the Millton Clos de Ste. Anne Pinot Noir, grown in the elevated, sloping Naboth's Vineyard, shows Santenay-like grace and tension. In Hawke's Bay, John Hancock's Trinity Hill High Country Pinot Noir can also surprise with its verve and 'pinosity'. Cooler sites, often at higher altitudes, are the key to success in these regions.

What is so attractive about New Zealand Pinot Noir? Its most obvious distinguishing feature as a drink is the vibrancy and intensity of the aromatics and primary fruit flavours. Pinot is not alone here — exuberant fruit characters are a trait common to all New Zealand wine. In Pinot's case an uninhibited burst of sweet floral scents followed by forward, often spiced, fruit flavours offers something deliciously approachable. These flavours are brightened by a firm acidity, often a fleshy mouthfeel. Depth and structure are evident in wines with more vine age from well-tended vineyards. Increasingly, New Zealand's better expressions compare well to reputable Burgundies.

Beyond this overview lies a more complex picture of emerging regional and, sometimes, subregional characters. There has been a premature eagerness to make much of New Zealand Pinot's varying regional accents. There are certainly differences emerging, but it is still difficult to correctly identify the region of origin of a group of New Zealand Pinots at a blind tasting. Viticulturist/winemaker experimentation and influence, together with the youth of the vines, help blur the boundaries. The exception is Central Otago Pinot, which in its short history has shown distinctive qualities.

It isn't possible to offer any explanation of what lies behind New Zealand's capacity to produce Pinot Noir of quality without entering the sometimes hazy world of 'terroir'.

Terroir is a French term (said by British wine writer Hugh Johnson to have originated in Burgundy) that has engendered much debate over recent years. Broadly speaking, it is the sum of all the environmental factors that make a wine turn out the way it does. Most definitions of terroir cite four key, contributing elements: the land (mainly soil and aspect), climate, grape variety and human involvement.

The notion that a wine's personality is shaped by the place in which it is grown, and by the people who grow it, is perfectly logical and easy to grasp. All living things are to a certain extent products of their environment. It's also easy to see the concept's appeal from a marketing perspective. Everyone likes to differentiate their product and that is exactly what terroir does. Terroir is a way of saying USP (unique selling proposition) in more poetic language.

But there is debate in regard to how the various factors impose themselves on the finished wine. Some say the actual taste of the soil can find its way into the glass, while others dismiss such claims as fanciful. There is also the question of the human factor. If a winemaker is insensitive to the natural environment and applies too heavy a hand or — worse — some preconceived recipe, the result will never be a *vin de terroir*. 'As long as you work in harmony with nature, you can find terroirs all over the world that produce great wines,' says Anne-Claude Leflaive of Burgundy's Domaine Leflaive. 'The role of the producer is to act as a kind of musician, interpreting the music of the terroirs.'

Right: Pinot plantings in the Gibbston Valley, Central Otago.

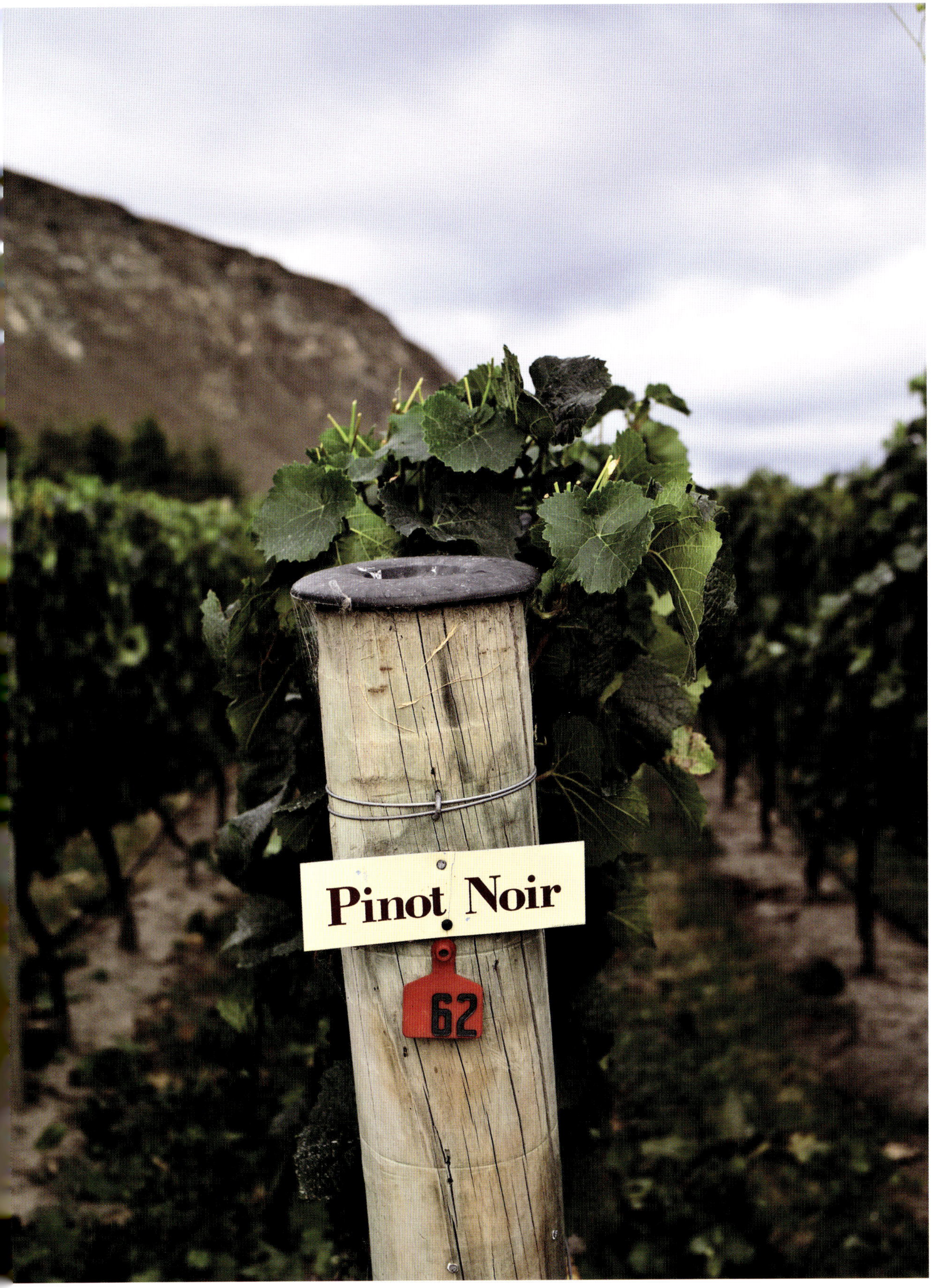
Pinot Noir
62

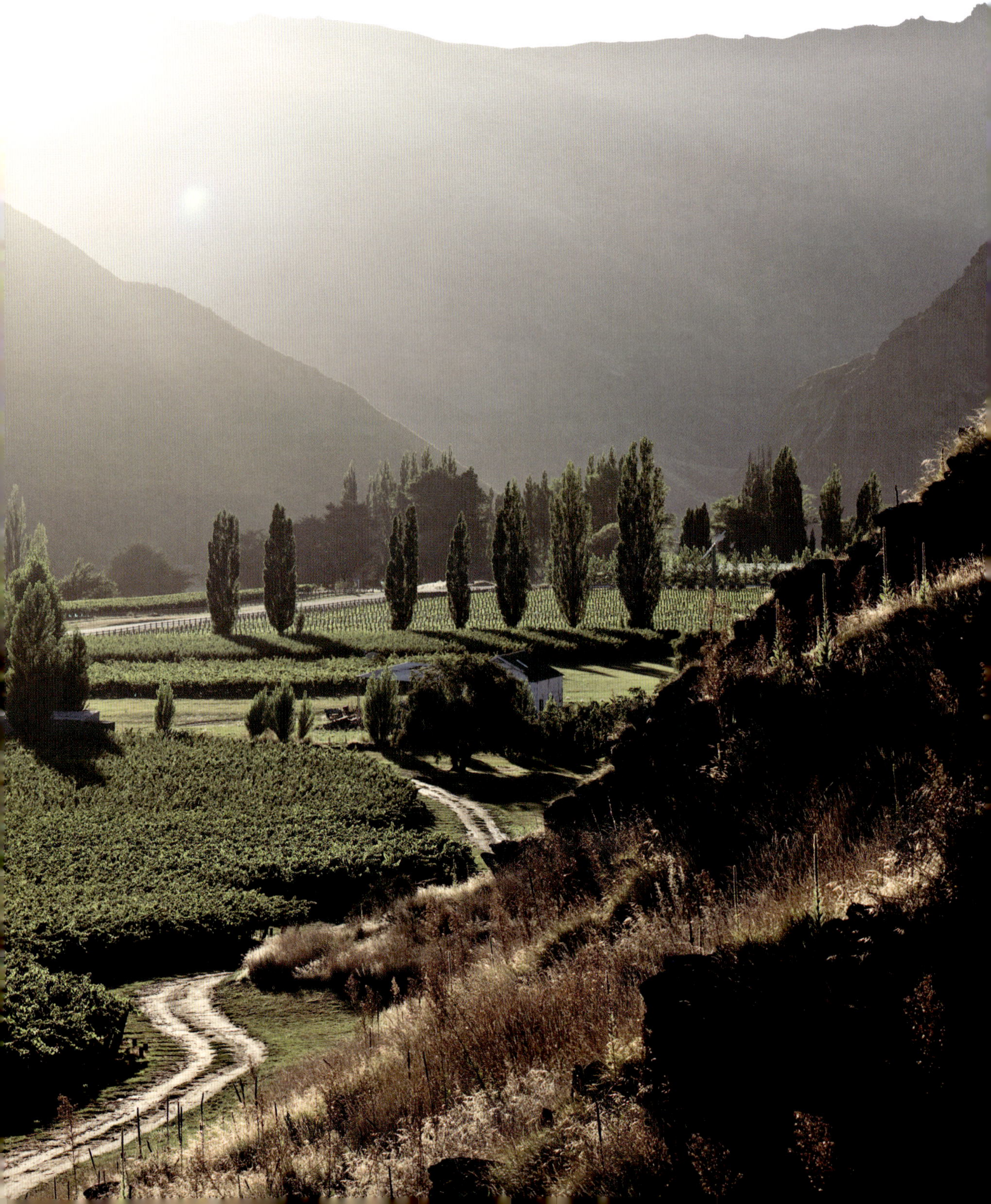

Pinot Noir, with its famous capacity to express its place of upbringing, is arguably the ultimate terroir grape. The Burgundians would say (and many have) that it's still too early to start talking about terroir in New Zealand. Our young vines with their undeveloped root systems are not yet capable of interpreting the voice of the land. And our winemakers, even those who have notched up a twentieth vintage, have so much more to learn about their vineyards.

That hasn't prevented the discussion from starting. For wherever there's good Pinot, there's a desire to find out what makes it so.

Looking at soil first, we can say that Pinot Noir seems comfortable growing in a range of soils in New Zealand. The free-draining gravels on the Martinborough terrace, the clay of the Moutere hills, the clay-cum-limestone of Pyramid Valley near Waipara, the sandy schist of Alexandra — all have produced splendid Pinots. This aligns with recent French studies that have found no connection between wine quality and specific soils.

In terms of assessing the precise influence of these soils on the character of the wine, a picture is starting to emerge. An obvious example is the fleshiness and roundness of wines grown in Southern Valley clay in Marlborough compared with their leaner, lighter neighbours grown on the alluvial gravels of the Wairau Valley floor. But there's so much more to learn in this area. Soil is slower than the other factors in the terroir equation to reveal its hand.

Emeritus Professor Warren Moran made the point, in his excellent paper on New Zealand's terroir delivered at the 2001 Wellington Pinot Noir conference, that over the early period of our Pinot story, 'climate has been more important than soils in ensuring wine quality for Pinot Noir'.

Terroir Aotearoa certainly offers the 'close call' climatic conditions the fringe-dwelling Pinot Noir needs to excel. New Zealand is a windswept archipelago located in the Roaring Forties. It is sometimes said, in jest, that the country doesn't have a climate — it just gets weather. There is a wild, dynamic, unpredictable aspect to the country's meteorology.

Most of the weather we do get comes from the west, and that's where the country's

Left: The Chard Farm home vineyard sits on a ledge in the Kawarau Gorge.
Following page: p95 The Clayvin Vineyard, one of Marlborough's key Pinot sites.

mountainous spine comes in. All our significant Pinot regions are sheltered from much of the rain arriving from the west by the mountains, giving the vineyards the requisite dryness (800mm rainfall a year is regarded as the upper limit). This effect also allows enough heat to take hold to ripen the grapes. Chilling nights provide the diurnal disparity in temperature that both promotes and protects the flavour development so necessary for Pinot. The fact our best Pinot vines are either under snow or in sight of it during winter is also telling. Pinot Noir likes a cool climate's coolest parts.

Which brings us to the role of *Homo sapiens*. For Warren Moran, this is terroir's driving force, but is too often downplayed: 'The soil exists but the terroir arrives when somebody makes an expressive wine from grapes grown in it. Without people and wine the word terroir would not exist.'

People are the key. It may be going a little far to describe New Zealand's leading Pinot producers as the Cistercians of the South Pacific, but there are similarities in the intelligence and dedication they have brought to their exploration of what Pinot Noir can achieve in this country. I've already alluded to the fact that Pinot attracts a certain type of person: inquisitive, uncompromising, delightfully obsessed. We seem to have a ready store of such characters in New Zealand.

There are profound differences between the rise of New Zealand Pinot Noir and that of Sauvignon Blanc. The latter piqued the interest of London's influential wine cognoscenti as a newborn, when no more than a handful of wineries existed in Marlborough. As a consequence, for more than two decades (until the bountiful, difficult 2008 vintage), demand outstripped supply.

Pinot Noir's early promise did not trip off the same kind of critical heatwave and instant popularity. Positive responses and wine show triumphs were certainly registered, but there was also slight regard expressed in some quarters. Robert Parker, the world's most influential critic, has dismissed New Zealand Pinots a number of times over the years as being 'green and vegetal'.

Belief in what the variety could achieve in New Zealand largely rested during the 1990s with a growing band of small producers and their equally small local followings. People were planting it and making it principally because they had themselves fallen under its spell. The motivation was personal rather than commercial.

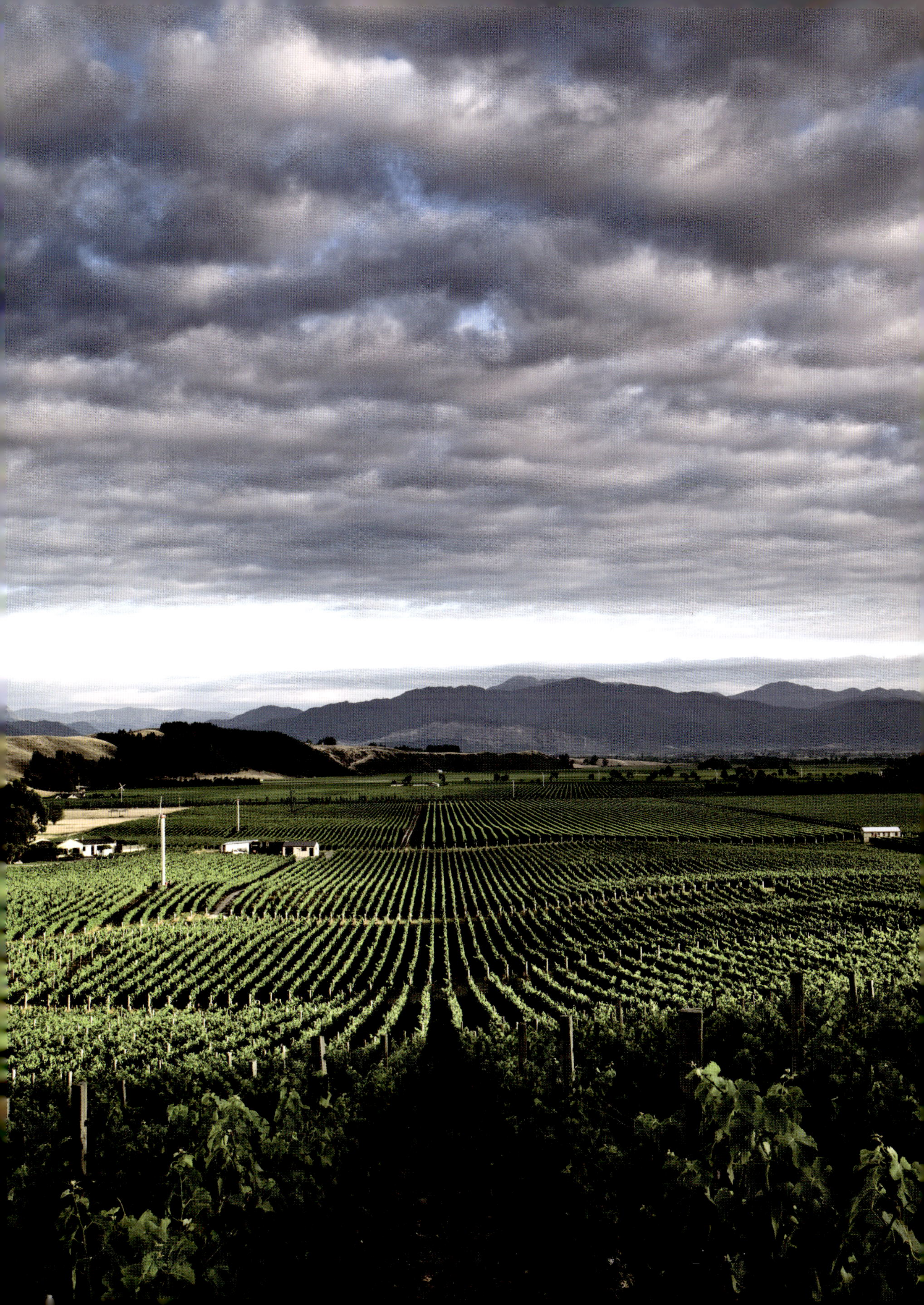

At the first Wellington Pinot Noir conference in 2001, there was a palpable current of unease concerning the amount of Pinot New Zealand was producing, when little hard thought had been given about how it would be sold.

The call had gone up at the event for a larger producer (read Montana) to produce an inexpensive, 'entry level' Pinot Noir in supermarket quantities. This would serve to introduce consumers in overseas markets to the joys of New Zealand Pinot, so they would theoretically become interested enough to move up to the costlier, more serious wines made by the boutique wineries. The parallel was made with Sauvignon Blanc where punters were able to develop a taste for the variety via the Montana standard Sauvignon and then aspire to Cloudy Bay.

Montana duly obliged with its release the following year of the Montana Marlborough Pinot Noir 2000, which retailed in New Zealand for $15.95 a bottle. A local wine writer called it 'the wine everyone in the world has been waiting for'. It was light and juicy in style and possessed good varietal definition for such an inexpensive Pinot.

It sold well and certainly put the 'New Zealand Pinot Noir' brand out in front of more consumers. But it never became a Sauvignon-like sensation. That's because it could really only offer a hint of how great Pinot could be, and also because the two varieties and their followings are just too different for one to expect that the success of Sauvignon would be replicated by Pinot. Simply put, one has fans, the other disciples.

And yet New Zealand Pinot Noir export sales began to rise steadily in the wake of the conference. Between 2000 and 2007, exports of New Zealand Pinot rose from 0.84 million litres to 5.88 million litres — a hike of 593%. The key markets, as with all New Zealand wine, were the UK, the US and Australia. Behind this lay a number of influences in addition to Montana's contribution. There were successful marketing events such as the one that Central Otago producers staged in London to showcase their breakthrough 2002 vintage. There was the dumb luck of the stunningly successful movie *Sideways*. And not least there was burgeoning quality, across the board.

The differences in tone between that 2001 Wellington conference and the ones held later in the decade were telling. The words of British wine merchant Jasper Morris seemed to have been heeded. Years earlier he had told the gathering, 'Great New Zealand Pinot Noir will come when people worry less and relax more. You need to go with the flow. You must be brave enough to do nothing.' In 2007 and 2010 there was a confidence that had been lacking in 2001. Comparisons with Burgundy were fewer, as were cringe-making discussion topics such as where New Zealand Pinot might rate on some mythical world league table. What many had known for some time was now commonly accepted: for between NZ$30 and $60, you'll have a superior Pinot experience buying a New Zealand wine rather than one from Burgundy at a similar price.

What hadn't changed was a sense of impatience. Buoyed by their progress, producers were eager to go further. The beneficial effects of vine age (older vines provide more complexity and structural depth) were just starting to be felt at some wineries. Speculating what vine age will do for their wines is a favourite pastime among New Zealand Pinot producers.

Even Robert Parker — or at least, his helper Neal Martin — has come round to New Zealand Pinot Noir. Martin was dispatched by Robert Parker.com to conduct a major report on New Zealand for the website in 2007. As a result, New Zealand wine entered Parker's orbit for the first time. Three Pinots received 95 points, the highest score Martin gave any New Zealand wine — one from Felton Road and two from Rippon Vineyard.

And yet it would be wrong to see the New Zealand Pinot Noir phenomenon as a friction-free, sunlit ascent to some promised land. Many in the industry are concerned with aspects of the fallout of the recent explosive growth.

'In my observation the best and most interesting Pinots in the various New Zealand regions still come mainly from the early producers,' says Tim Finn of Neudorf. 'Those who have come later are perhaps driven by different desires.'

Homogeneity of style is one issue. The success of the early Central Otago expressions — dark, dense, voluptuous — has prompted many others, both within the region and outside it, to try to make their Pinot Noir in that image, regardless of site and microclimate. People often refer to these wines as being recipe-driven, rather than true expressions of terroir.

Allied to this is a belief in some quarters that big is better. At a dinner in Auckland in 2008, Georg Riedel, head of the internationally renowned eponymous glassware company, held up a glass of New Zealand Pinot and said to the winemakers present: 'What are you New Zealanders doing with your Pinot Noir? It should never be black like this.' There were nods of agreement around the table. Big, dark, oaky, over-extracted wines that thunder across the palate are not unusual, their winemakers mistakenly equating power with quality. It is a stage some New Zealand winemakers seem to have to go through before they realise they are doing the variety a disservice.

'I began trying to make the biggest Pinots I could — a red wine for red wine drinkers,' says Andrew Greenhough of Greenhough Vineyard in Nelson. 'Then at the 2000 Southern Pinot Noir Workshop there were so many big, masculine wines I realised we'd gone to extremes in our desire to make a more substantial wine. I went away from that workshop thinking about getting more delicacy and layers in my Pinot, accepting the variety for what it was. As a result I've become more aware of the vintage and the vineyard … I know it's a cliché, but winemaking is done in the vineyard.'

Another concern is the looming spectre of commodification. Heightened consumer interest has led to large-scale plantings, production short-cuts and a rising tide of what the industry calls 'entry level' Pinot. We're seeing more and more Pinot Noir on shop shelves selling for under $20 a bottle. The base of the New Zealand Pinot Noir pyramid has widened considerably. There are fears this dilution of quality is putting the hard-won reputation built by the early, boutique producers at risk.

'There's a danger it will go down the Sauvignon Blanc track,' says Neil McCallum of Wairarapa winery Dry River. 'Sauvignon Blanc is farmed industrially — nothing has to be hand-done. But you can't make good Pinot without spending money on viticulture. With Pinot Noir we have the opportunity to make sure we have the equivalent of a grand cru and premier cru classification in place. The New Zealand wine industry can only develop successfully if there is, at its core, an increasing number of quality producers who are noticed worldwide and who can promote the image of a high quality on behalf of the industry as a whole.'

The pattern is a familiar one: where a dedicated few have followed a dream and achieved great success, commerce inevitably follows. But, as has been pointed out, there is plenty of ordinary Pinot Noir produced in Burgundy as well, yet Burgundy's great domaines have maintained their greatness.

There are certainly no signs of quality dilution or compromise coming from our best Pinot producers. Quite the opposite. More and more of them are producing truly fascinating wines from special parcels or single vineyards as their vines age and offer more distinct, complex expressions of place.

Speaking at the close of the Pinot Noir 2010 conference in Wellington, Oz Clarke underlined the importance of not veering from that path. The problem of oversupply that had beset New Zealand Sauvignon Blanc from the 2008 vintage was degrading the New Zealand wine brand in the eyes of British consumers. 'Pinot Noir must now lead the fight … for the sake of New Zealand's hard-won reputation,' he exhorted the gathering.

The land is, and always has been, New Zealand's provider. With our finest Pinot Noir, it has provided at a rare level of sophistication. Few other products that are grown, crafted and packaged in New Zealand are so coveted by the world's epicurean cognoscenti.

And yet Pinot Noir has given New Zealand much more than a red wine to make us proud. The terroir conversation essentially began in this country because of the importance placed on the concept by Pinot producers. Their hunger to know more about sites and soils, about viticulture and vinification practices, and about what great wine really is, has been felt across the wine industry and beyond. Many producers of Hawke's Bay Syrah, an exciting new wine style, have acknowledged a debt to Pinot.

The exploratory bent and respect for the land is at the heart of what our Pinot vignerons have achieved. It has enabled them to take Burgundy's delicate, demanding, revered red grape and allow it to tell the truth.

....

Following pages: p100 Biodynamic preparation at Millton Vineyard, Gisborne; p101 Most Pinot producers hand-harvest their fruit.

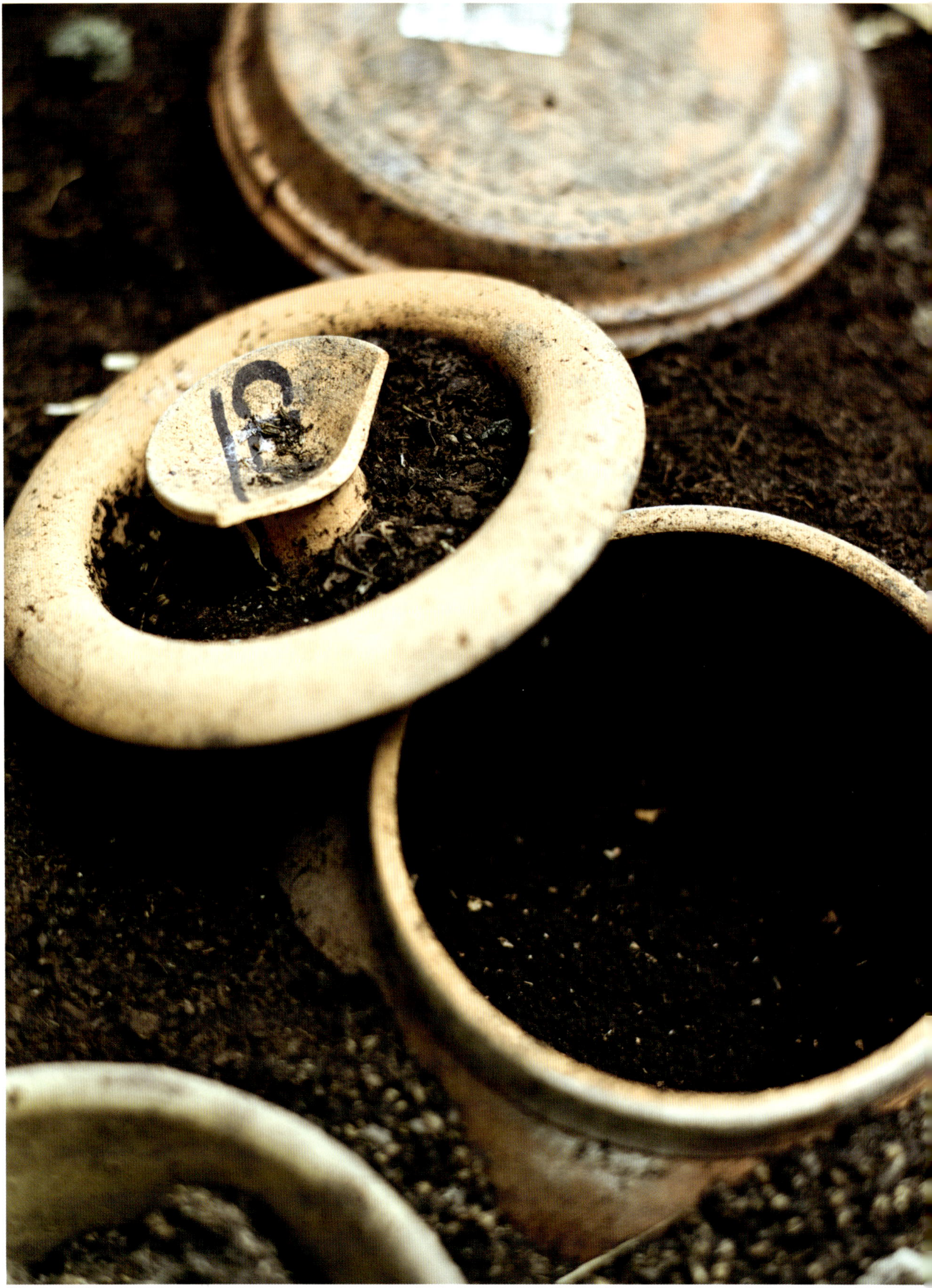

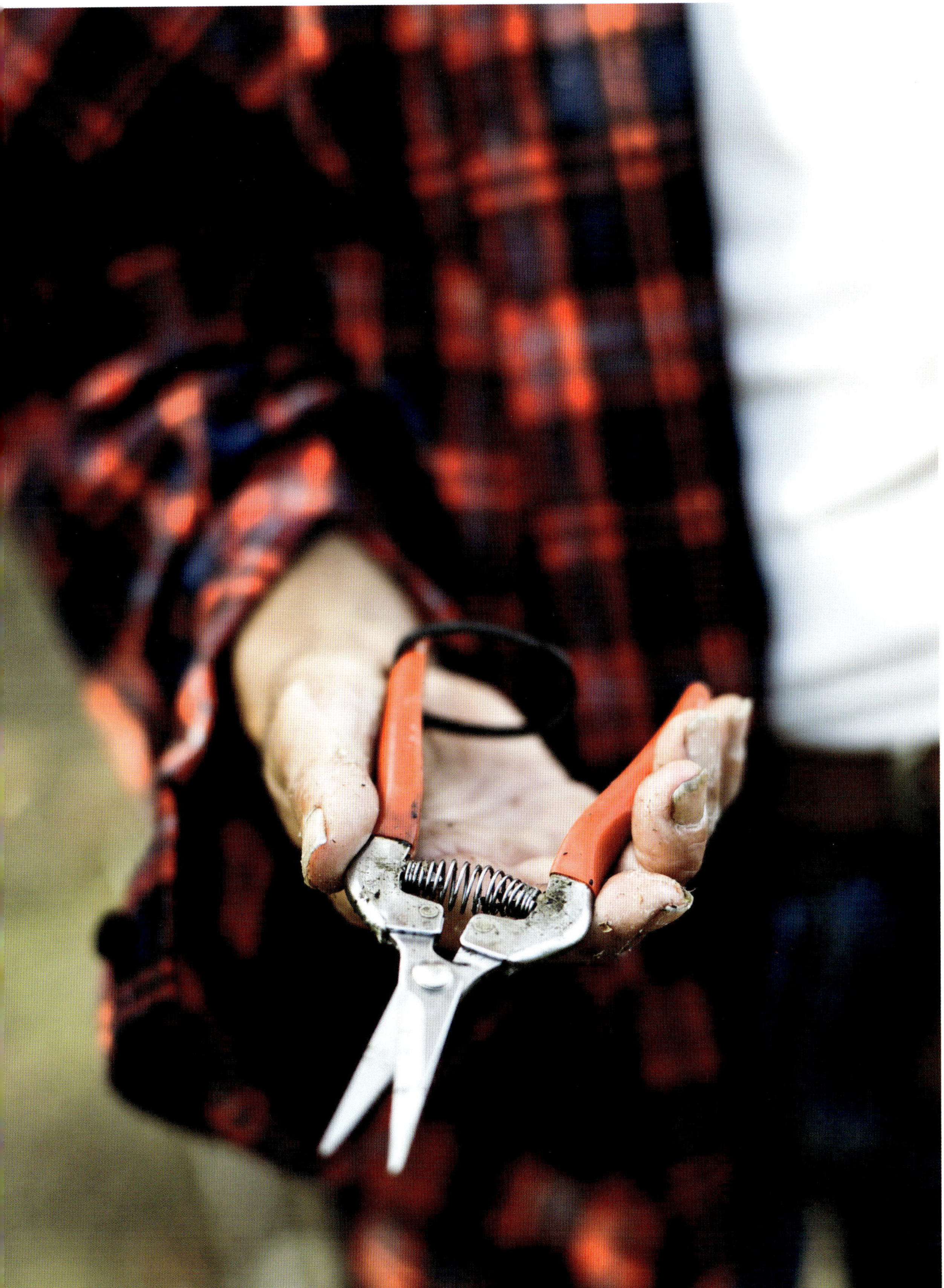

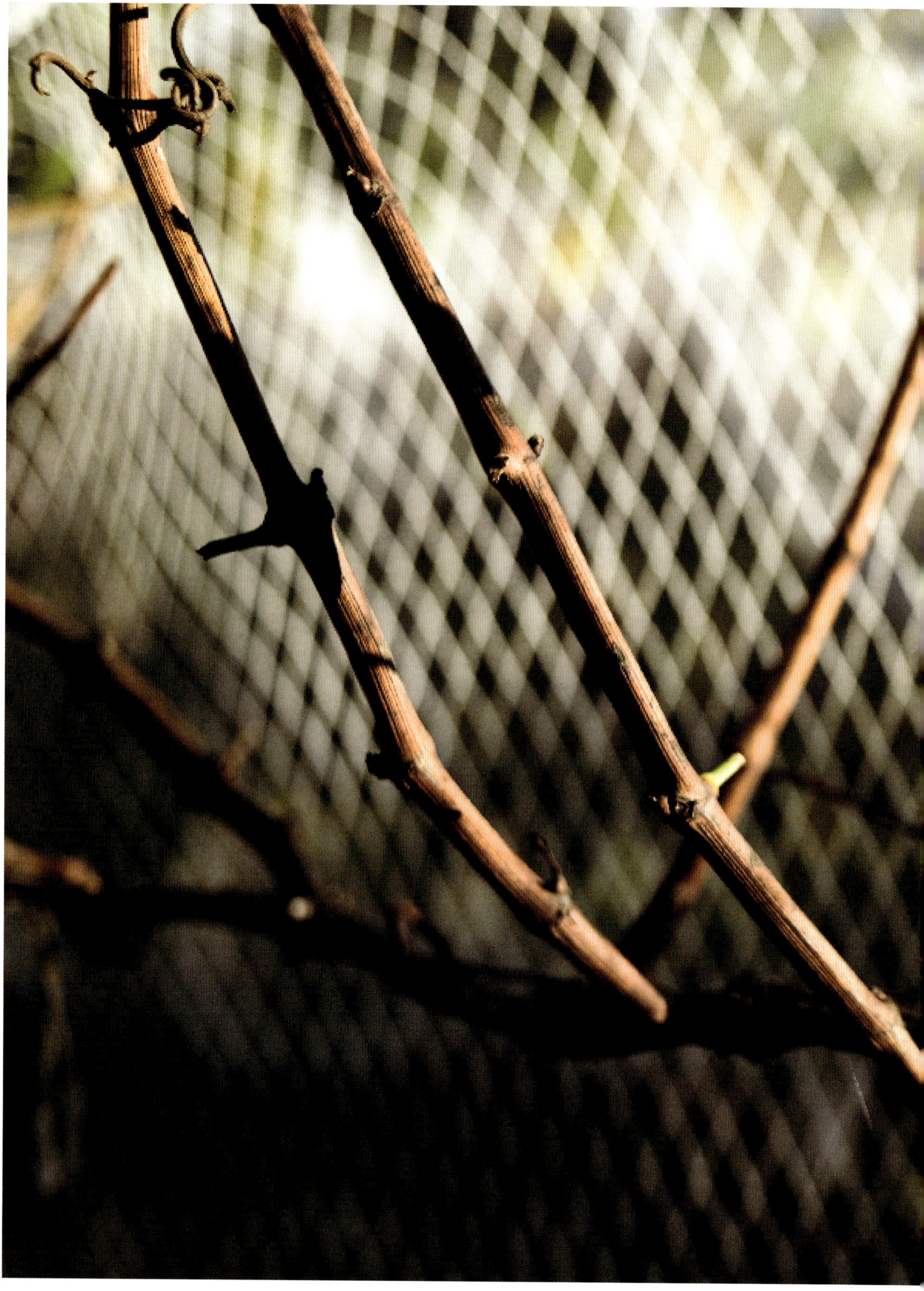

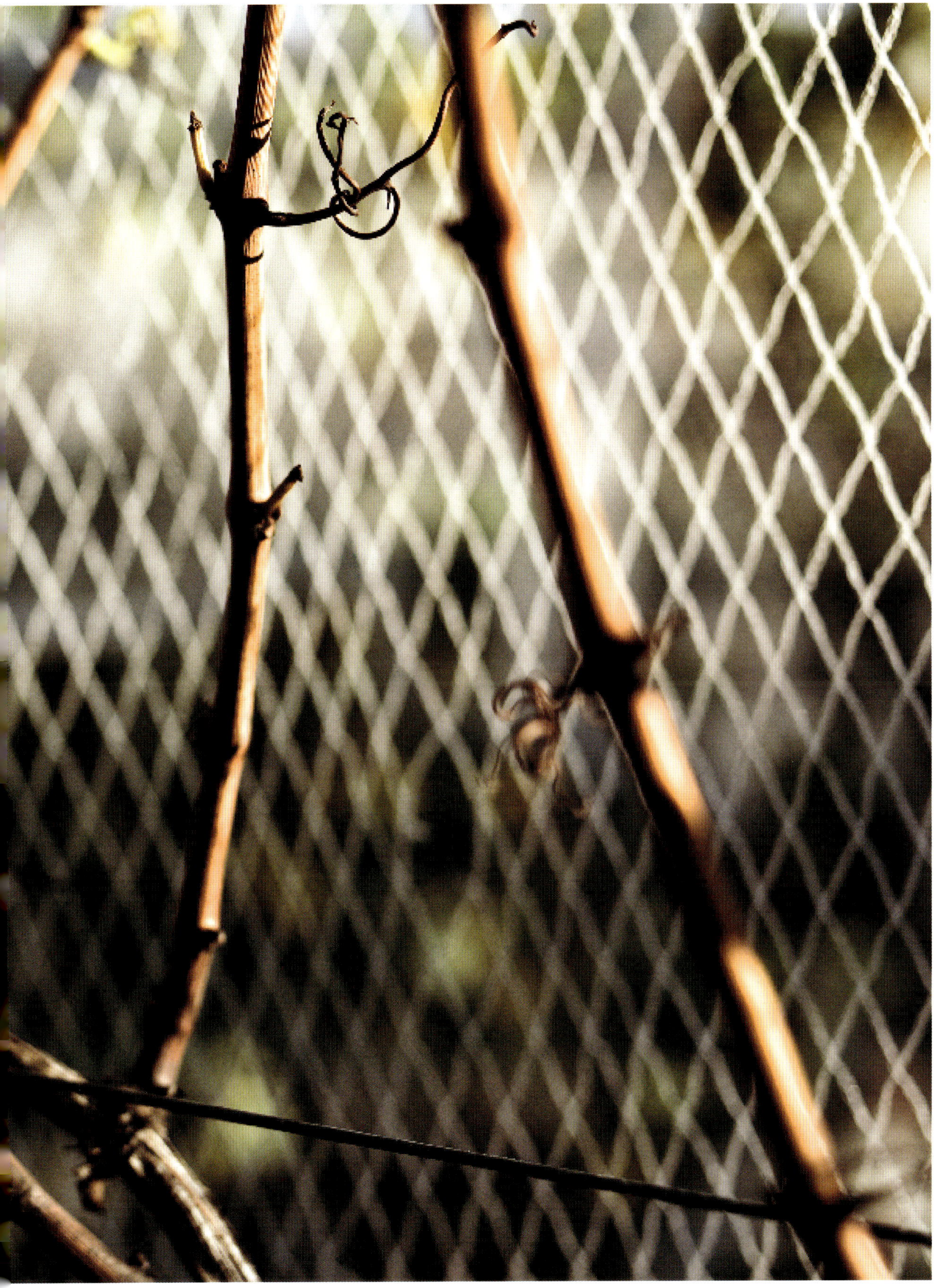

SEVEN. WAIRARAPA

Travelling north from Wellington on State Highway 2 and crossing the Rimutaka Range is the most common way to enter the Wairarapa. Very often it is a journey of two seasons — the cloud and drizzle on the west of the divide give way to an east-side offering of blue sky, warmth and sun. Wellingtonians enjoy escaping to this 'other country' on weekends. It is also a place that vines — Pinot, in particular — find remarkably hospitable.

A family thread links the two eras of Wairarapa Pinot Noir. Derek Milne, the soil scientist whose report reignited winegrowing in the region in the 1970s and who helped found Martinborough Vineyards, is married to a great-great-niece of William Beetham. With his French wife Marie Zélie Hermanze Frère, Beetham was the first to plant Pinot Noir in the Wairarapa in the 1890s and often marvelled at how well the grape performed. The connection was celebrated when Martinborough Vineyards produced its first limited-release Marie Zélie Pinot Noir with the 2003 vintage.

Beetham's early enthusiasm was more than justified by the success of Pinot's twentieth-century return to the region. By the end of the 1980s, the small South Wairarapa town of Martinborough had become New Zealand's Pinot Noir *chef lieu*. The first of the new wave of Pinot producers — Martinborough Vineyards, Ata Rangi and Dry River — succeeded in making exciting wine, thanks mainly to Milne's scientific analysis having led them to an excellent growing environment. Other factors came into play as they went on to found an influential Pinot Noir-producing culture. These Martinborough first-footers were inquisitive, collaborative, quality-driven and painstaking; their wineries were small, hands-on operations concentrated in a small area. Not least, the capital city was very close, very interested and very supportive from the beginning.

Growth in the Wairarapa continued as it began. In Martinborough, small wineries and their vineyards filled spaces in and around the town, completely transforming it. Only Palliser Estate and later Craggy Range in the Te Muna valley brought dimensions that could be considered beyond boutique. During the 1990s, more family-owned wineries began to sprout, away from Martinborough, up the low-rainfall corridor through Gladstone to Masterton. The two towns have now given their names to two Wairarapa subregions from which we are already seeing some very promising Pinot Noir.

While the strong development of these subregions is exciting, Martinborough remains the Wairarapa's vinous epicentre. It accounts for two-thirds of the wine made in the region and is home to its most prestigious wineries. As early as 1986, Martinborough producers began making attempts at creating an appellation system to both protect and enhance the reputation of the wine that grew within a prescribed area. It was not a smooth process and the appellation's scope and shape changed several times before it was decided to register a defined area under the Geographical Indication (GI) Act. The boundaries were settled upon in 2003, making Martinborough the first New Zealand wine region to establish such an appellation. The GI's rectangular area, bounded by the Ruamahanga River in the west and the famous 'three canoes' ridgeline on the eastern side, incorporates all the vineyards in and around the town, along with the more recent plantings on Te Muna Road, to the east of Martinborough. Te Muna's vineyards are slightly more elevated and cooler than those closer to the town, though with similar free-draining gravel soils.

Despite its global reputation, the Wairarapa accounts for only 3% of New Zealand's vineyard area. What's more, the low yields that are a feature of the region mean it contributes just 1% of the nation's wine production. Nevertheless, it is home to 10% of New Zealand's Pinot Noir plantings.

The Wairarapa occupies the southeastern corner of the North Island, a territory consisting of spectacular coastline abutting the Pacific Ocean to the east and Cook Strait to the south, and a large plain that is open to the south and flanked by hill country.

It is on this plain that the Wairarapa's grapes are grown. Visually it is not unlike Marlborough's Wairau Plain — a wide river flat, with dry, low, rolling pastoral hills on the eastern side, while opposite are dark, bush-clad peaks that close the door on much of the precipitation carried by the prevailing northwest wind flow. In terms of latitude, there is actually not a lot of difference between the Wairarapa and Marlborough. The Wairarapa is the northernmost of the recognised Pinot Noir regions, and when you consider wine styles as well, it's tempting to describe it as the North Island's South Island wine region.

The river that runs almost the entire length of the plain is the Ruamahanga. Its changing course over many tens of thousands of years has gifted the region deposits of deep alluvial gravel, typically covered with a thin layer of silty loam or loess. Most of the vineyards are planted on these ancient, free-draining river terraces in which there is also an element of limestone. The hills along the Wairarapa's eastern side were part of an old seabed and are host to large deposits of limestone. Slippage down to the plain over the millennia has added varying quantities of this calcareous material to the gravels.

Local vignerons like to call the climate 'semi-maritime'. Strong summer heat, peaking at 32–34°C, is tempered by the cooling influence of the nearby coastline. Temperatures can drop to 10°C at night. Martinborough township is the driest place in the North Island, with an average rainfall of 700–800mm a year. The prevailing nor'wester is mostly a dry wind, which helps prevent botrytis taking hold.

Helping to extend the growing season are autumns the French would call *doux*. They are reliably mild and dry, lowering anxiety levels and often allowing growers to harvest precisely when they choose. Autumn generosity has come to the rescue of many a less-than-satisfactory Wairarapa summer.

But the place is not without its challenges. Cool, windy weather in spring and early summer often affects flowering, resulting in naturally low yields. Spring frosts have become a regular threat. In 2007, many producers took in only 10% of their usual Pinot Noir harvest, courtesy of a severe November frost. The chilling, wet southerlies that flay Wellington are also felt by the Wairarapa.

Over the past decade there have been almost as many disappointing vintages in the Wairarapa as favourable ones. And yet, when the region's best practitioners are able to get it right, there are arguably few more fascinating expressions of New Zealand Pinot Noir. The wines are elegant and layered, notable for spicy, incense-like aromatics, fine structure and enviable depth.

Following pages: Ata Rangi vines on the Martinborough Terrace.

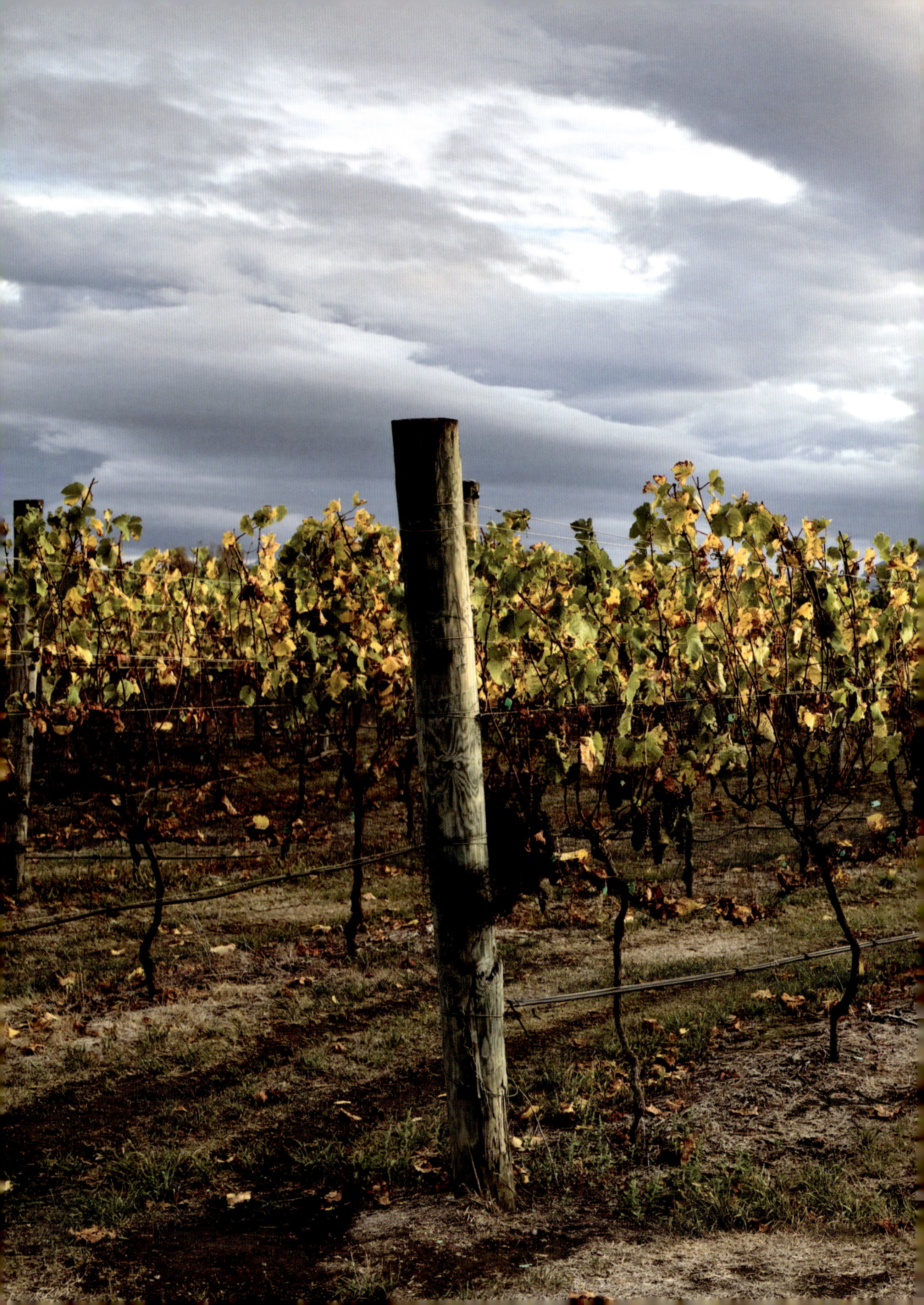

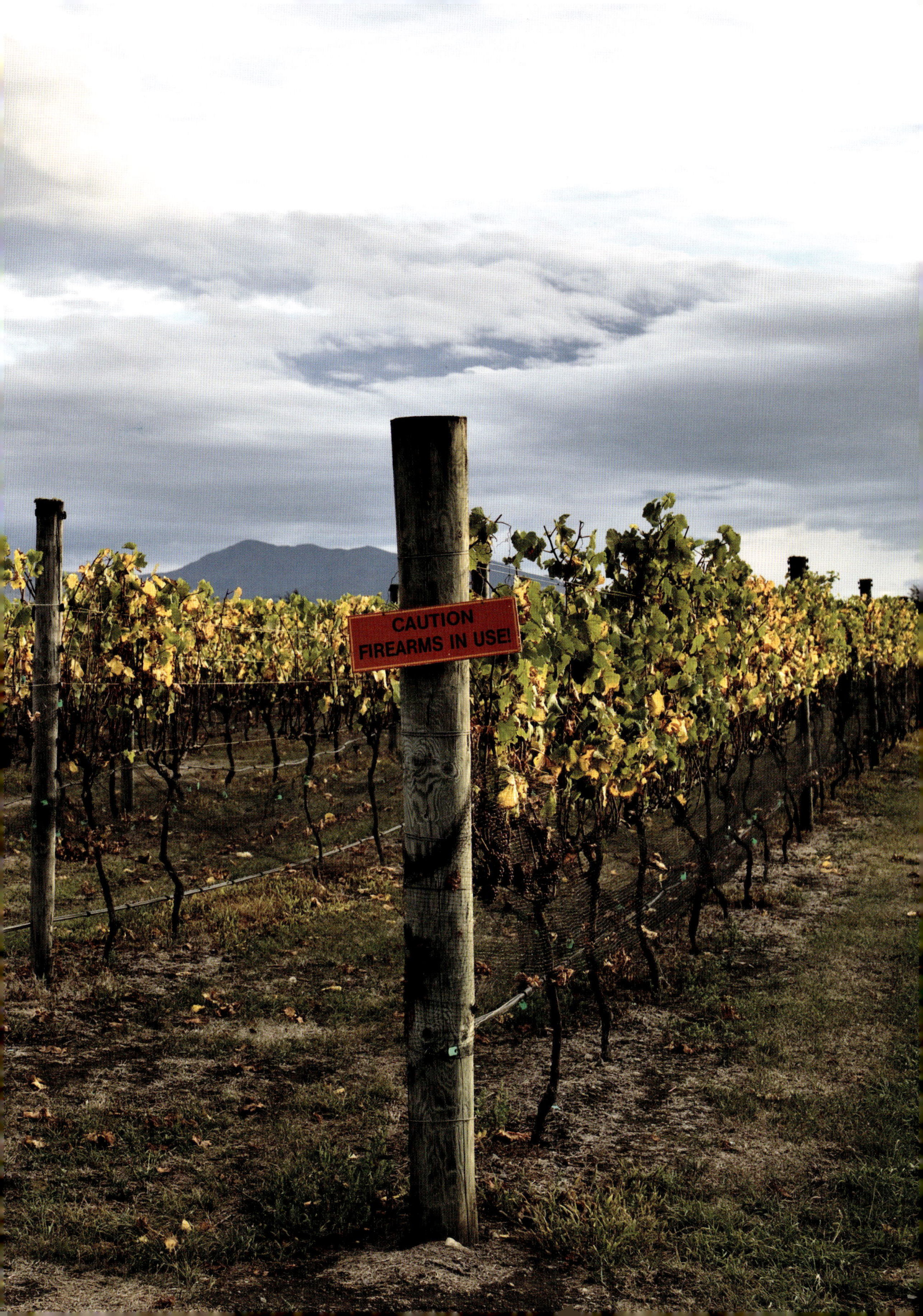
CAUTION
FIREARMS IN USE!

KEY PRODUCERS

ATA RANGI

The winery created by rugby-loving former dairy farmer Clive Paton produces one of the thoroughbreds of New Zealand Pinot Noir. Ata Rangi is on everybody's perennial shortlist of top Pinots in this country and is one of a select group to have achieved international renown.

The charms of Pinot Noir revealed themselves to Paton via a bottle of Chambolle-Musigny in 1982. 'It really turned my head,' he says. 'It was fantastic — that classic yin and yan thing; strong and powerful, yet with a classy, feminine side. It was a wine that made me think.' His vineyard on Martinborough's Puruatanga Road was two years old at the time, with Pinot one of several red varieties planted. After blending them all for two or three years, he produced a varietal Pinot Noir for the first time in 1985. He and Ata Rangi have never looked back: 'That first wine was the one that really cemented in my mind what Martinborough could do with Pinot Noir.' The Ata Rangi 1985 was made using the now-famous Abel clone, discussed in Chapter 4. Paton was the first to discover the clone's attributes, hence it is often called the Ata Rangi clone.

The early years were tough. Vegetables were grown between the vine rows to help with cash flow; wine was bottled and sold as quickly as possible for the same reason. Paton was joined early on in the business by his sister Alison, and then by Phyll Pattie, herself a trained winemaker, who became his partner 'in life and business'. Family involvement is an important part of the Ata Rangi approach; it seems unlikely the winery will ever stray from family ownership.

Ata Rangi became the first New Zealand winery to take out the prestigious Bouchard Finlayson Trophy for the top Pinot Noir at the 1994 International Wine and Spirit Competition in London. Then it became the only New Zealand winery to repeat the feat when it won again a few years later.

The Ata Rangi Pinot Noir has always been made from fruit grown at different

Right: Clive Paton, Ata Rangi.

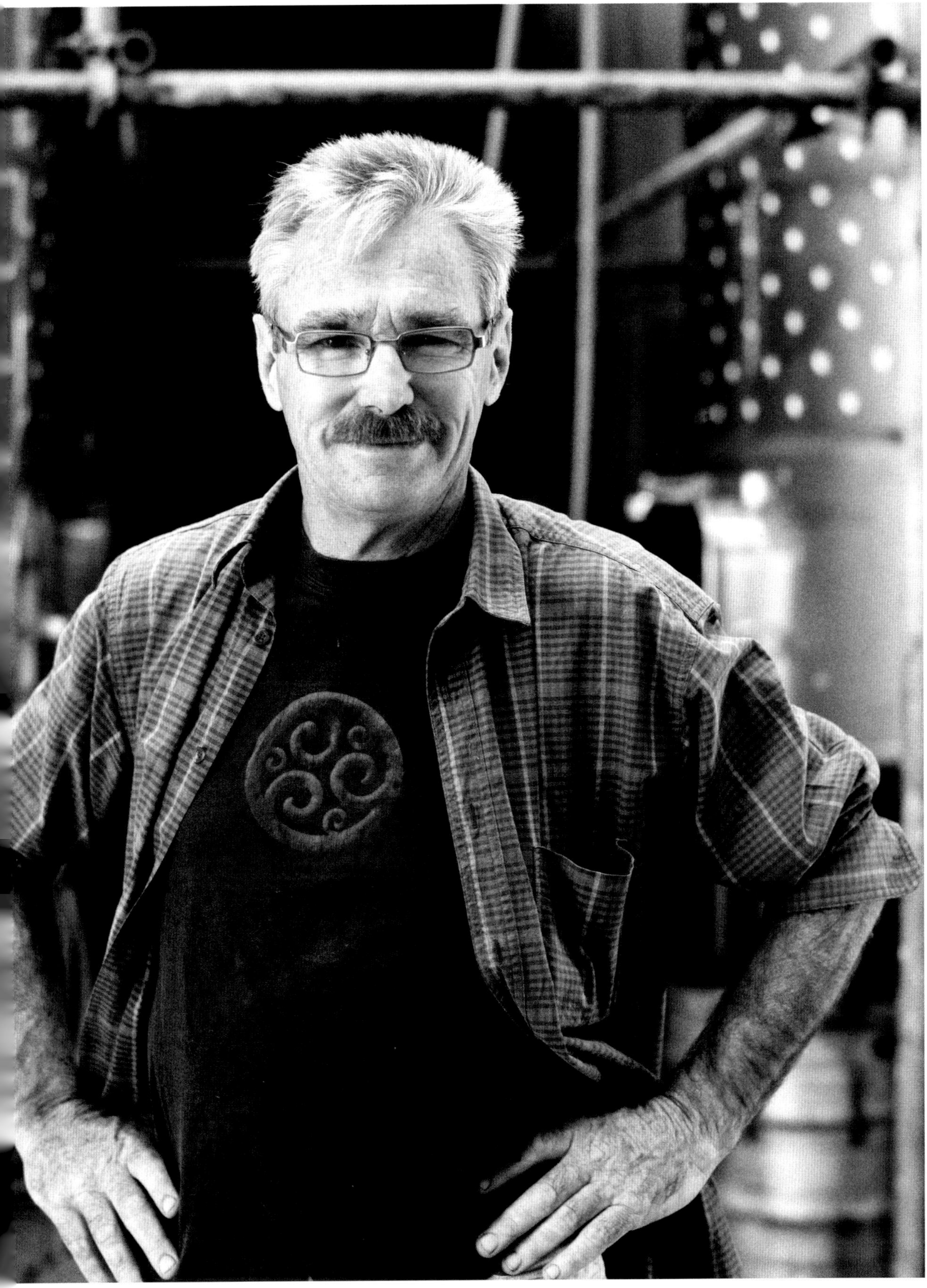

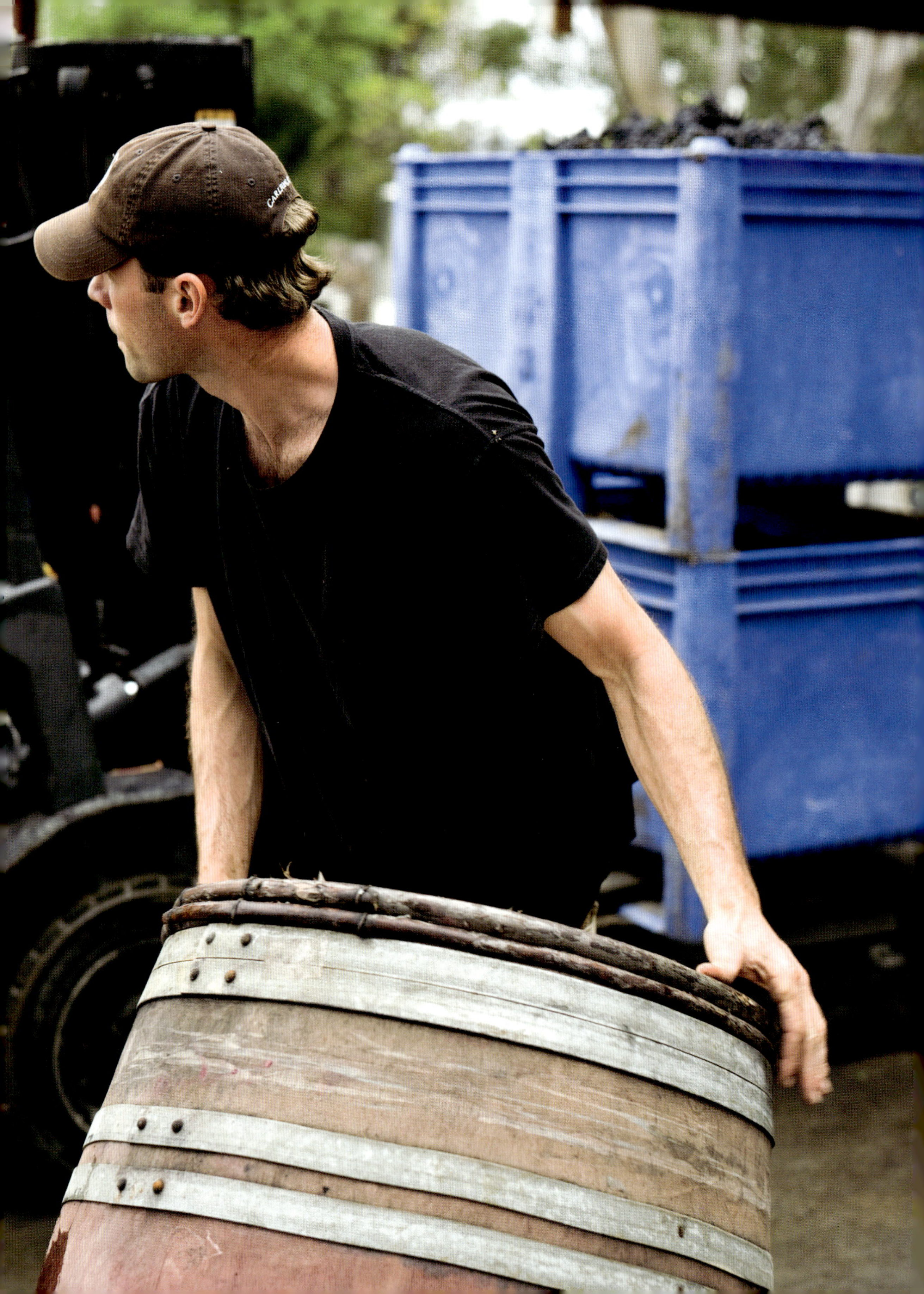

sites around Martinborough. Today the winery takes in Pinot grapes from 75 different plots, including a number of blocks that are leased by the winery or belong to contract growers. Some of the vines are approaching 30 years old.

The aim has always been to create a perfumed, refined, silken-textured wine; Paton often invokes Musigny as his guiding light. Ata Rangi has also always taken great pride in the longevity of its Pinot, with good reason. In 2003, the winery held a vertical tasting featuring the previous 10 vintages of its Pinot Noir. The oldest wine at the tasting, the 1994, was remarkable — still performing with sweet, spiced, velvety aplomb.

Helen Masters, who began her winemaking career as cellar rat at Ata Rangi in 1990, took over as head winemaker in 2005. Under her stewardship, the Pinots have been as aromatically impressive as ever, and a little more robust.

'We get the best perfumes from our Abel clone fruit,' Masters says. 'Clones UCD5 and 10/5 in our older vineyards also perform well. But Martinborough struggles with the Dijon clones — they're in too much of a rush to ripen.'

In 2004, a young-vine Pinot Noir called 'Crimson' was released for the first time. It's a wine made for immediate drinking. Its creation was in part inspired by Clive Paton's love of native flora — proceeds from the sale of the wine go to 'Project Crimson', a planting and protection programme for New Zealand's red-flowering pohutukawa and rata trees.

For the first time ever, a single-vineyard wine was made in 2007. Don and Carole McCrone, owners of a Pinot vineyard in Oregon that supplies fruit to legendary winemaker Ken Wright, became 'bi-hemispherical' when they bought and planted a plot near the Ata Rangi winery. They realised their goal of having single-vineyard wines from both countries with the Ata Rangi McCrone Vineyard Pinot Noir 2007 — a dense wine with a distinctive blend of sweet and savoury flavours.

The applause was long and hard when Ata Rangi (along with Central Otago's Felton Road) received a Tipuranga Teitei o Aotearoa award at the Pinot Noir 2010 conference in Wellington. The award was created to honour the top tier of New Zealand Pinot producers.

Following pages: Harvest time at Ata Rangi.

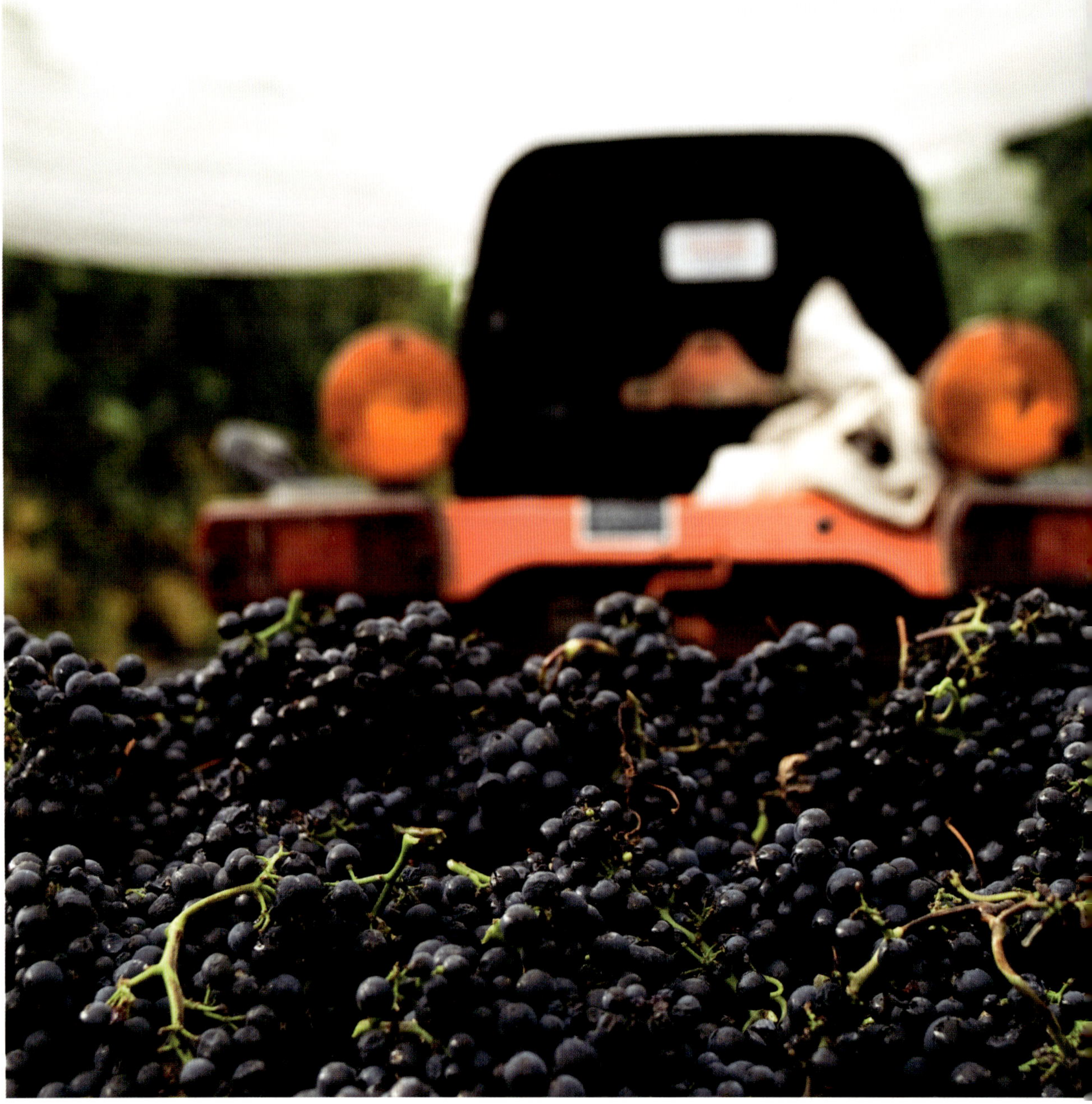

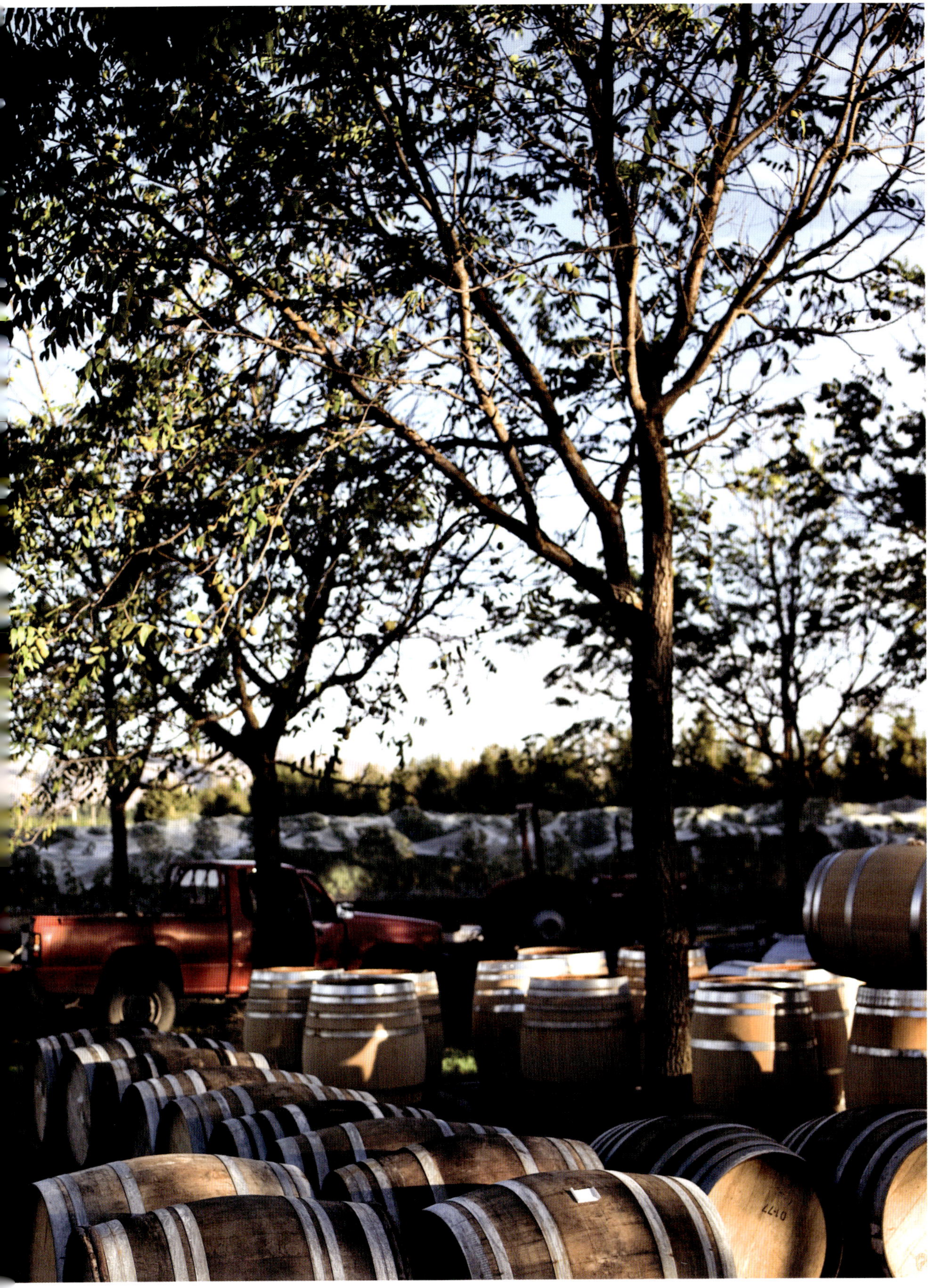

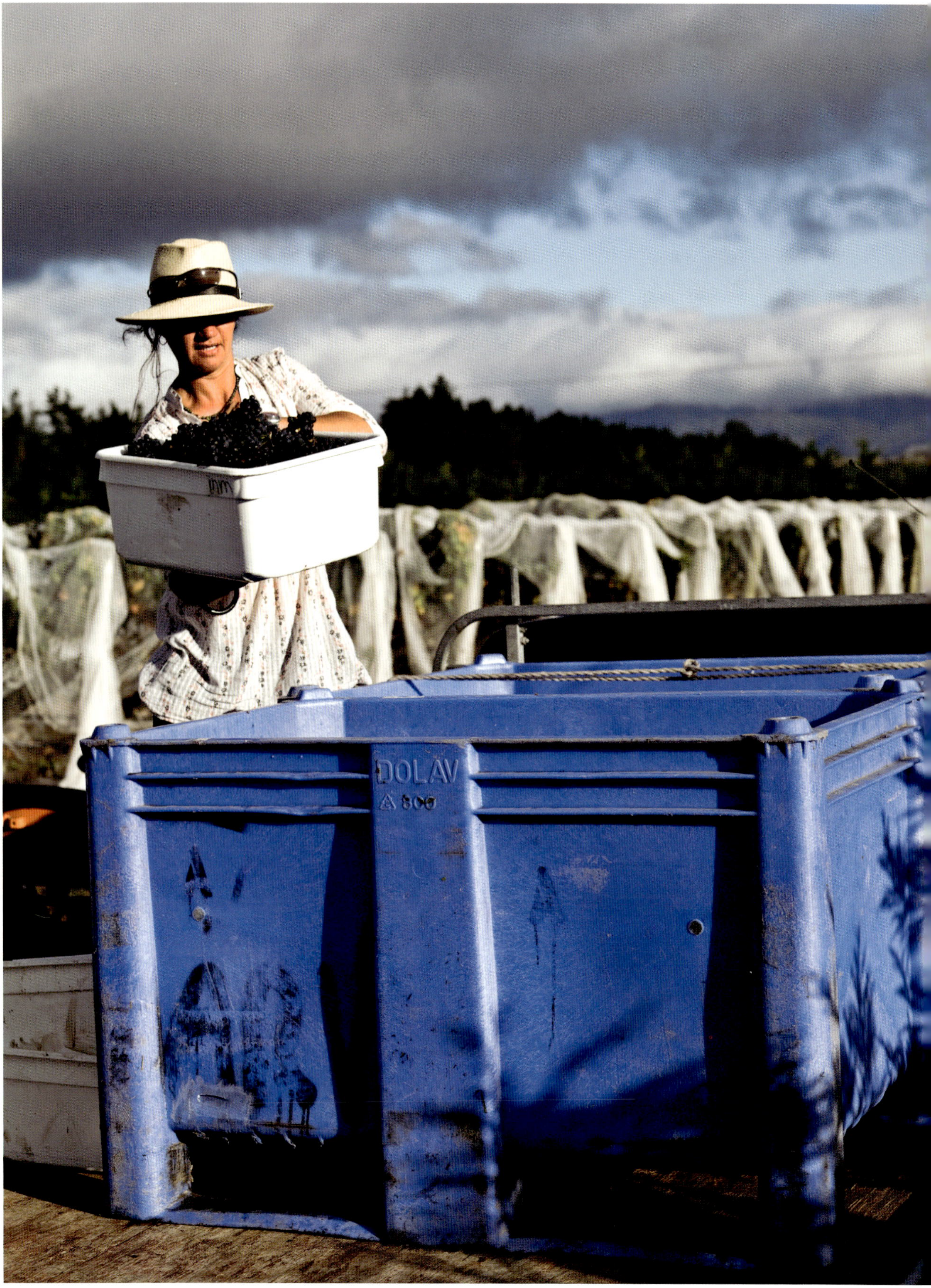
DOLAV
800

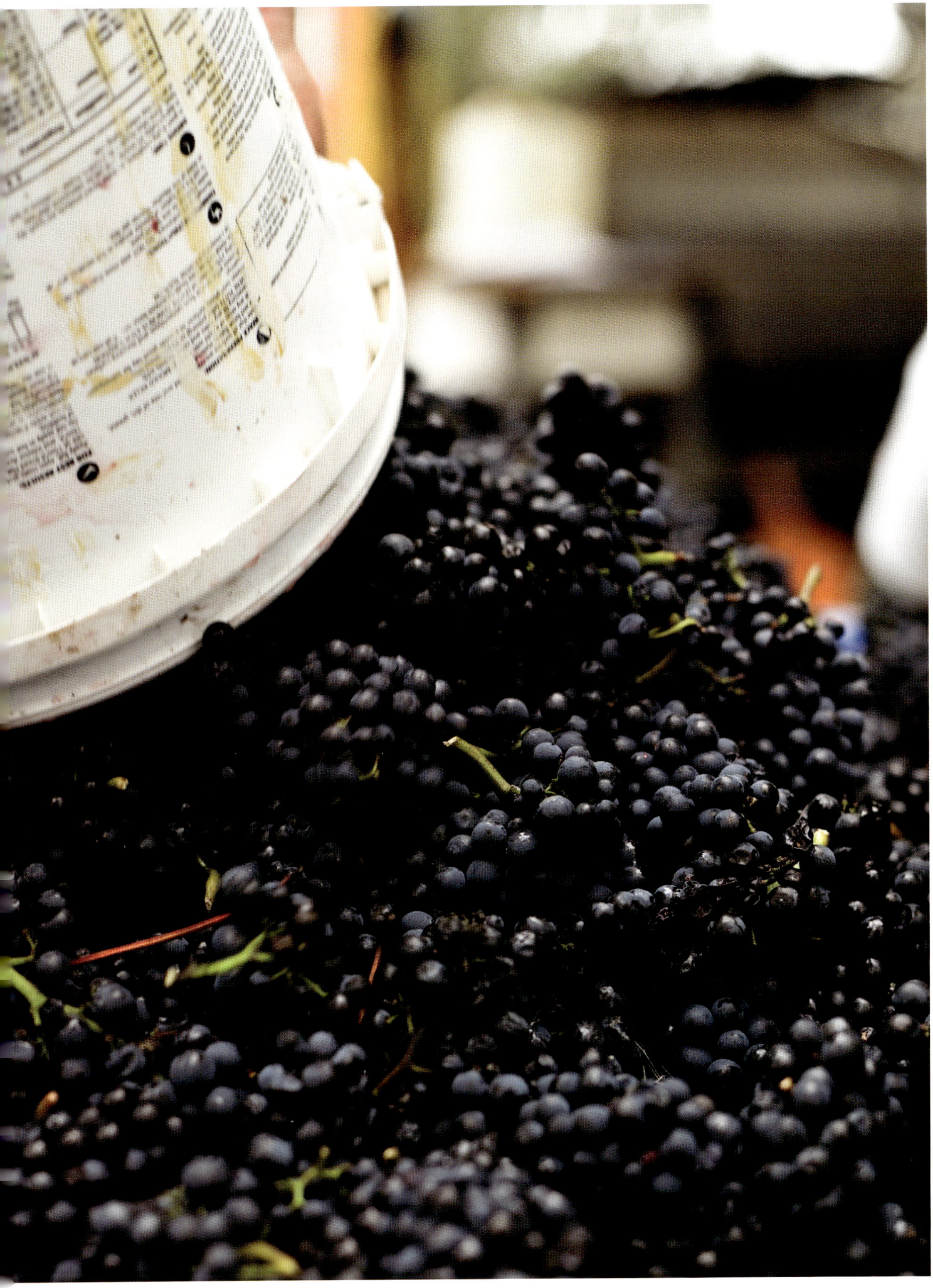

CRAGGY RANGE

'Single vineyard, single minded', Craggy Range's founding concept, was always going to have particular relevance for famously site-expressive Pinot Noir.

The idea of a winery that specialised in making single-vineyard wines from chosen sites around the country was something Steve Smith (former group viticulturist for Villa Maria and a Master of Wine) had harboured for some time. In 1998 he met Australia-based multi-millionaire Terry Peabody, who had no vinous expertise but a skeletal ambition to create a quality-driven, family-owned wine enterprise. Each ticked the other's empty boxes and Craggy Range was born.

Craggy Range HQ is on the Waimarama Road in Hawke's Bay, where its Giants Winery and Terrôir restaurant set new standards of scale and flair for the New Zealand wine industry. Its vineyard holdings extend from the Bay to Central Otago, each site carefully selected to host a certain variety.

'Pinot has been the variety we have spent the most time planning and researching where we wanted to go,' says Smith, Craggy Range's director of wine. 'Our search was about two things: quality on a regular basis, and character/style/*typicité* — in other words, which wines in which regions did we love the most?'

Smith has always admitted to a strong fondness for Pinot from Martinborough, and the first of Craggy's Pinot sites was the 33-hectare Te Muna Road Vineyard. Te Muna is situated to the east of the township and, as an extension of the same river terrace on which the Martinborough vineyards are planted, it benefits from similar alluvial gravels.

The first Craggy Range Te Muna Road Pinot Noir was produced in 2002, made by the late Doug Wisor, Craggy's young Pinot-engrossed American winemaker who was killed in a kite-surfing accident in 2004. Planted in Pommard, Abel and a selection of Dijon clones, the wine has come to display tension, austerity and depth — what Steve Smith likes to call an 'intellectual' quality. One prestige release has been made with fruit from the vineyard, the Aroha Pinot Noir 2006.

Soon after establishing Te Muna, Craggy Range's Pinot attentions turned south. The label now makes three single-vineyard Pinot Noirs from Central Otago and another from the Waitaki Valley. The Craggy Range Calvert Vineyard Pinot Noir made its début in 2006. The arrangement is unique: three producers (Craggy Range, Felton Road and

Pyramid Valley) share the crop to each make a single-vineyard Calvert Pinot. Viticulture is handled biodynamically by Felton Road's Gareth King. Craggy Range's Calvert Pinots have a bright, plump black fruit presence tethered to a silken texture.

Just 500 metres to the west of the Calvert vineyard is the source of the fruit for the Craggy Range Bannockburn Sluicings Vineyard Pinot Noir, first produced in 2007. Despite sharing many features with the Calvert, the Bannockburn Sluicings Pinot is quite different, with keen floral aromatics and often displaying the dried herb character very typical of the subregion.

The last of the Central Otago trio is the Craggy Range Zebra Vineyard Pinot Noir, first produced in 2007. The vineyard is located in Bendigo, on the alluvial rocky soils of the Bendigo Station Terrace. Bendigo's warmth helps produce a wine with vibrant fruit and structural elegance.

The Craggy Range vineyard in the Waitaki Valley, called the Otago Station Vineyard, was planted between 2004 and 2006 and has yet to produce a commercial harvest. The company describes its Waitaki project as 'interesting, pioneering viticulture'.

All the wines are made in Hawke's Bay. Adrian Baker, who looks after the making of Craggy's Pinots, says the distance between the vineyards and the winery is not an issue:

> The grapes from all our vineyards are picked into small, open-sided crates, immediately refrigerated and then transported to the winery in refrigerated trucks. They arrive in perfect condition. We see it as being more important to have the vineyard in the right place and our winemakers together in one state-of-the-art facility.

Baker adds that while winemaking technique is adapted to each vineyard, there is an overriding Craggy Range dictum. 'We want to make wines with complexity of fruit and maturation characters. Each wine must have a serious tight-knit, slightly savoury texture that is vital to the production of fine Pinot. Plush, open, fruity, sweet-edged wines are not our style.'

DRY RIVER

Despite coming to Pinot Noir a little later than its fellow Martinborough pioneers, Dry River was quick to make a name for itself with the variety. In fact, through subtle brand presentation (the curiously lipped bottles are a talking point), rarity (the winery has never produced more than 1000 cases of Pinot a year) and an unerring dedication to quality, Dry River's has arguably become the cult Martinborough Pinot.

Neil McCallum and his wife, Dawn, bought land to establish the winery in 1979. It was a bold move. McCallum, who has a doctorate from Oxford, was in mid-career as a DSIR research scientist. But at Oxford he had come under the spell of great wine, and the urge to act on the report by his friend and fellow-scientist Derek Milne, which declared Martinborough to be a viticulturally chosen spot, was too great to resist. McCallum's attention to detail was clear from the start:

> Before I started making wine, I sat down in a scientific library and I read every research paper that had ever been written in English on making wine. The development of the vineyard work has been done on the same basis. Many of the ways in which these developments have gone have meant we are different from many other operations. We have simply tried, in a systematic manner, to pursue quality and do it in a logical fashion.

These words, however, belie another important strand of McCallum's personality: a belief in, and enjoyment of, the unquantifiable romance of wine. When it comes to assessing wine, he prefers 'to replace science with poetry. Not the rhyme, but the subtlety and complexity of expression, the innuendo and evocativeness, which the scale of 1 to 10 can never express.'

In 2003, the McCallums sold Dry River to an American pair: Californian viticulturist Reg Oliver and Wall Street businessman Julian Robertson, who has a number of other business interests in New Zealand, including the spectacular Cape Kidnappers golf course. McCallum remains central to the winemaking at Dry River, though the day-to-day winemaking duties are now handled by the capable Katy (Poppy) Hammond, whose husband Shayne Hammond is the Dry River viticulturist.

Dry River makes one Pinot Noir annually, from fruit grown on three estate-owned vineyards: Dry River, Craighall and Lovat, all of which are located on the alluvial gravels of the Martinborough Terrace. About 80% of the vines are clone UCD5 (which McCallum introduced to Martinborough), with the remainder split between 10/5 and Dijon clones. McCallum has reported he is 'least impressed by the Dijon clones which we feel ripen too early under our conditions'. Average vine age is 20 years old.

McCallum doesn't irrigate, cropping levels are held to below 6 tonnes per hectare, and harvest decisions are based on phenolic ripeness. There can be up to 15% whole bunches in each ferment, depending on the year, after a cold soak lasting five to seven days. Both the primary and malolactic ferments are winemaker-induced. Time in barrel is generally a year, with the proportion of new oak at 25%. McCallum:

> We want to fully express the terroir, and winemaking decisions are made on the basis of minimum damage or change to the flavours as harvested. Other than terroir, we feel good Pinot should be long-lived to allow development. Farming and making wines for this requires a tannin structure which, inevitably, will highlight the differences in climate for these vineyards when compared to the Old World continental counterpart. It is important to allow the wines to express this rather than adapt winemaking to produce look-alikes of Burgundy.

The Dry River Pinot Noir is big and masculine in style, deeply coloured and complex, with a serious, savoury edge to the fruit. Several years ago British wine writer Jancis Robinson confessed that Dry River and Ata Rangi Pinot Noirs were the only New World wines she bought by the case.

Following page: Dr Neil McCallum, Dry River.

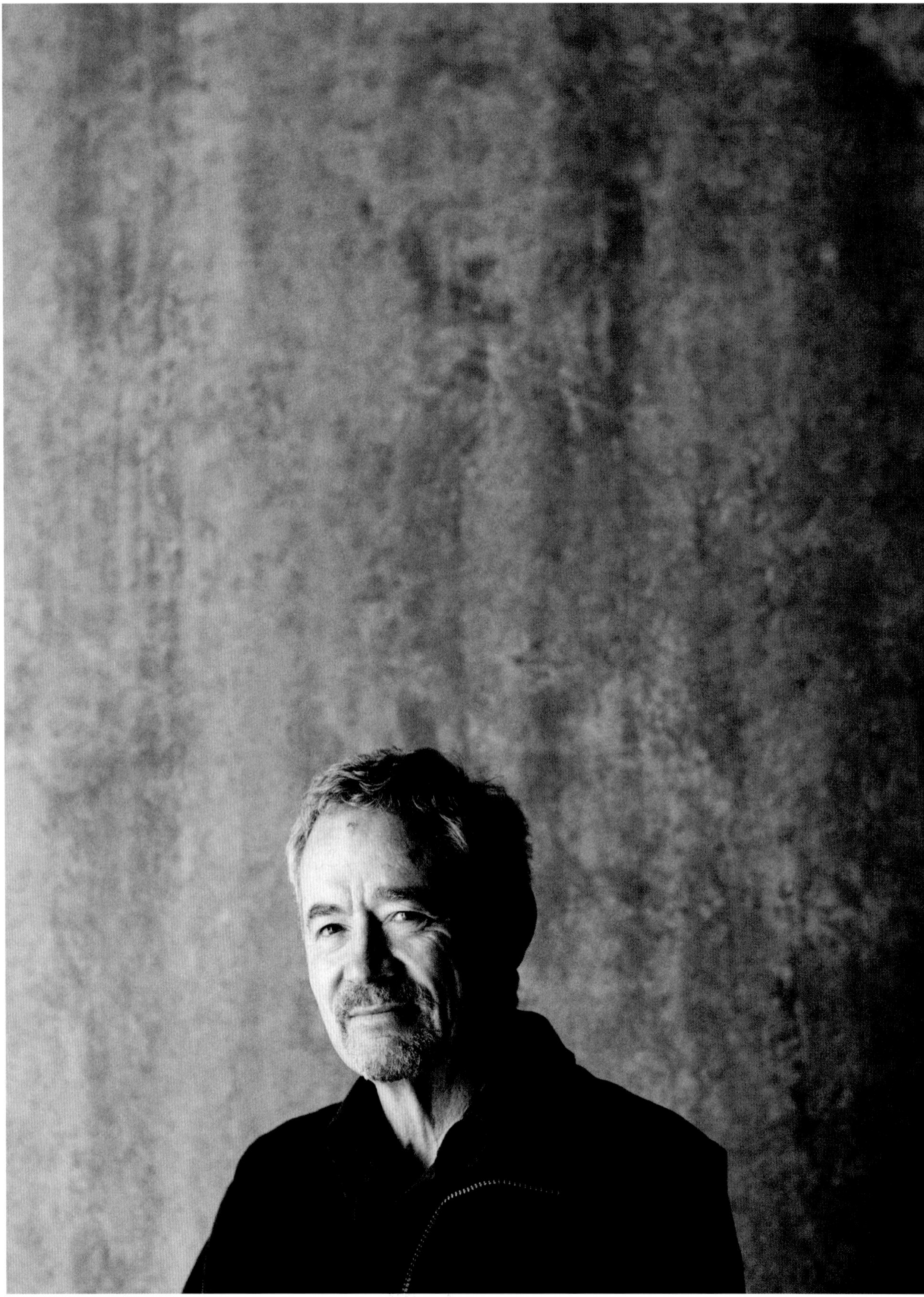

WE HAVE SIMPLY TRIED ... TO PURSUE QUALITY AND DO IT IN A LOGICAL FASHION.

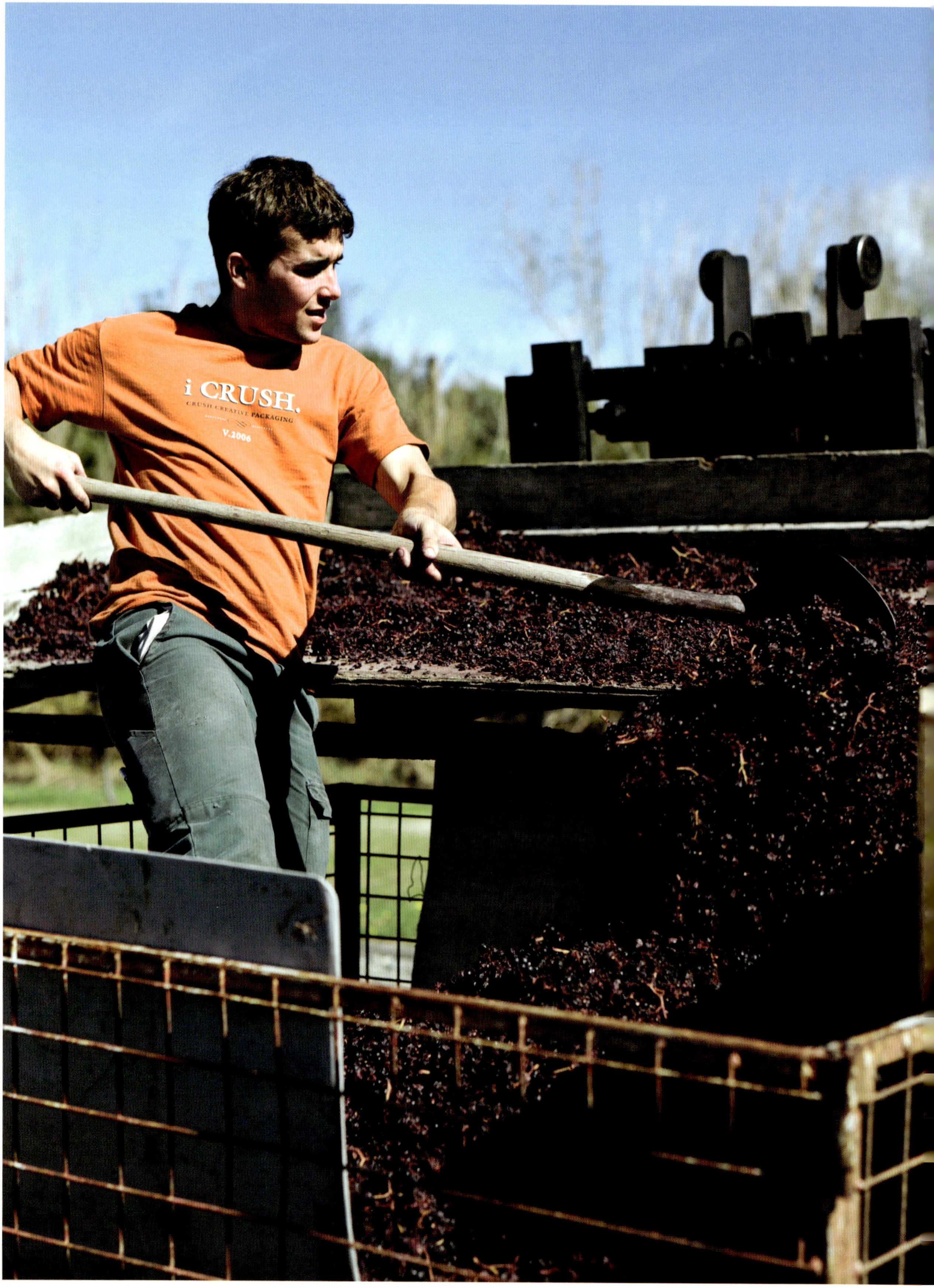
i CRUSH.
CRUSH CREATIVE PACKAGING
V.2006

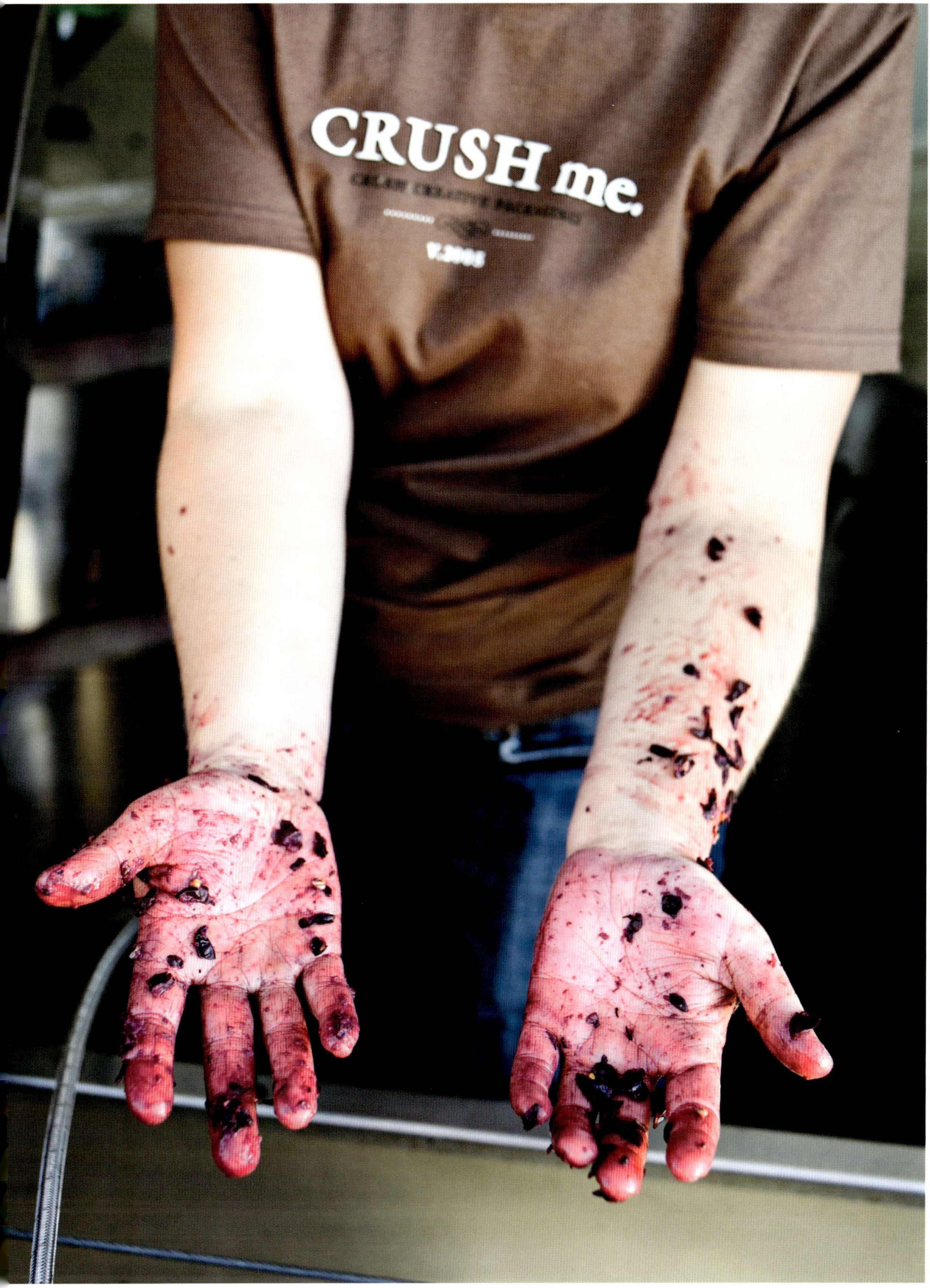
CRUSH me.

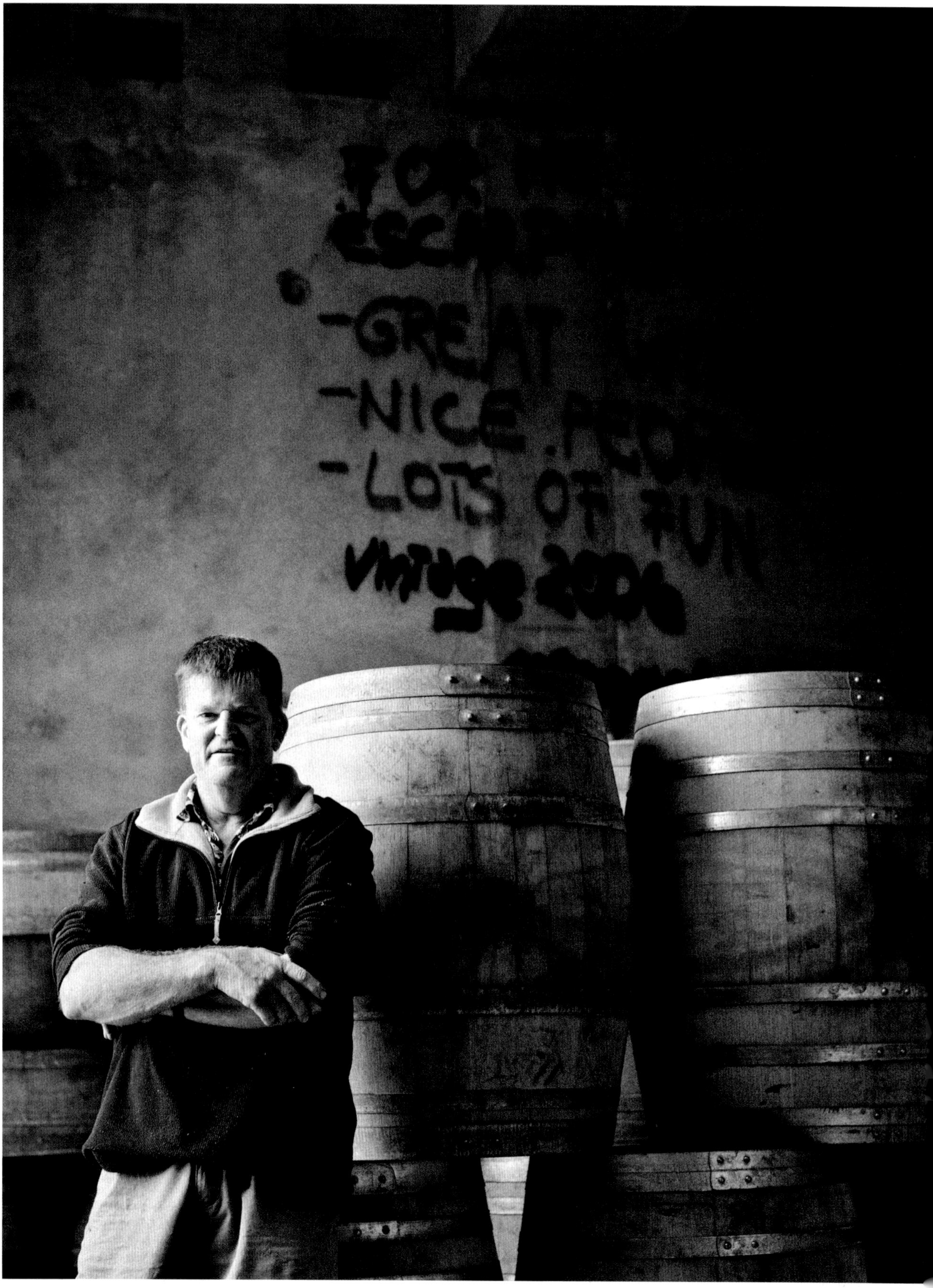
-NICE
-LOTS OF FUN
Vintage 2006

ESCARPMENT VINEYARD

New Zealand's original 'Mr Pinot' is Larry McKenna, founder and director of Escarpment Vineyard. His personal Pinot Noir journey mirrors that of the country as a whole. It didn't have its genesis in Burgundy or Oregon or elsewhere, although those places were later to become formative. His passion took hold in New Zealand, as every new vintage he oversaw with Martinborough Vineyard's infant vines not only revealed the grape's fondness for its new home, but also pointed to the treasures that lay ahead for the committed explorer.

Hopelessly infected, he has been on the 'Grail trail' ever since. More than that, McKenna has been a leader in fostering the culture of quality and knowledge-sharing that helped propel New Zealand's small Pinot-growing fraternity onto the world stage in the 1990s.

After 14 vintages at Martinborough Vineyard, McKenna left to do his own thing. He and his wife, Sue, established Escarpment in 1999 with another Australian couple, Robert and Mem Kirby who came on board as business partners.

After a nationwide site search, McKenna settled on reasonably familiar territory: the Te Muna valley to the east of Martinborough township, a big attraction being Te Muna's free-draining river terrace soils. McKenna's move to the area tripped off a wave of planting by other producers, which included Craggy Range's large neighbouring vineyard.

For the first few vintages, Escarpment's Pinot production was confined to the Escarpment Pinot Noir, a vibrant expression made mostly from Te Muna fruit, with the addition of some grapes sourced from other Martinborough sites.

With the 2003 vintage, McKenna's plans to 'set new directions for Pinot' began to be revealed when he released the single-vineyard Escarpment Kupe Pinot Noir. Produced with the fruit of close-planted (1 metre x 1.5 metre) Abel clone vines on their own roots at the Te Muna vineyard, it is McKenna's homage to Burgundy, a complex, visceral, savoury and compelling wine.

This was followed by the release of a raft of single-vineyard wines with the 2006 vintage. Calling them the 'Insight Series', McKenna said it was 'time to start defining

Left: Larry McKenna, Escarpment Vineyard.

the Martinborough terroir with wines that allow for individual vineyard expression from exceptional sites utilising some of the oldest vines in the district'.

The three new single-vineyard wines joining Kupe — Te Rehua, Moana and Voyager — were all from leased or contract sites on the Martinborough Terrace, close to the township. (Moana and Voyager have recently been rechristened Kiwa and Pahi respectively.) The wines are indeed individuals and represent a bold step forward, for McKenna and the region. Voyager, or Pahi, offers a particularly intriguing glimpse into Martinborough terroir. It is made with fruit from a single clone — 10/5, planted on its own roots — and vines that are 25 years old. The result is a wine that enters the palate without fanfare, but blossoms into a complex amalgam of sweet fruit and black tea notes, and shows rare structural finesse.

The 2006 vintage also saw Escarpment produce an entry-level Pinot for the first time, called 'The Edge'. It is described by the winery as a 'New Zealand Pinot Noir, but most likely Martinborough'. It is unwooded and offers a very good Pinot experience for the price.

In early 2010, McKenna released his 2008 'Insight Series' wines. The traits of each site are even more marked than in the 2006 offerings, confirming for McKenna that the future is all about single-vineyard Pinot Noir:

> The Old World consumer is more interested in the place a wine comes from than the brand or producer. As New World purchasers, we are driven firstly by brand, then district and price. That will change … genuine consumers will eventually become more interested in district and vineyard. But that type of demand will only be driven off the back of wine districts and producers creating Pinots which truly reflect their provenance. As producers we have to allow the vineyards to firstly mature, then understand the fruit we are given and faithfully allow the wines to be expressions of themselves.

V
TONNELLERIE
REMOND
MTL+
ABEL
C.P.
4012
QUALITE
CTB
CERTIFIEE

MARTINBOROUGH VINEYARD

The release of the Martinborough Vineyard Pinot Noir 1984 ended a Pinot drought in the Wairarapa that had lasted over 70 years. Derek Milne and a small group of founding partners — his brother Duncan, Duncan's wife, Claire Campbell, and Russell and Sue Schultz — formed Martinborough Vineyard and began planting the variety in 1980.

While the group was not ignorant of the success William Beetham and others had had with Pinot in the Wairarapa at the turn of the nineteenth century, their Pinot compulsion was driven by Milne's own research and the success Danny Schuster had been having with the variety at Lincoln and St Helena. Milne was also aware that the rewards for getting it right would make any early sacrifice worthwhile.

'The best Pinot Noir grown in Burgundy is the most expensive red wine in the world,' he says. 'It is also the most difficult grape to grow elsewhere in the world. I said that if we can grow it well and make wine approaching red Burgundy in quality, we're quids in.'

They planted various clones obtained from Danny Schuster. Most fell by the wayside, though the 10/5 'coloured up beautifully' from the start, says Milne. The right place had the right vine. The right person arrived in the form of Larry McKenna, in 1986. Although McKenna was a trained winemaker — Martinborough's first — his experience with Pinot was extremely limited. But it didn't take him long to fall under the spell of Pinot and the challenge it presented.

McKenna stayed at Martinborough Vineyard until 2000, winning multiple awards for his Pinots along the way. He has had worthy successors: first Claire Mulholland and then Paul Mason, who took over in 2007.

The Martinborough Vineyard Pinot Noir is the winery's centrepiece Pinot. With 24 vintages to its name, it is one of New Zealand Pinot's elder statesmen, a wine that is starting to evoke terroir as strongly as any. Over recent vintages it has been a sturdy wine, built to last a decade at least, beautifully layered and elegant, a savoury seam balancing the sweet fruit flavours.

Contributing to the wine are 10/5 and Abel vines on their own roots and more than 20 years old, along with younger UCD5 and Dijon clone plantings, all grown on

Right: Paul Mason, Martinborough Vineyard.

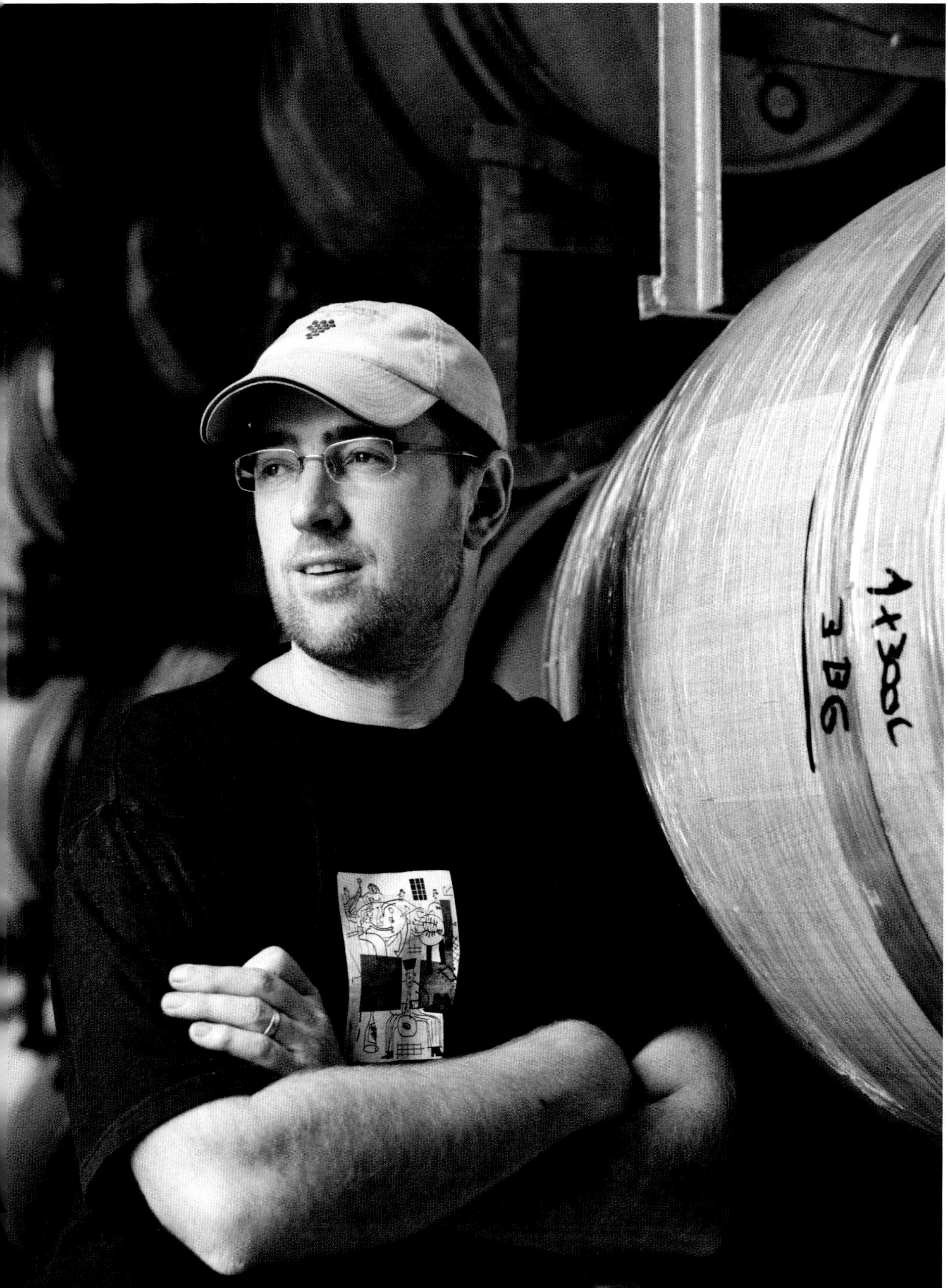

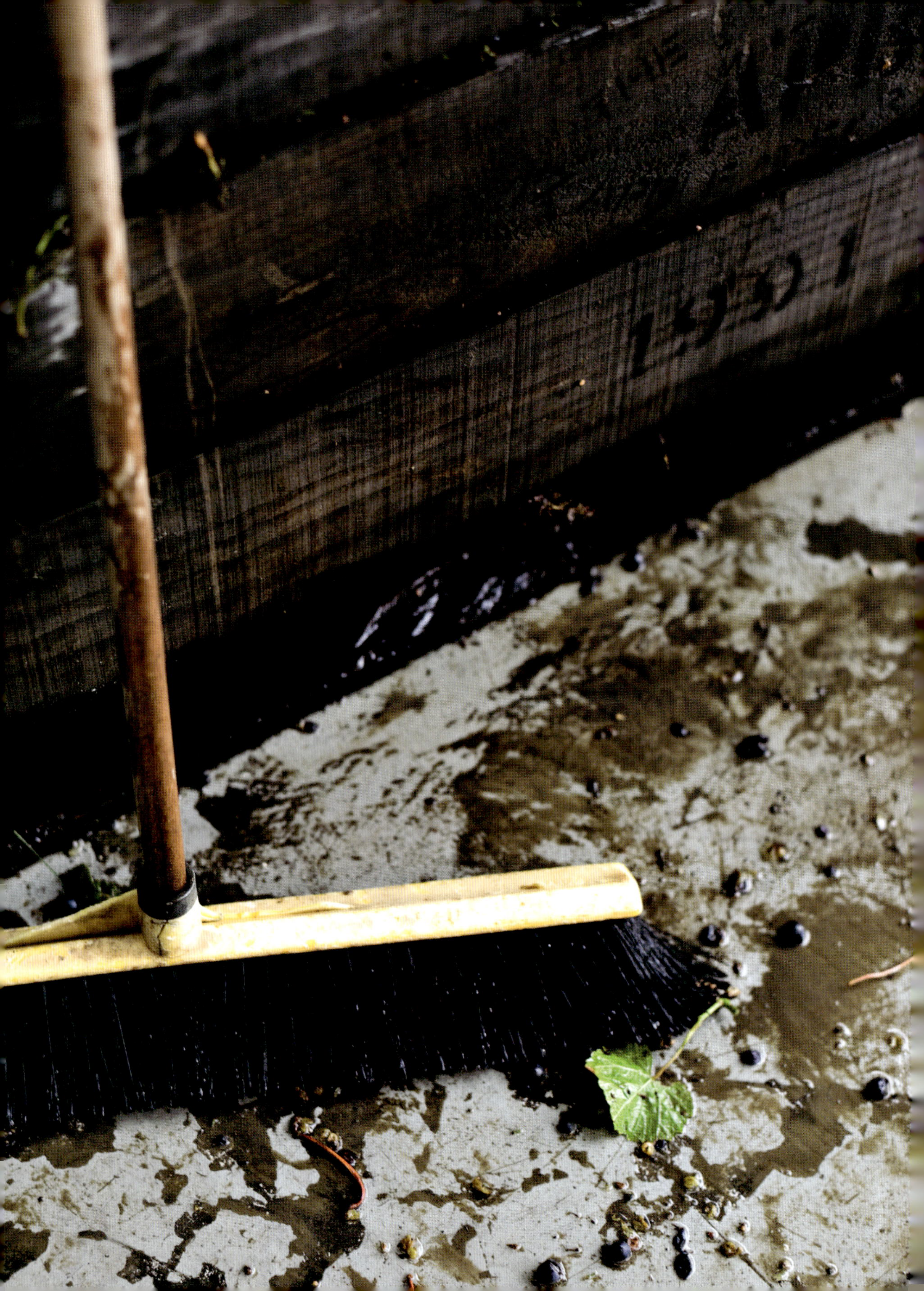

the Martinborough Terrace. Pains are taken in the vineyard to manage the canopy, yield and soil health. 'Our aim for our soils is to make them "biologically and fungi abundant",' says Mason. In the winery the approach is hands-off: natural ferments, no fining, no filtration.

A Martinborough Vineyard Reserve Pinot was produced in exceptional vintages during the 1990s. The reserve label was discontinued in 1999, but in 2003 the winery released the stellar Martinborough Vineyard Marie Zélie Pinot Noir. Priced at $170 a bottle, only 737 bottles were produced, made with fruit from a small parcel of the older vines. A follow-up, the Marie Zélie 2006, was released in late 2009.

The winery produces three other Pinots: the perennially excellent-value Martinborough Vineyard Te Tera Pinot Noir; the single-vineyard Burnt Spur Pinot Noir, grown 'off-Terrace' on heavier soils; and the Russian Jack Pinot Noir, inexpensive and fruit-forward.

Paul Mason wouldn't want to be making Pinot anywhere else in New Zealand:

> I believe this region has the greatest potential to make world-class Pinot Noir. For me, Martinborough achieves better physiological ripeness than other New Zealand regions. This shows as fruit tannin or density coming through in the mid- to back palate. This can be balanced against tannins coming from oak use and, along with ripe fruit, brings the wines into harmony. Our Pinots often shy away from being fruit driven and instead tend towards spicy and more savoury characters as they age, and this is a good thing.

PALLISER ESTATE

Palliser Bay is the large dent in the eastern coastline at the base of the North Island, and Cape Palliser marks both the bay's eastern end and the island's southernmost extremity. Palliser Estate winery is closely associated with another historic local name: Riddiford.

Richard Riddiford, managing director of the Martinborough winery, is a member of a farming family whose roots in the region reach back to 1846. Riddiford was working in marketing in Wellington when he became an investor in Palliser Estate in 1988. The winery had been founded four years earlier by local accountant Wyatt Creech (who went on to become a cabinet minister), himself a descendant of nineteenth-century gentleman farmer and pioneer winegrower William Beetham.

Riddiford brought his marketing acumen to Palliser Estate — and to New Zealand Pinot Noir as a whole — when he became managing director of the company in 1991. Sheep may have been the original Riddiford mainstay, but he was quick to see the exciting potential in this new land use. He has served on a number of industry bodies and was instrumental in setting up the inaugural Wellington Pinot Noir conference, an event that has become a successful triennial showcase.

Palliser Estate grew steadily through the 1990s to become the largest winery in the Wairarapa, its cellar door a popular regional attraction. The winery currently produces 15,000 cases of Pinot in a normal year, though Riddiford pertinently adds 'whatever a normal year is'.

Arriving at Palliser Estate a few months before Riddiford was winemaker Allan Johnson. He has carved a name for himself as a skilled exponent of aromatic varieties, particularly Riesling, yet his Pinot is consistently among the region's finest.

The Palliser Estate Pinot Noir is the winery's flagship Pinot. It is made from fruit drawn from a number of the winery's own vineyards on the Martinborough Terrace. Twenty-year-old 10/5 vines make a key contribution, along with more recent plantings of Dijon clones 115, 667 and 777.

In the vineyard, Johnson follows a 'treat them mean, keep them keen' line on irrigation: 'After fruit set, we manage soil moisture levels carefully to put the vines

Right: Allan Johnson, Palliser Estate.

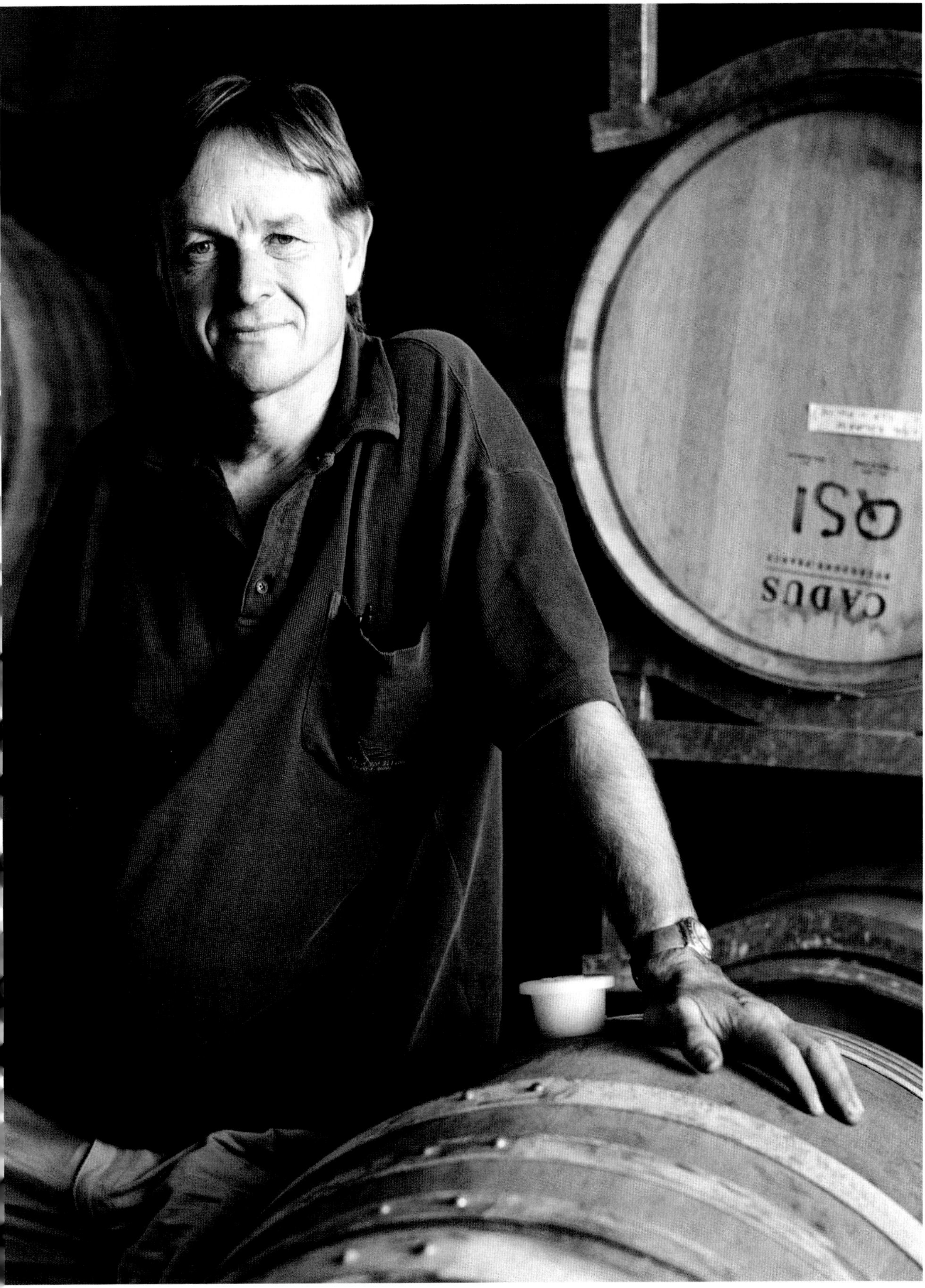
CADUS

07 PALLISER CH WILD

N90

under some stress. We don't want the vines defoliating, but some stress encourages earlier ripening, which allows us to harvest when we want. We tend to pick late … I look for a balance of maturity of skins, seeds and flavours.'

The Palliser Estate Pinot tends to be feminine and graceful in style, with plenty of savoury complexity.

The winery was one of the first serious Pinot producers in New Zealand to release a second-tier label. The Pencarrow Pinot Noir's first vintage was 2000. 'The aim was to introduce the varietal to a new class of people,' Riddiford says. 'A lot of consumers get put off by the words "Pinot Noir". They see it as complicated and expensive. We wanted to show them something that was neither.'

In exceptional years, a rare, premium Pinot is made in small quantities from selected barrels. Only four have been made so far: The Great Bear 2002, The Great George 2005, The Great Harry 2006 and The Great Walter 2008.

VOSS ESTATE

From careers in fisheries research, Gary Voss and Annette Atkins (married in 2003) plunged into the uncertain seas of vineyard ownership and wine production in 1988. Voss had spent a year studying oenology in Australia and gained a couple of vintages' experience before he and Atkins bought a vineyard site on Martinborough's Puruatanga Road, on the stretch between Ata Rangi and Dry River.

Voss's family arrived to pitch in and help the pair plant their original vineyard. In those days, everything — even the hole digging — was done manually. They split the vineyard between Pinot Noir, Cabernet Sauvignon and Chardonnay.

'If I was to do it all again I would plant 100% Pinot from day one,' says Voss. 'The Cabernet was a marketing ploy … the distributors told us that Bordeaux blends were popular and that everyone had to have one of some sort.'

Cabernet Sauvignon no longer figures in the Voss portfolio. The annual output of 1000 cases of Pinot Noir now makes up 70% of this boutique winery's production. Voss Estate makes a single Pinot Noir from fruit grown in three estate-owned vineyards. Contract-grown fruit is used occasionally, 'but only if we get to control it right from pruning'.

In 2004, the original vineyard on Puruatanga Road was sold to Tirohanga, a Martinborough newcomer, although Voss Estate continues to obtain fruit from the site. A vineyard on Todds Road, also on the Martinborough Terrace, was bought in 1999, and later the nearby Kotinga Vineyard.

There is a spread of clones across the vineyards, with the oldest vines on the Puruatanga Road site being Abel, 10/5 and UCD5. Yields are kept to 3–4 tonnes per hectare. Voss says the vineyards have complementary qualities: the free-draining, bony gravels of Puruatanga Road give backbone to the wine, while the presence of clay at Todds Road provides flesh.

In the winery, every care is made to produce a wine of delicacy and balance.

'There's a fine line between concentration and over-ripeness', declares Voss.

In the old days, we welcomed those hot years which gave us inky black wines. But the Pinots I came to enjoy most were from average to cooler years, which had more nuance and interesting flavours, and also kept well.

Now we're careful not to over-extract in the winery. We've backed off from plunging four to five times a day and we're continually tasting the ferment to look for balance. We've also backed off on new oak. Oak tannins give those punchy vanilla flavours but they fade. The fruit tannins from whole bunches are broader and give the wine more longevity.

A combination of experience (Voss and Atkins have close to 20 Martinborough vintages under their belts), vine age, skill and reasonable pricing has made the Voss Estate Pinot excellent value over the years. There are no plans to grow bigger. A recently contemplated change was to produce a single-vineyard wine from old-vine fruit, but in the end, Voss decided the market was not yet sophisticated enough.

EIGHT. NELSON

Nelson has always had a distinctly different feel from other New Zealand towns and regions of similar population density. It has an open, interested, relatively tolerant and surprisingly cosmopolitan heart. This spirit, combined with a gentle, sunny climate and much natural beauty, has made the Nelson region a haven for people seeking what is often loosely described as an 'alternative lifestyle'. It was host to a number of communes in the back-to-the-land days of the 1960s and 1970s, and the local art and craft scene has always been lively.

Which all sounds very Pinot-esque. Indeed, if you were to match a New Zealand region to a grape, going on nothing but general perceptions of the culture and personality of each, you'd instantly pair Nelson with Pinot Noir. A cursory look at the rolling landscape of Upper Moutere would confirm the coupling — in appearance, it's the most Burgundian of any of our Pinot regions.

A cluster of German colonists in the early 1840s tried and failed to establish vineyards and wine production in Upper Moutere. They fled to South Australia, where they found the Barossa Valley far more hospitable. Since then, at various times, a number of small winegrowing operations in the Nelson region have come and gone.

The first known planting of Pinot Noir in the Nelson region was in 1975 at the original Seifried vineyard in Upper Moutere. At the time, Hermann Seifried thought he had planted Gamay, but it later proved to be the Pinot clone UCD22. This became the basis of the early Nelson Pinots, pioneered by a small group of wineries — principally Glovers, Greenhough, Ruby Bay and Neudorf — in the early 1990s.

Comparatively, like its much larger neighbour Marlborough, Nelson was slow to join the Pinot movement. The surge came between 2001 and 2007, when the area planted in Pinot Noir increased from 33 to 188 hectares, at the cost of many an unprofitable apple orchard. By any measure that's an impressive rate of growth, but the second figure does indicate how tiny Nelson's Pinot Noir presence remains overall. Nelson's share is only 4% of the variety's national plantings.

Despite its size, the region's suitability for growing Pinot is today unquestioned. Some Nelson Pinots rank among the country's best, and it's not uncommon to hear Nelson being talked of as the country's most underrated Pinot region. Recent developments such as the Woollaston Estates winery building in Upper Moutere have enhanced the region's profile in regard to producing Pinot. This remarkable, four-level, gravity-fed facility was designed by Oregon architects to cater specifically for Pinot Noir.

The region is blessed climatically, largely thanks to the presence of mountain ranges on three sides — west, south and east. Nelson sits in a pocket, protected from many of the weather systems that buffet other parts of the country.

An abundance of sunshine is the source of much local pride. It is common for Nelson to lead the country in sunshine hours, with about 2400 hours per year. Summer daytime temperatures usually peak at 25°C (although they can top 30°C) and drop to 14°C at night. Average annual rainfall is roughly 900mm, with the driest months being January, February and March. This rainfall pattern suits early-ripening Pinot; it is rare for Nelson Pinot crops to suffer from wet weather.

The region can be divided into two quite distinct subregions: the Waimea Plains and the Moutere Hills. They share similar climates, but differ markedly in soils. The majority of the region's wine is grown on the plains, although most of the recent growth in Pinot planting has been occurring in the Moutere Hills.

The Waimea Plains spread out horizontally to the south and west of the town of Richmond. The soils are not unlike those in Marlborough's Wairau Valley — free-draining loams (variously sandy, silty or clay) over a gravel base. These soils tend to suit aromatic varieties better than Pinot; the best Pinot producers on the plains are those where the soils have a higher clay component, which adds weight.

The low, rolling hills around the old German settlement of Upper Moutere have sandy topsoil over clay subsoil, below which are the deep weathered gravels of an ancient river system. Well-drained, north-facing slopes are features of the better sites.

They may share their clay soils with other regions around New Zealand, but Pinots from the Moutere Hills are developing an interesting profile all of their own. There is a savoury, visceral underlay to many of them. Neudorf winemaker John Kavanagh describes this as a 'dried blood character', an apt descriptor. The best wines from the Waimea Plains are more feminine, with lively aromatics.

MERCUREY FRANCE

"HOPE PN 06'

20/7/07

SB

KEY PRODUCERS

GREENHOUGH

'Am I a Pinotphile?' Andrew Greenhough ponders the question.

'Yes. In terms of what I drink and what provides the most challenge and interest in winemaking … yes, Pinot covers all those bases.

'With Pinot, I slowly allowed what I did to become more complex. The more you know, the more intricate the web becomes. If you're genuinely interested in something, it's the complexity that sustains your interest. That's what makes Pinot so satisfying and keeps so many people interested. It's never boring. It has its disappointments. I wish I could get on with the next vintage right now.'

Andrew Greenhough and his partner, Jenny Wheeler, arrived in Nelson from Auckland in 1990. They bought what was the old Ranzau winery and vineyard near Hope, on the eastern edge of the Waimea Plains. The 1.5-hectare site was planted mainly in Müller-Thurgau, but included little pockets of Pinot Noir, Riesling, Cabernet Sauvignon and Pinot Blanc.

'We just played with that for the first few years,' Greenhough said. 'Then we slowly removed what we didn't want, so the Müller-Thurgau and Cabernet went. Since then we've acquired little bits of land from adjoining properties, planting them as we went.'

After feeling his way through the early years, his knowledge strengthening through experience and attendance at the early Southern Pinot Noir Workshops, Greenhough looks back at 1997 as the breakthrough vintage.

'It was the first year I consciously labelled the wine Hope Vineyard, as I felt it was reflecting this site; 1997 wasn't necessarily the best Hope Pinot, but it was a good one. By the mid-1990s we had more clones to play with and we'd had a chance to look at them over a few vintages.'

Since that vintage, the Greenhough Hope Vineyard Pinot Noir has been the winery's flagship Pinot. The Hope Vineyard itself is nestled into an elevated terrace

Right: Andrew Greenhough, Greenhough.

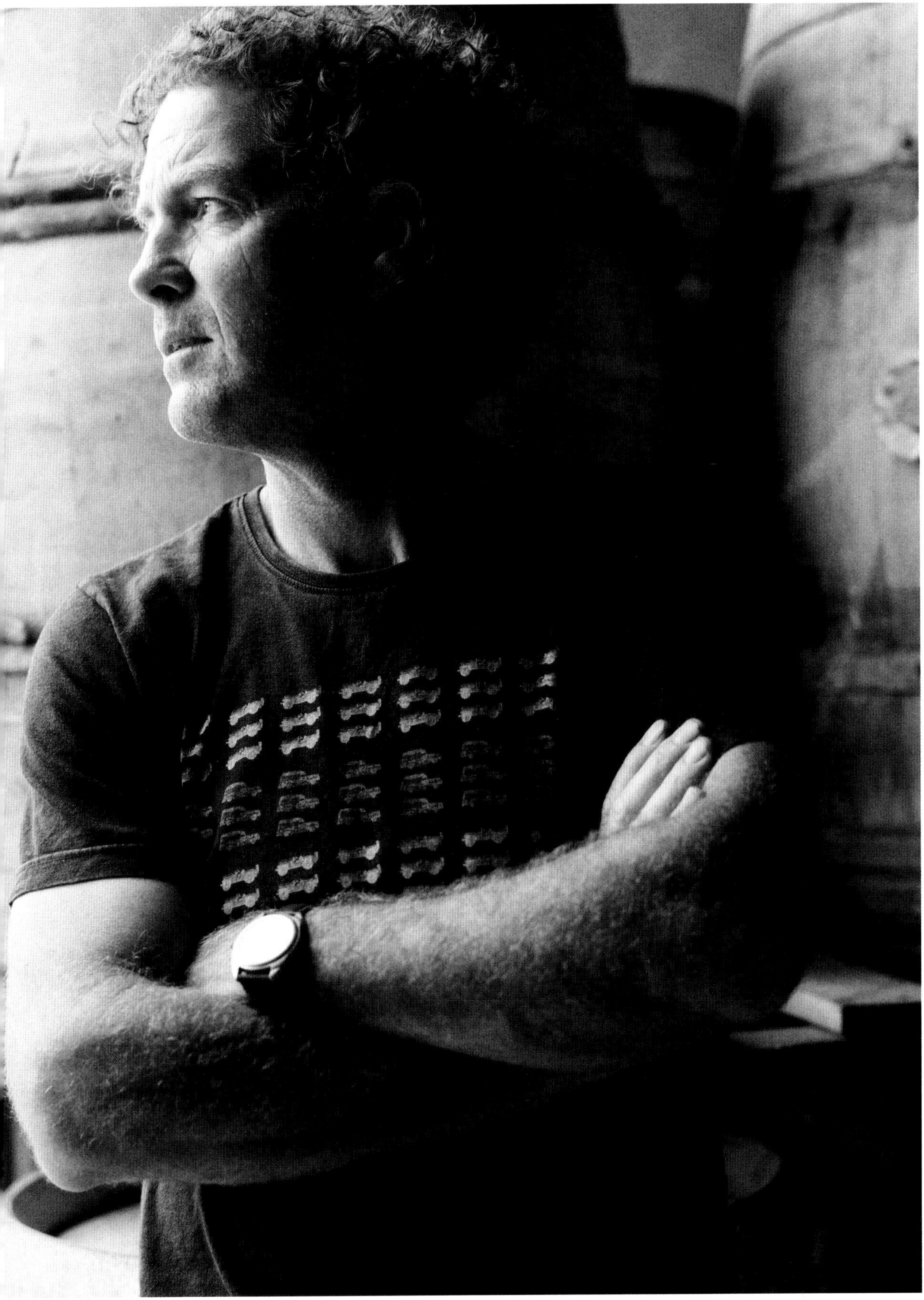

of the southeastern Waimea Plains beneath the foothills of the Barnicoat Range. The proximity of the hills adds to the clay content of the soils, making them an interesting mix of clay-rich loam and river gravels. A spread of clones planted between 1976 and 2007 includes 10/5, Pommard, a number of Dijon clones and the most recent addition, Abel. Greenhough says the 1976 plantings of UCD22 (the clone originally thought to be a form of Gamay) produce very attractive fruit if cropped low. 'Though I wouldn't plant any more of it,' he adds. The vineyard is being managed organically, with a view to obtaining full BioGro certification for the 2011 vintage.

Leading up to harvest, Greenhough is 'looking for flavours without hint of greenness and acids that are no longer harsh. The juice has a weight and fullness at a certain point, and the flavours should also remain primary and fresh. Underripe is green, while overripe is dull and lacks brightness.'

The Hope Vineyard Pinot is typically matured in French oak (both 228- and 500-litre barrels, 30% of which are new) for 12 months. The result is a wine that, in good vintages, shows rare finesse, depth and intrigue, demonstrating a light but deft touch.

The winery makes a second Pinot, the Greenhough Nelson Pinot Noir. It was first made in 1991 as a wine produced entirely from fruit grown on the home vineyard. It has since expanded to include fruit from sites in both Hope and the Moutere Hills (the Barrand Vineyard). These are vineyards Greenhough established and now leases from the property owners.

This wine is made with younger vines and a shorter period of skin contact to produce a more fruit-driven, approachable style of Pinot.

'I imagine my philosophy on Pinot Noir is similar to many who make and love to drink the variety,' Greenhough says.

I have an imagined ideal which I strive to make but have yet to fully achieve. Vintage plays such a big part in shaping the wine and there are always factors beyond control. Often there are elements of the ideal wine successfully captured but others missing. Sometimes it's too big and structured, sometimes not generous enough. Sometimes the wine shows too much dark fruit and not enough of the red cherry and berryfruits which to me should be predominant if the wine is to have elegance and brightness. This is an essential character of the ideal Pinot and is associated with the 'right' aromatics. There should be a sweet centre of fruit to the wine which provides generosity, but this should give way to other more savoury, flora and spice influences. Tannins and acidity should provide structure that is supple and enveloping. Balance is the key and not always easy to achieve.

Following pages: pp148-9 The Barraud Vineyard, Greenhough's site in the Moutere Hills; pp150-1 Andrew Greenhough in the Greenhough Hope Vineyard, close to harvest.

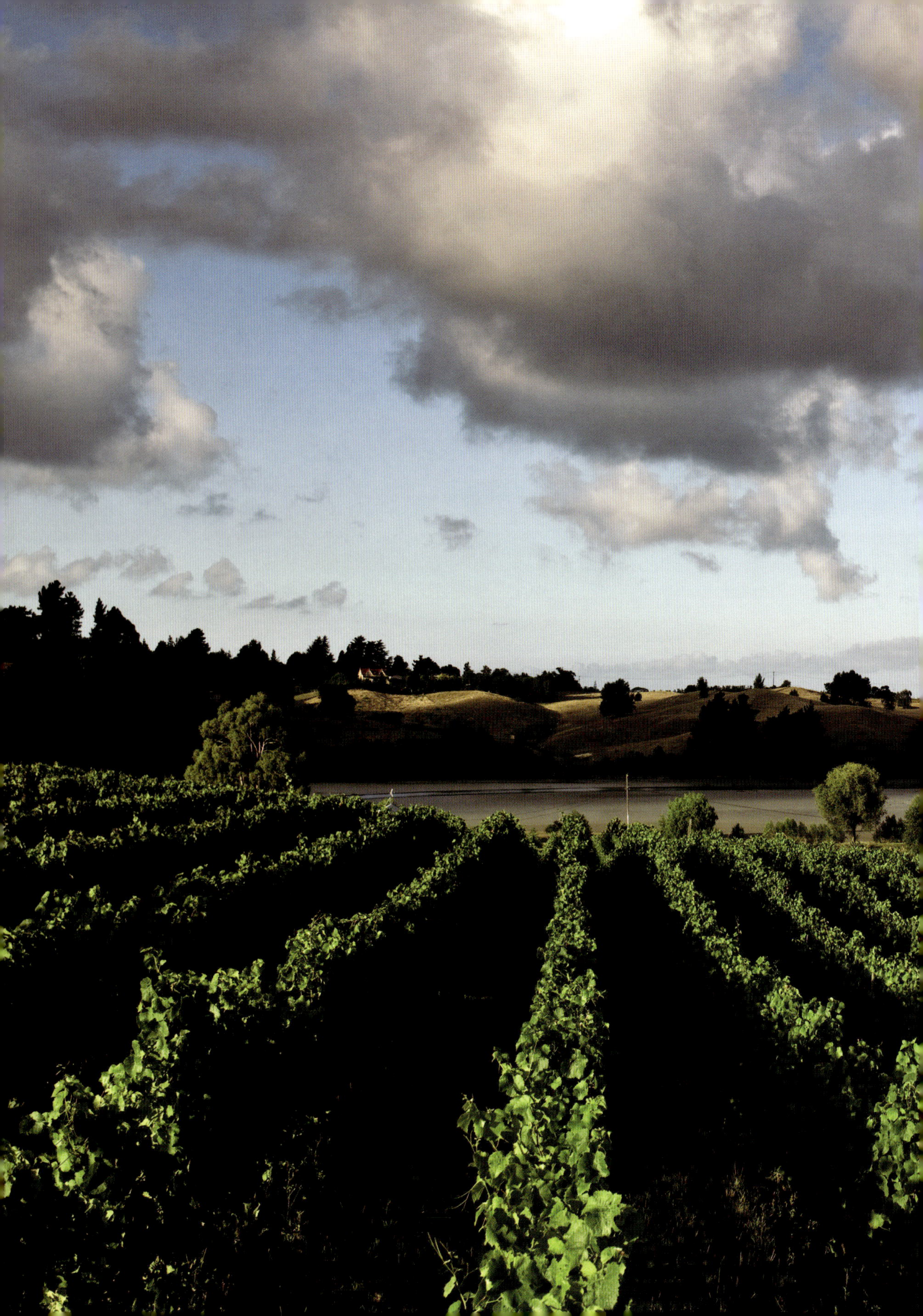

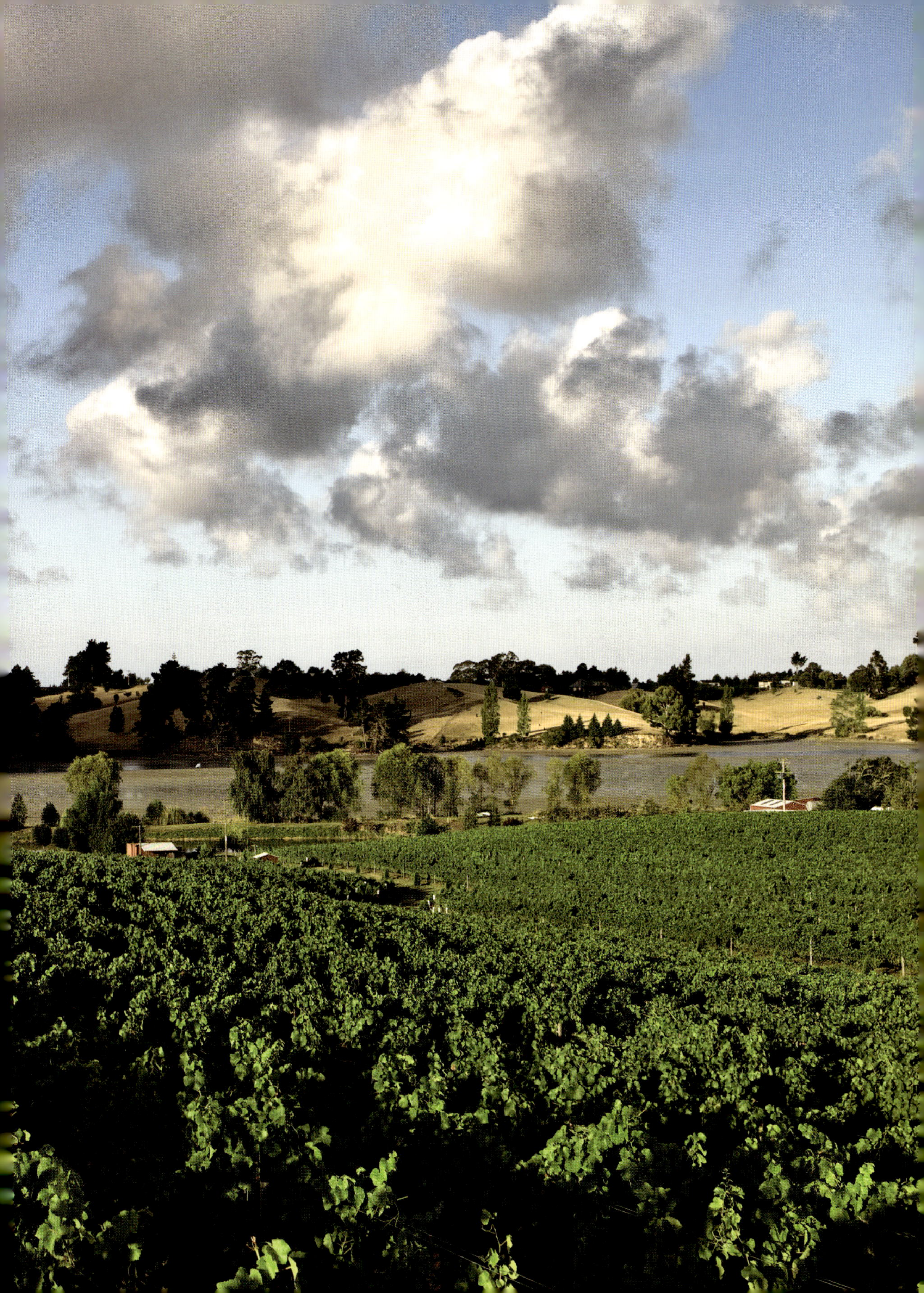

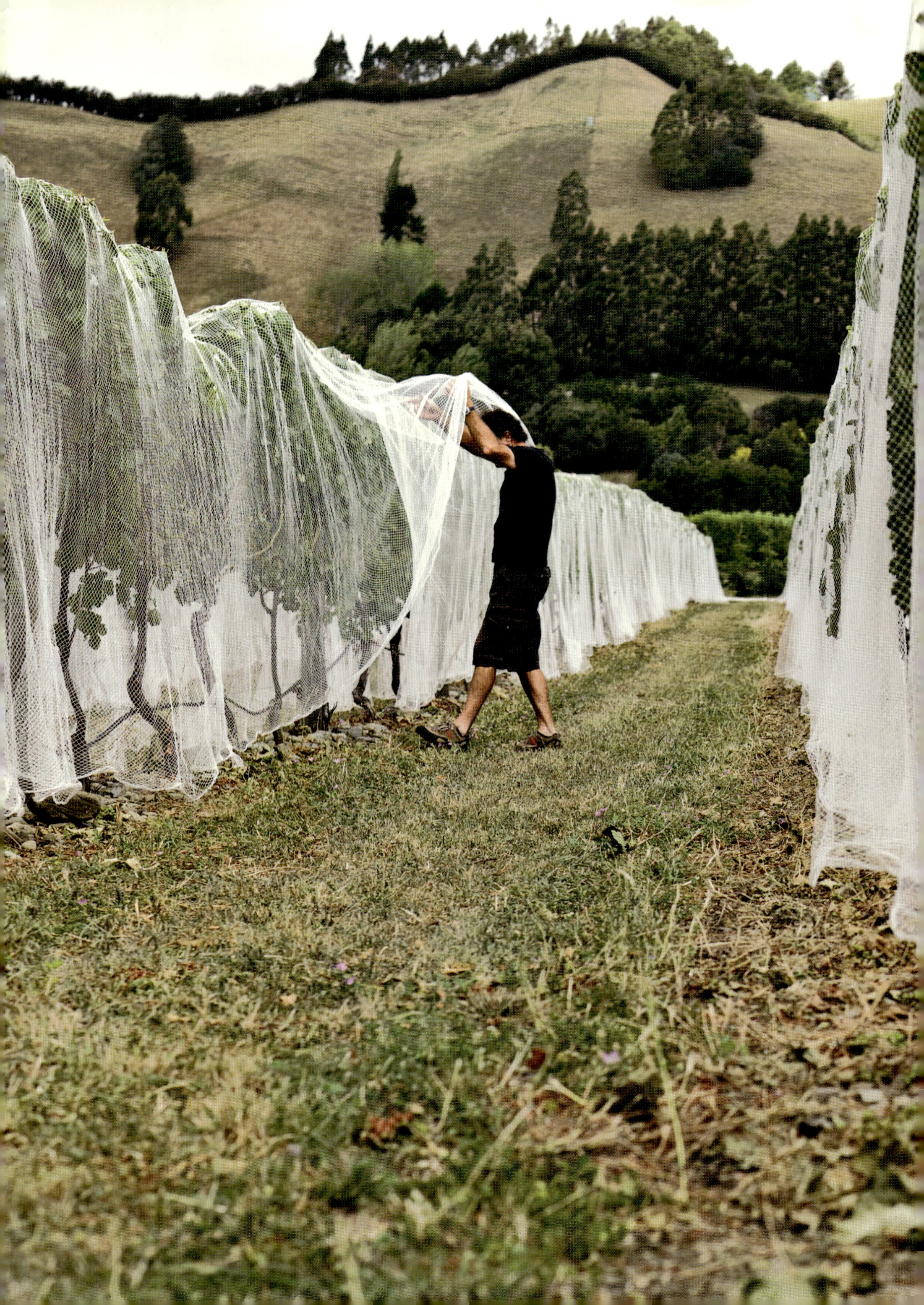

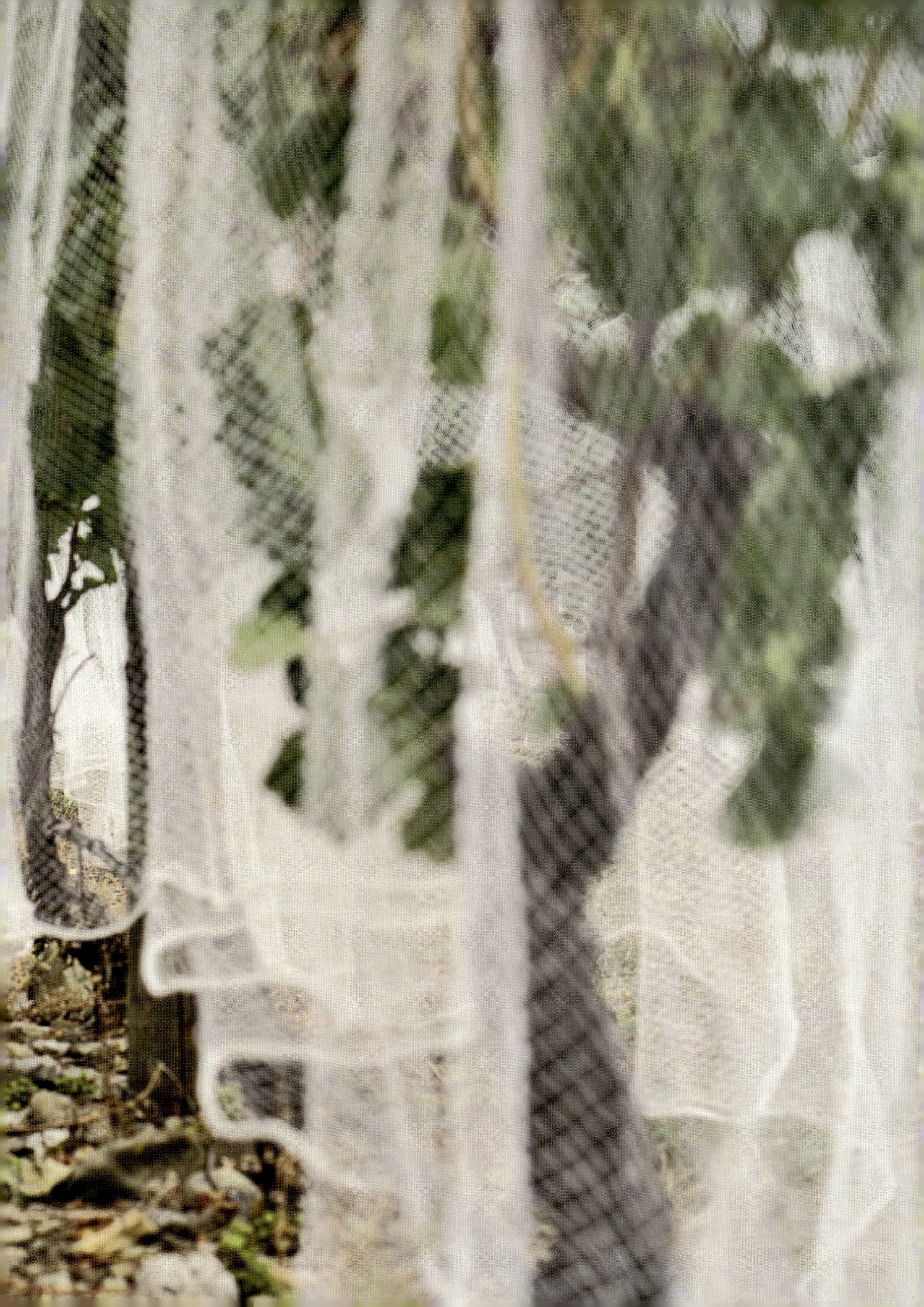

NEUDORF

If Neudorf was the only Nelson winery to make Pinot Noir, the region could still claim significance on the New Zealand Pinot map, such has been (and continues to be) this exceptional winery's contribution. Its devotion to craft is backed up by a distinguished line of elegant wines produced across three decades.

Tim Finn has a master's degree in science; his wife, Judy, has a background in journalism. They bought what had been an old hippie commune on Neudorf Road in the Moutere Hills in the late 1970s and began planting vines. A number of varieties were tried and rejected, among them Merlot, Cabernet Sauvignon, Chenin Blanc and the inevitable (for that era) Müller-Thurgau. Although the début Neudorf Pinot was produced in 1988, Finn regards his first serious attempts with the variety as coming in the early 1990s. He credits this to the Southern Pinot Noir Workshops, at which he was present from the beginning.

Neudorf produces two Pinot Noirs annually, and a third in some vintages. The Neudorf Moutere Pinot Noir is the Pinot that has been made since 1988. The fruit is sourced from two Moutere Hills sites — the Home and Pomona vineyards. Both occupy north-facing slopes, gently rolling as is typical of the area. Finn describes the soils as 'sandy/clay gravels — a glacial deposit formed during the last Ice Age from ancient river gravel bound with a sandy, generally draining clay. This is overlaid by a modest layer of topsoil, initially deficient in many elements.' The first Pinot vines planted on the Home Vineyard were clone UCD22, soon followed by 10/5. These vines are now close to 30 years old and contribute enviable depth and structure to the wine. The Pomona Vineyard is 10 minutes away from Neudorf's home base. It is leased to Neudorf and contributes fruit from 10/5 vines of similar age to those of the Home Vineyard. The Neudorf Moutere Pinot Noir showcases what the Moutere Hills subregion can achieve; it is a refined wine, and treads an interesting path between restrained fruit charm and visceral, earthy austerity.

The Neudorf Tom's Block Pinot Noir took over in 2005 where the Neudorf Nelson Pinot Noir left off. A proportion of the Nelson Pinot had been made with fruit sourced

Right: Tim Finn, Neudorf.

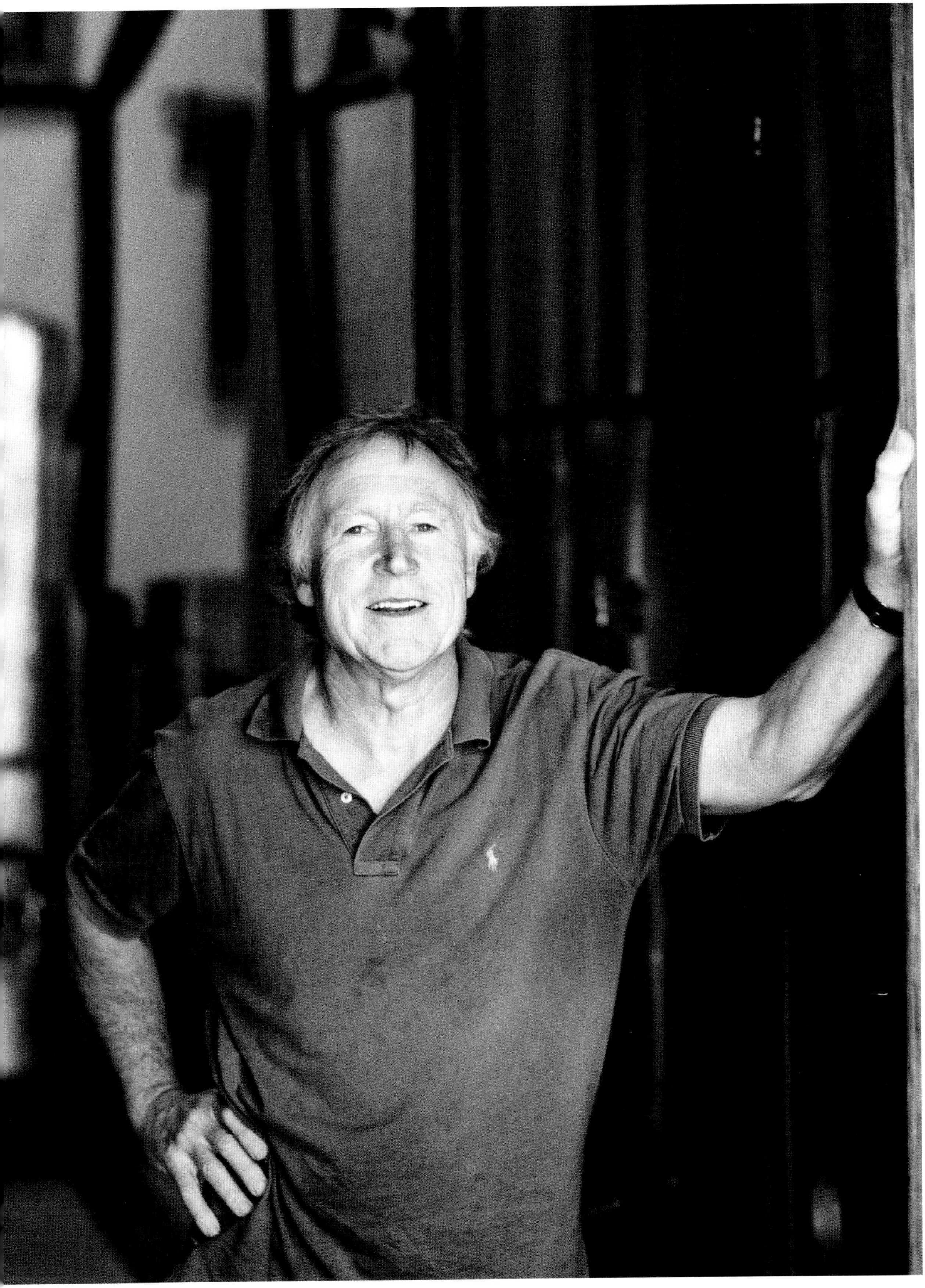

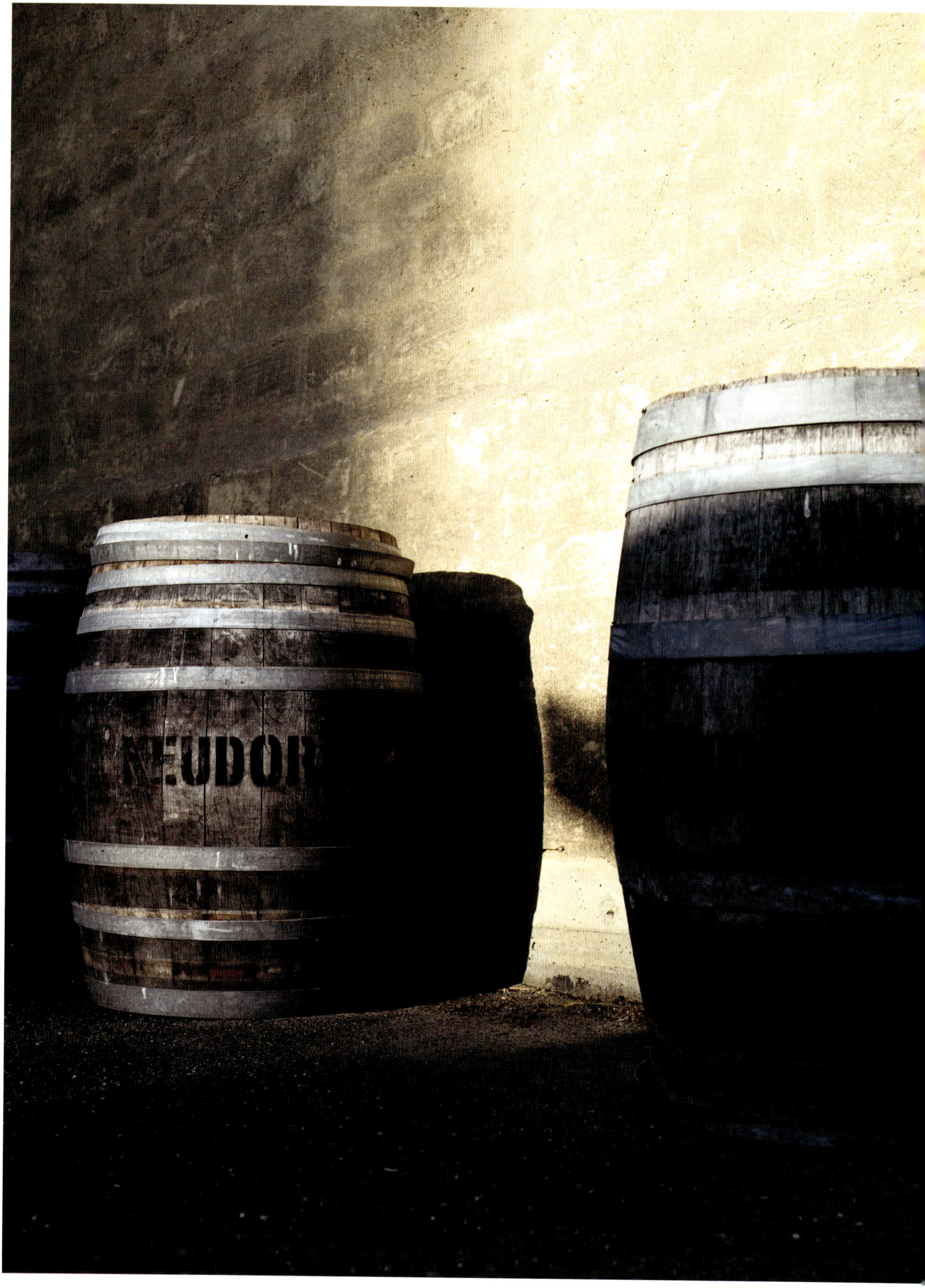
NEUDOR

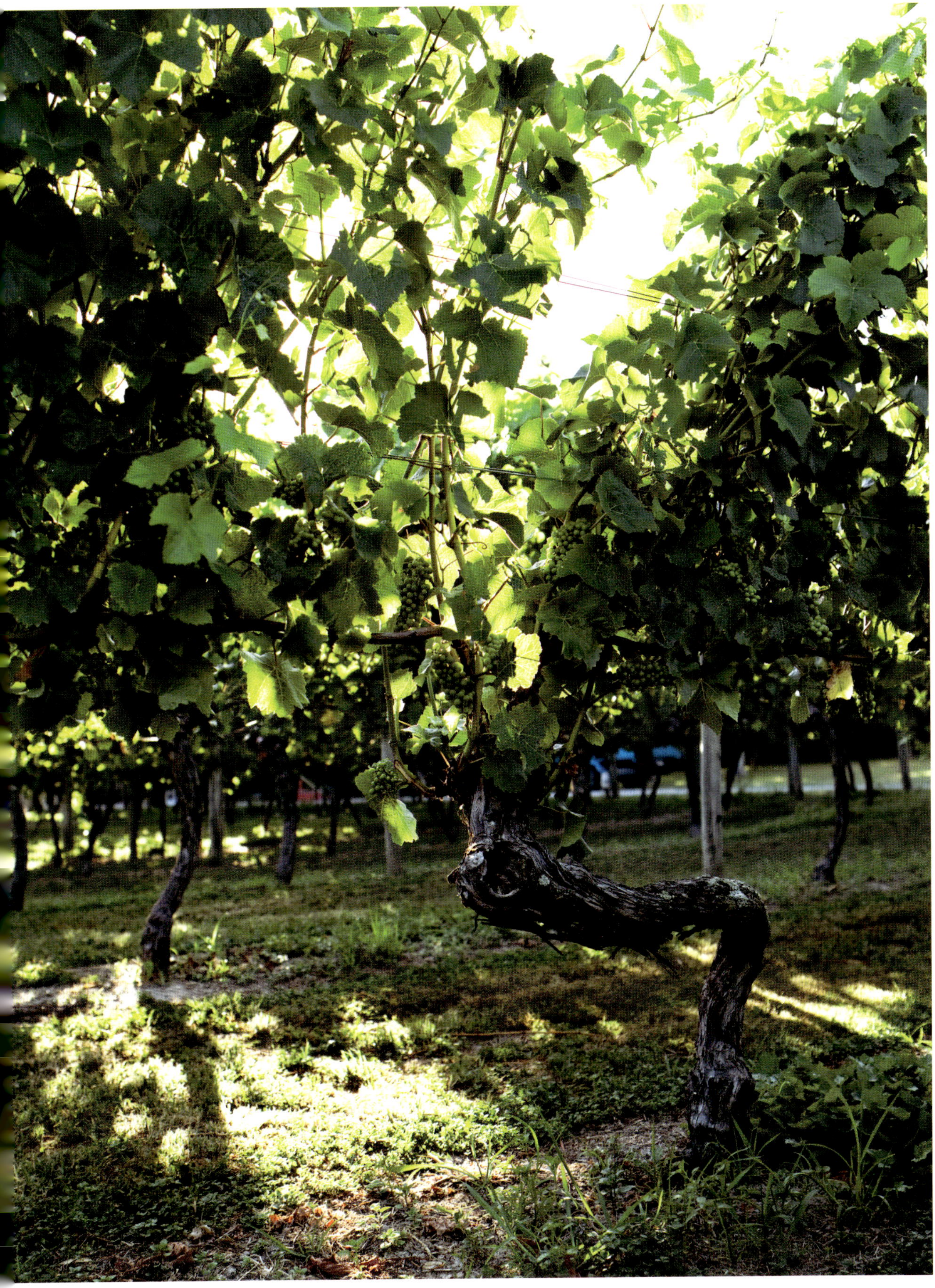

from a Neudorf vineyard at Brightwater on the Waimea Plains. It was decided to rechristen the wine when the fruit from a 4-hectare block adjoining the Home Vineyard came on stream. This block, planted in three clones (UCD5, 667 and 777), was sold to the Finns by Tom Francis, hence the wine's name. The Tom's Block Pinot currently consists of 85% Moutere-grown fruit, with the remainder coming from Brightwater. The Waimea Plains influence comes through in the wine's lighter, sweeter edge.

A small planting of 15-year-old UCD5 vines forms the core of the Neudorf Moutere Home Vineyard Pinot Noir. It is only made in some years and the quantity never exceeds 120 cases. It is a dark, yet ethereal expression. Finn regards UCD5 as being particularly suited to the Moutere clay/gravel soils, 'producing wines with silken phenolics and great finesse'.

Neudorf's vines are not irrigated and the crops are thinned to achieve yields of 5 tonnes per hectare. Winemaking is traditional in style and labour-intensive. All fruit is hand-picked, bunches are hand-sorted, destemmed to small open tanks and macerated for about six days before the natural ferment takes hold. The tanks are hand-plunged up to three times daily. Time in oak (between 20 and 35% new) ranges from 11 to 15 months, and the wines are usually bottled unfined and unfiltered.

Tim Finn sums up Neudorf's approach this way:

> As a starting point we expect our wines to exhibit the characteristics of good wines everywhere — balance, ripe flavours and tannins, sufficient concentration, and length. These result from astute sustainable viticultural practices and on-site experience spanning three decades. But our aims go beyond that and incorporate our personal preferences. We admire and aim for wines that have the complexity to be intellectually stimulating, wines that reflect their sense of place and characters other than ripe fruit, wines that have elegance and some transparency. These are the characteristics that can move Pinot Noir beyond good to great.

Right: Tim and Judy Finn established Neudorf on the site of an old hippy commune.

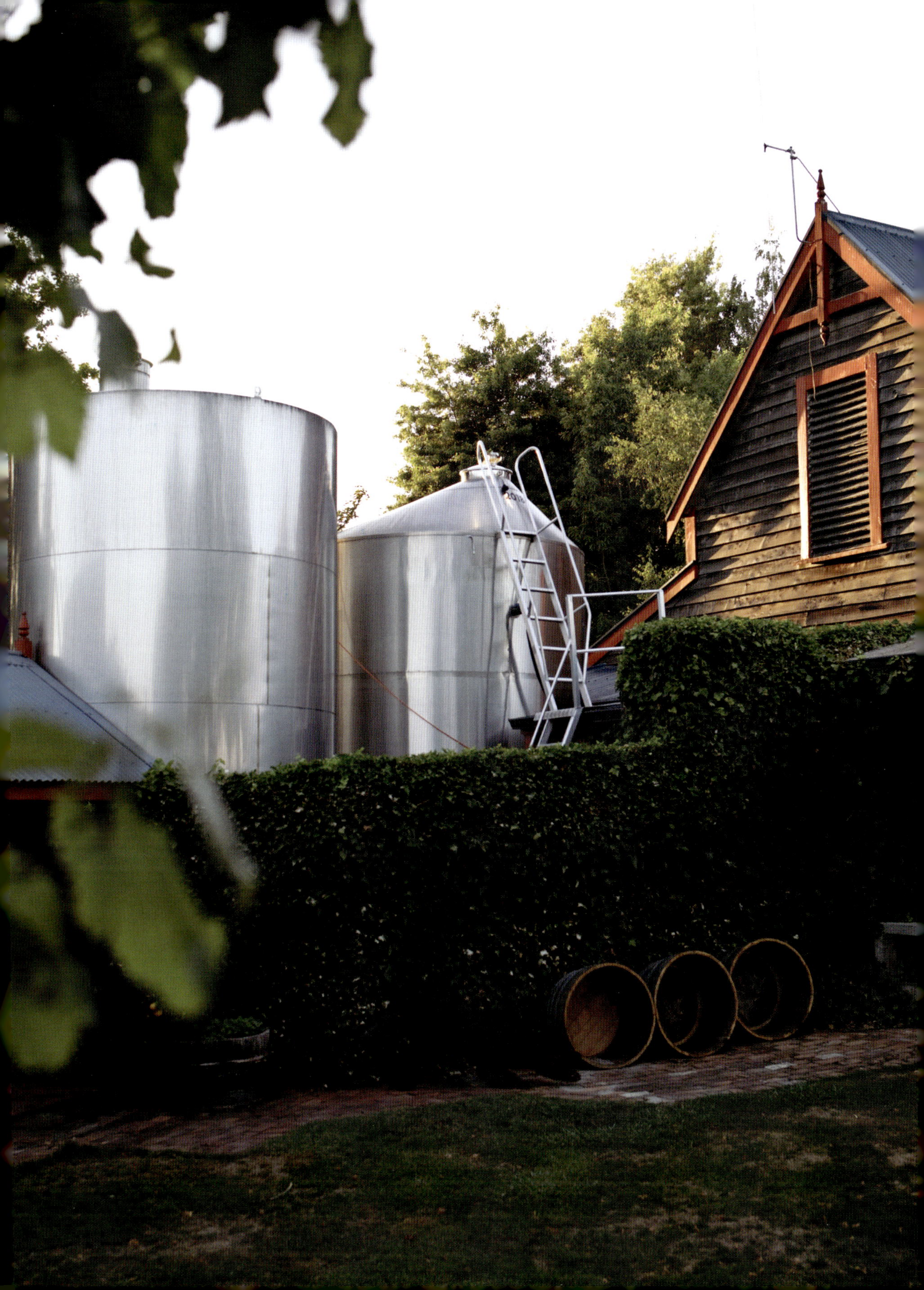

NINE. MARLBOROUGH

Marlborough is to New Zealand wine what rugby is to New Zealand sport: the big presence. And Sauvignon Blanc is to Marlborough what the All Blacks are to New Zealand rugby: the global face — and cash cow — of that presence.

Grassy, racy Marlborough Sauvignon Blanc, a wine style that redefined the grape variety for consumers around the globe, is easy to make in quantity, easy for consumers to understand and — until very recently, at least — has been easy to sell. Marlborough has happily prospered from the facility that underpins 'Sauvignonomics'. However, such giddy success with one variety has made it difficult for others — Pinot Noir included — to establish a strong regional base.

The first Pinot Noir vines were planted in Marlborough at the same time as Sauvignon Blanc, when Frank Yukich of Montana made his now-famous initial foray to the region in 1973. This was 100 years after farm manager David Herd started the region's first commercial winery, Auntsfield, in the Ben Morven Valley. The present-day Auntsfield winery occupies the same site and, in a tribute to Herd's pioneering endeavours, a dash of his 1905 brown Muscat was added to Auntsfield's Heritage Pinot Noir 2005.

Yukich and his viticulturist Wayne Thomas favoured the Swiss Pinot Noir clone Bachtobel (also preferred by Nick Nobilo at Huapai in the 1970s), and in 1975 the company made the decision to plant 20 hectares of Bachtobel on the Fairhall Estate vineyard on the southern side of the Wairau Valley. In 1979, the year of the company's first commercial vintage of Sauvignon Blanc, it also released Marlborough's first-ever Pinot.

'Pinot Noir, the noble grape of the Burgundy region, grows to perfection in Marlborough', declared the label of the Montana Marlborough Pinot Noir 1979. That brazen expression of confidence in the region-variety pairing proved premature — at least in regard to the making of still Pinot Noir. The wine reappeared only twice over the next 17 years, in 1983 and 1989. Lindauer, Montana's inexpensive and popular méthode traditionelle, became the favoured destination for the company's Marlborough Bachtobel grapes.

The Sauvignon tide spread out across the Wairau Valley during the 1980s. Viticulturist Mike Eaton remembers swimming against it in 1990 when he returned inspired from France and developed the Clayvin Vineyard on the clay-dominant north-facing slopes of the Brancott Valley. In association with Swiss vigneron Georg Fromm, Eaton close-planted predominantly Pinot Noir. He recalls:

> The biggest companies actually offered the greatest support. They helped out with ideas and equipment to get us over some of the earlier obstacles of the close row spacings. Generally, though, most locals saw us as a threat to the trophy status of their flat Sauvignon Blanc-dominated vineyards. We were a challenge in the form of red grapes (God forbid), high density (how on earth will you get the Massey Ferguson down those pathetic rows?) and hillside planting (outright heresy). One valuer (who himself grew Müller-Thurgau at 25 tonnes per hectare) predicted we'd go bust inside five years. My, how things change.

ONE VALUER PREDICTED WE'D GO BUST INSIDE FIVE YEARS. MY, HOW THINGS CHANGE.

The breakthrough year for Marlborough Pinot is generally considered to be 1994. Mike and Robyn Tiller, founders of Isabel Estate, had planted a 14-hectare vineyard in Pinot Noir in the lower Omaka Valley in 1987. The Tillers were alert to the suitability of the site's soils — dense clay laced with some calcareous elements — especially after discussions with visiting Burgundy experts Jasper Morris and Remington Norman. They planted clones drawn from a variety of sources, including some 10/5 from Ruby Bay Vineyard in Nelson and Pommard supplied by Danny Schuster. Vine density was a revolutionary (for the time) 2 metres x 1 metre.

The vineyard's first few vintages were absorbed into the Cloudy Bay sparkling wine Pelorus. That changed in 1994, an excellent Pinot vintage, when the harvest was shared among a number of interested winemakers — Rudi Bauer at Giesen, the Donaldsons at Pegasus Bay and Hätsch Kalberer at Fromm. Kalberer also used some of the fruit to make a first Pinot for Isabel Estate.

The prescience of the Tillers' planting soon became apparent when some of these wines began collecting gold medals and trophies. Ironically, the one that is perhaps remembered most fondly, the Fromm Winery's La Strada Pinot Noir Reserve 1994, wasn't entered into shows because of company policy. That didn't prevent it finding its way into high-profile tastings in Australia and Europe where it was more than once mistaken for a great young Burgundy.

'It helped put Marlborough [Pinot Noir] on the map,' says its maker Kalberer. 'People were saying there's nothing coming out of this region, then here was this concentrated wine that gave so much pleasure and was very much at the level of the Martinborough Pinots at that time.'

Despite this flurry, the Pinot uptake in Marlborough was relatively slow. In the 1998 edition of Michael Cooper's comprehensive *Buyer's Guide*, there were still only 13 Marlborough Pinots listed, just three more than Hawke's Bay. That year of drought (1998), which was unkind to the region's Sauvignon Blanc harvest, spawned another Marlborough Pinot to make waves. The Wither Hills Pinot Noir 1998 swept all before it at the 1999 Air New Zealand Wine Awards, taking Champion Wine of the Show.

Serious growth began from 2000. In that year, Marlborough had 527.8 hectares planted in Pinot Noir, more than 20% of which was used to make sparkling wine. Over the ensuing decade, this sparkling-wine share receded and overall Pinot plantings took off. By 2009, Marlborough had 2000 hectares of Pinot Noir in the ground. This is far more than any other region, and in fact accounts for close to half the country's Pinot Noir output.

Marlborough's late charge received impetus when, after taking a wait-and-see approach, the large companies Montana and Villa Maria both decided to expand their still Pinot operations. Marlborough was the chosen region because of its remarkable consistency from vintage to vintage. Montana's expansion was impressive. In just a few years the company become one of the top three producers of still Pinot Noir in the world.

Pinot's gains certainly did not come at the expense of Sauvignon Blanc, which over the same period increased in area more than five-fold.

Marlborough was all about one valley — the mighty Wairau — during its early years as a wine region. The Wairau Valley is a large flood plain, running west to east, with the river of the same name flowing along its northern side to meet the sea at Cloudy Bay.

The valley is a favoured spot viticulturally. The dark, bush-clad Richmond Ranges form a northern wall that keeps at bay much of the precipitation borne by the prevailing nor'westerly winds. Along the southern side the Wither Hills, and beyond them the higher Kaikoura mountains, shelter the valley from the cold stab of the southerlies, so often the bane of Waipara and the Wairarapa during flowering in the spring.

Inside this protective fence is one of the country's driest and sunniest climates. While the growing season is warm, it is not overly so, thanks in part to cooling summer sea breezes from the east. As with all the country's Pinot regions, the diurnal shift is pronounced. Marlborough growers like to point to the region's low mean temperature of the warmest month — 17.4°C, which is 2.2°C cooler than Dijon in Burgundy — to illustrate Marlborough's long, unrushed ripening season, so important for the retention of fruit characters.

There is nothing monolithic about the Wairau terroir. The valley's northern side is both warmer and wetter than its southern side. Of particular significance to Pinot Noir, the soils also vary markedly. The free-draining gravels near the river give way to higher levels of clay as you move south.

When interest in Pinot began to rise in the 1990s, many wineries planted the grape on the gravels alongside their Sauvignon Blanc vines. These soils provide aromatic lift and a lean, bony texture — great for the Marlborough style of Sauvignon; not so great for Pinot. The wines tended to be simple and lacking in warmth, weight and generosity. Not helping was a pervasive Sauvignon viticultural mentality with its high-yield emphasis, along with a continued reliance on the older clones, particularly 10/5.

From the mid 1990s it was apparent that the low-vigour clay soils on the Wairau's southern side delivered Pinots with deeper colour and better mid-palate texture. Marlborough's upswing in Pinot planting over the past decade has been marked by a southward drift in order to take advantage of these conditions. In concert with improved clonal material and viticultural practices, what became known as 'southern valley flesh' raised the quality bar for Marlborough Pinot.

The southern valleys are a trio of smaller valleys — the Brancott, the Omaka and the Waihopai — that run perpendicular to the southern edge of the Wairau. Their north-facing, clay-rich slopes today provide most of the region's key sites for Pinot Noir.

Marlborough's other main valley, the Awatere, has also shown remarkable growth over the past decade. Running roughly parallel to the Wairau on the southern side of the Wither Hills, the Awatere is altogether a tougher grape-growing environment. It is cooler and drier than the Wairau and gets raked by the nor'westerly winds during the summer. The frost danger is high in many parts of the valley. The soils are predominantly stony alluvial, although there are pockets of deeper wind-blown soils.

There is plenty of Pinot planted in the Awatere, but the valley has established more of a reputation for its zesty, acidic expressions of Sauvignon Blanc and Pinot Gris than for Pinot Noir. Awatere Pinots lean towards lighter, vibrant, slightly herbal styles, although there are exceptions.

It has become something of a cliché to place the style of the best examples of Pinot from Marlborough somewhere between Central Otago and Martinborough, an amalgam of lushness and spicy depth. As with all clichés, it has achieved that status because it is apt and worth repeating. The pristine fruit that defines the region's other varietals is certainly present in its Pinot Noir.

Pinot Noir will always be a Marlborough subset, but it's a subset that is becoming more assured with every passing vintage.

KEY PRODUCERS

CHURTON

Shropshire-born Sam Weaver became smitten with red Burgundy during his years working as a fine-wine buyer in London. 'I always found good Burgundy more interesting than Bordeaux,' he says. 'I particularly remember a magnum of 1959 Clos-de-Bèze from Armand Rousseau that was opened amongst some very fine old Bordeaux. Its aromatic dimensions, the peacock's tail finish, its voluptuousness and femininity … alongside it the Bordeaux seemed very austere. That's become a vision for me, a perception of what is possible. The trouble is, what's possible with Pinot is almost infinite.'

Sam and his wife, Mandy, arrived in New Zealand in 1988. Sam's first job was fine-wine consultant at New Zealand Wines and Spirits. He remembers doing an assessment of all the local Pinot Noirs on the chain's shelves at that time.

'I recommended we kick all bar three of them out,' he recalls.

He'd always harboured a desire to make wine, and that began in 1989 at Hunters. By 1999 he was chief winemaker at the Corbans-owned Stoneleigh winery in Marlborough.

The Weavers launched the Churton label with a small amount of Sauvignon Blanc made with grower fruit in 1997. Encouraged, and eager to plant Pinot on a Marlborough hillside, they bought their Waihopai Slopes site on the northern side of the Waihopai Valley in 1999. Sam describes it as 'a wonderful block of land that had a great feel about it'.

It is certainly a singular site: 200 metres above sea level, gently undulating, mostly northeast facing. The Weavers eschewed earthmoving and used varying row orientations to make the most of each aspect, although sun exposure has been treated with care. 'Lowering exposure to that intense cooking afternoon sun makes a big difference for us,' says Weaver. Water has been a challenge. Weaver was hoping the high water retention of his clay loess soils would allow him to forgo irrigation, but he was forced to backtrack when the young vines struggled during rainless periods and now irrigates 'sparingly'.

The planting was not rushed. Weaver wanted to observe each block for a couple of years and learn from it before proceeding to the next. There are now five blocks planted in Pinot Noir — Abyss, Bowl, Dog's Leg, Shoulder and Clod — the last planted in 2009. The clonal mix is varied, though the Dijon clones have a majority.

'We pick by block or area, rather than by clone,' Weaver explains. 'All our ferments are multi-clonal. In terms of quality parameters, clones are way down the list. I rate site, soil, water management and vine density all higher.'

Weaver's desire to make a wine reflecting this special site has led Churton towards biodynamics. The vineyard is now largely managed biodynamically, and the aim is for the entire vineyard eventually to receive BioGro organic certification.

In the winery, all the fruit is destemmed. Says Weaver: 'I don't see Marlborough stems as being a positive influence. I look for fruit tannins and seed-derived tannins, achieved from long macerations.' Indeed, the warm, post-ferment maceration lasts three to four weeks, which Weaver says is crucial 'for integration and back palate focus rather than just sweet fruit'. The exposure to new oak is usually minimal (20–25%).

The first Pinot produced by the winery was the Churton Pinot Noir 1999. There has been a Churton Pinot Noir every year since, while a portion of fruit from the Abyss block was used to make a limited-release (150 cases) Abyss Pinot Noir in 2008. 'Those mature vines in Abyss are providing another dimension … great breadth and length, the fruit is joyful with something more intellectual,' Weaver says.

Weaver most assuredly doesn't make fruit bombs. Churton Pinots are typically lightish in colour, firm but fine in structure, and offer a beguiling array of flavours, from berries to lower, earthy notes. They are intense, but undemonstratively so; they keep asking something of you each time you return to the glass.

Following pages: p166 Churton's elevated vineyard on the northern side of Waihopai Valley; p167 Looking across the Wairau Valley to the Richmond Range from Cloudy Bay's Barracks Vineyard.

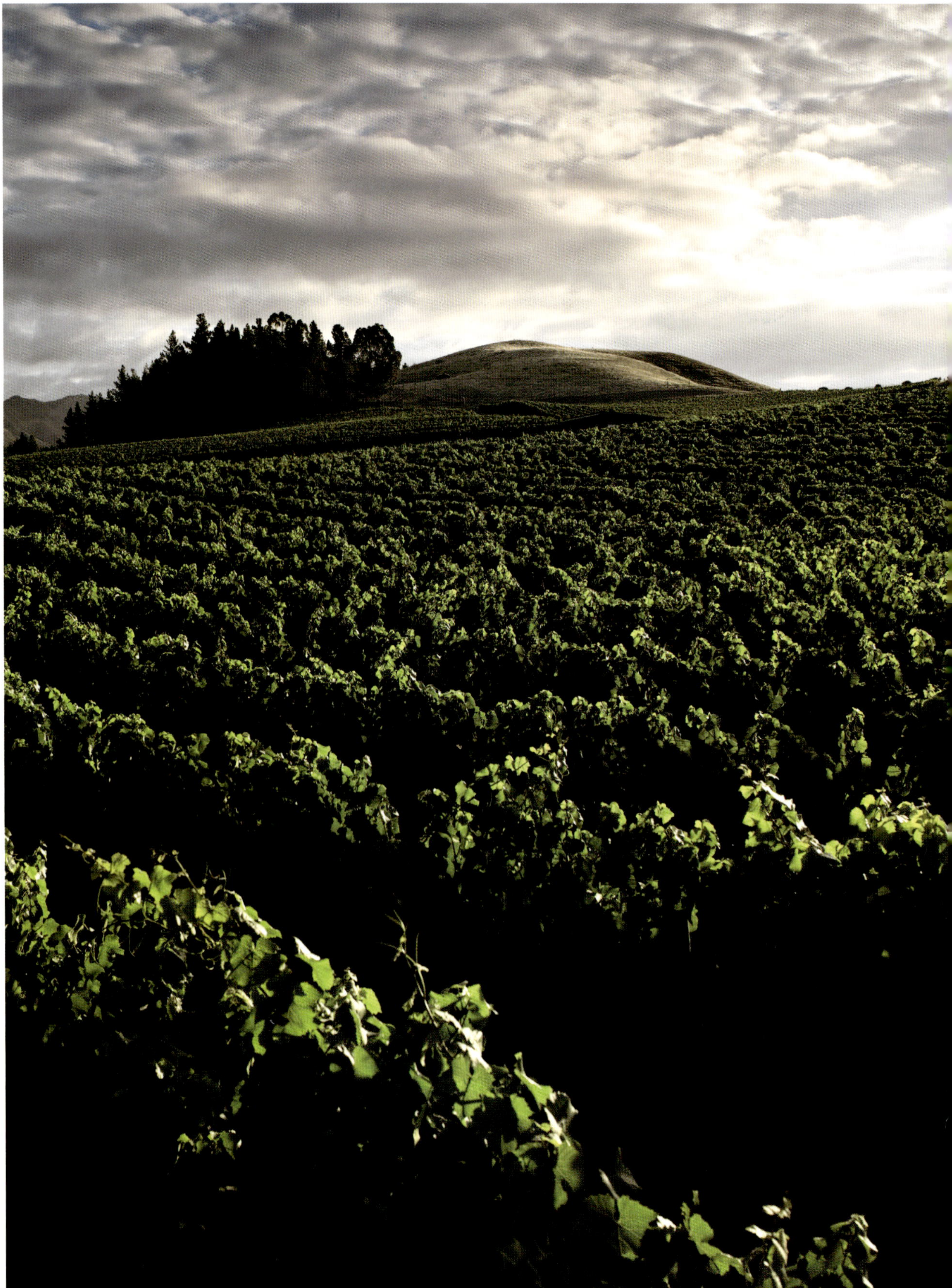

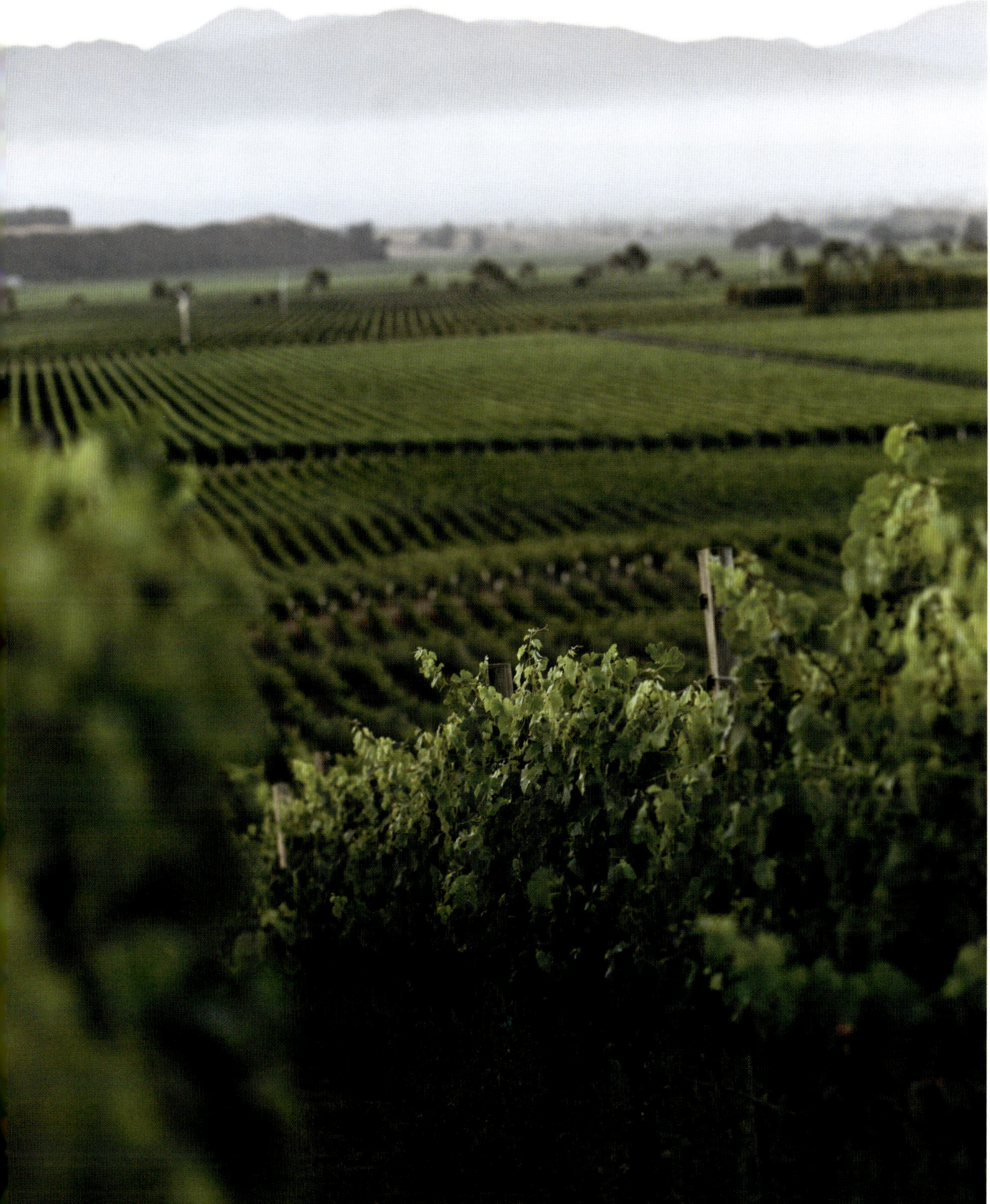

CLOUDY BAY

That most revered and potent of all Marlborough Sauvignon Blanc labels, Cloudy Bay, began producing still Pinot remarkably early. The first vintage was 1989, four years after founding winemaker Kevin Judd had produced the winery's inaugural Sauvignon.

As with most of the early Pinot production in Marlborough, there was a work-in-progress quality to Cloudy Bay's first attempts. The desire was there, says Judd, but lack of knowledge and unsatisfactory clones on alluvial gravel sites (whose Pinot fruit was more suited to making Pelorus, the winery's sparkling wine) held things back. Some poor vintages in the early 1990s were also a factor in Cloudy Bay only producing a varietal Pinot at every vintage from 1996.

There was a reliance on grower-supplied fruit during the early years. Again, this was not unique to Cloudy Bay. Marlborough has always been a region where contract growers have played a significant role. This did present challenges with Pinot, as many growers had become accustomed to the high yields and low labour demands of Sauvignon Blanc and the early sparkling-wine-designated Pinot plantings. Cloudy Bay, says Judd, worked around the high-yield mindset by purchasing its Pinot fruit on a per hectare basis, thus ensuring grower acceptance of the winery's yield expectations.

Winemaker James Healy arrived in 1991 and gave Cloudy Bay's 'Pinot project' an injection of fresh enthusiasm and talent. The variety's stock began to rise steadily while those of Cabernet Sauvignon and Merlot, for which winery founder David Hohnen had had early hopes, dipped. (Hohnen's proprietal association with Cloudy Bay ended in 2001. The company is now completely owned by French luxury goods group LVMH Moët Hennessy–Louis Vuitton.)

A leap forward was taken in 1994 when Cloudy Bay obtained the 45-hectare Mustang Vineyard on the heavier soils of the Brancott Valley. The move marked the beginning of a series of 'southside' acquisitions, with a site in the Omaka Valley joining the fold a couple of years later.

In Kevin Judd's estimation, 'respectable' Cloudy Bay Pinot began to emerge from the 1996 vintage.

Emblematic of Cloudy Bay's strengthened commitment to Pinot Noir over recent years is the 34-hectare, sloping Barracks Vineyard, a heat-trap in the Omaka Valley that was planted in 2004. It is given over entirely to Pinot, planted with Pommard, Abel and a number of Dijon clones.

Judd left Cloudy Bay in 2008. Overseeing Pinot production is current chief winemaker Nick Lane. His stated goals are lofty: 'We want to make the Cloudy Bay Pinot Noir as successful as our Sauvignon Blanc. We feel that we've only scratched the surface in terms of potential.'

Cloudy Bay makes a single Pinot Noir, a wine which has achieved a poise and delicacy over recent vintages. The arrival of fruit from the Barracks Vineyard has been influential and it now accounts for over 50% of the blend. The grower component is just over 20%. In 2008, a small quantity of a single-vineyard release using fruit from the Mustang Vineyard was produced.

Lane comments:

In terms of style, we want to make elegant and structured Pinot Noir with all the abundant fruit characters our climate affords us, but with a distinct savoury edge. Over the past five years we have been in a phase of maximum extraction, but as the vines age we are starting to see the vineyard give us more concentration than we necessarily need. This allows us to 'finesse' the extraction, fine-tuning the tannins so we finish with a silky wine with ageing potential.

Another one of our goals is to make Pinot that will not only age, but improve over an 8- to 10-year period.

Since 1999, the winery has also annually held the popular event Pinot at Cloudy Bay. Consisting of a tasting of Pinots from around the world followed by a long lunch, it attracts Pinot lovers from around the country and beyond.

DOG POINT

Taking its name from a hill which was home base to a pack of feral canines that roamed the district over 100 years ago, Dog Point Vineyard was established in 2002 by two couples: Ivan and Margaret Sutherland, and James and Wendy Healy.

Ivan Sutherland is a member of one of Marlborough's great sporting families. He was an Olympic rower who won bronze at Montreal in 1976, while his brother Allan was an All Black of note. Growing grapes, he decided, was a more profitable (and more fun) use of Marlborough land than the sweetcorn, peas and various stock of his parents' farm. And rowing trips to Europe in the late 1970s helped convince Sutherland that classic varieties made better economic sense than Müller-Thurgau. In 1981 he planted Riesling and Chardonnay; a couple of years later he added Sauvignon Blanc and Pinot Noir to the portfolio. Those first Pinot plantings, clone 10/5 planted on rootstock, are now some of the oldest Pinot vines in Marlborough. Sutherland went from grower to company man when he joined the Cloudy Bay team in 1986.

James Healy is a Rotorua boy who signed up with Corbans in 1978 after leaving university. He held the position of quality-control manager when he was lured south to take a winemaking position at Cloudy Bay in 1990.

The pair met, worked and learned together at Cloudy Bay. They still remember the first Pinot they were involved with at the winery. 'It was the '89,' recalls Ivan. 'There were a few balls-ups. We eventually sold it through the rowing club as a fundraiser.'

By the time of their departure, Sutherland was Cloudy Bay's chief viticulturist and Healy the oenologist. This gave Dog Point a wealth of Marlborough experience right from the start. Equally significant were a number of well-established vineyards on fine sites owned by the Sutherlands that would serve the new label.

Confining themselves to three of the varieties that Sutherland put faith in back in the early 1980s — Pinot Noir, Chardonnay and Sauvignon Blanc — and tightly focusing on quality above all else, Dog Point quickly became a respected Marlborough player.

From the début 2002 vintage, the Pinot Noir has exhibited a layered, poised, joyful personality, fleshy but firm in texture. That inaugural wine still has loads of sweet, earthy charm eight years on. Later vintages show consistent clovey, anise-like fragrances and excellent depth. Sutherland describes the Dog Point style as 'more Vosne than Pommard'. Roughly 3500 cases of Dog Point Pinot are produced annually.

Healy states:

> For me, there have been two significant things since we began. In the winery we've learned that less is more; as we've grown more confident, we no longer agonise over every little thing. The second thing is the evolution in the vineyard. Part of that is vine age — the average age of our Pinot vines is now 17 years, and that definitely helps the structure of the wines. Then there are changes we've made, like lowering the crop load to one bunch per shoot from 2007.

The Dog Point Pinot is made from fruit grown at the Dog Point Vineyard. The 100-hectare site straddles the hump between the Omaka and Brancott valleys, and runs out into the Brancott.

Says Sutherland:

> Sixty percent of the vines would be the Dijon clones, which we like. After that there's Pommard, Abel and a small portion of those old 10/5 vines. Our vines are mainly on clay, where you get a slow evolution of fruit development and where I believe it's easier to achieve vine balance.

Dog Point crops at 4.5 tonnes per hectare and the vines are typically planted just over a metre apart, with 2 metres between the rows. There is a gradual move to organic viticulture, which Sutherland emphasises is 'not a marketing thing … we're doing it because we think it's environmentally responsible'.

In the winery the grapes are hand-sorted, fermented using indigenous yeasts and the ferments are hand-plunged. The wine is treated to 45% new oak and is bottled unfined and unfiltered.

'Making a single-vineyard Pinot Noir is a long-term goal,' says Healy.

'It's not something we're going to rush into,' Sutherland adds. 'We're a couple of old-fashioned bastards … that's why we close the wine with corks.'

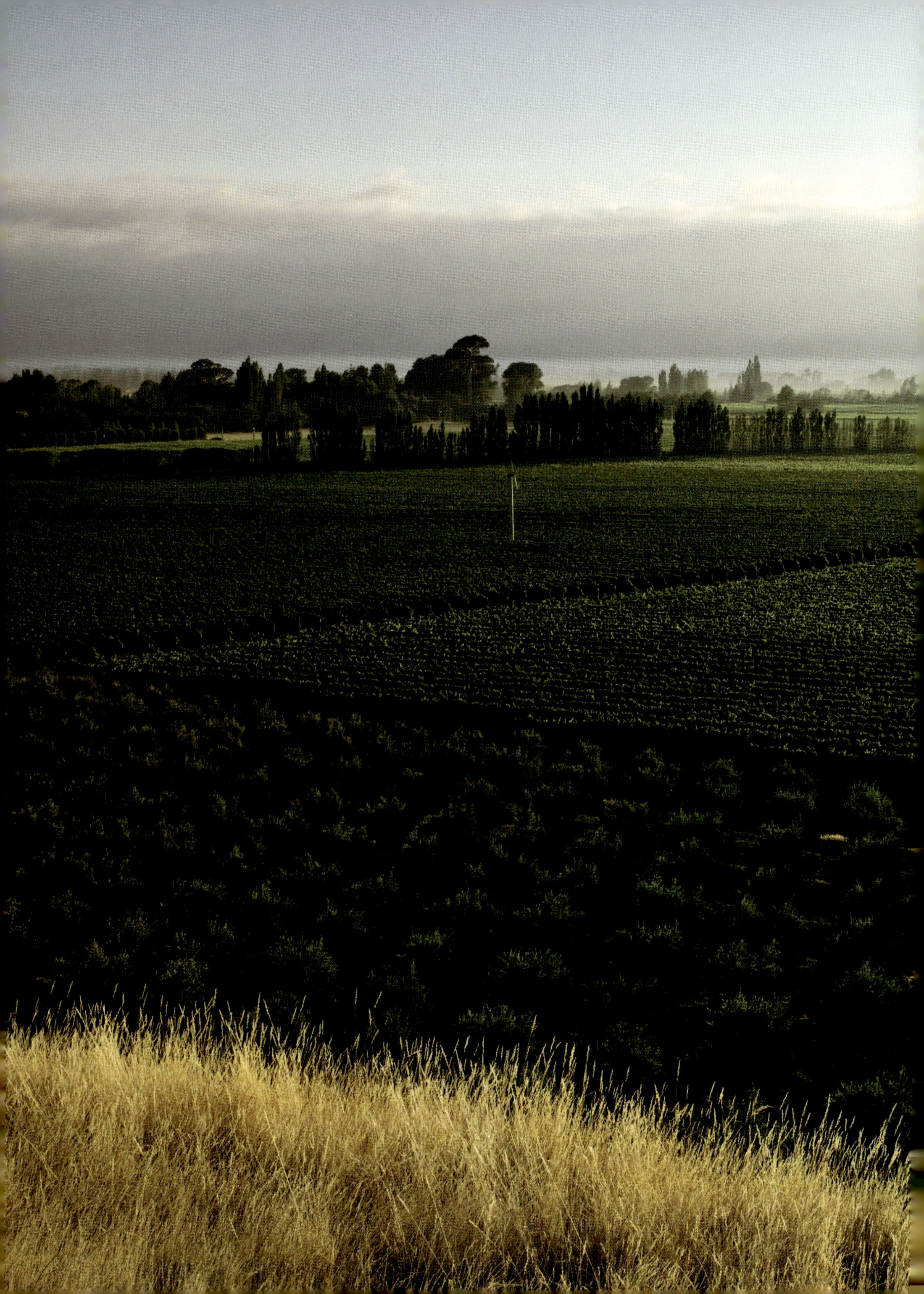

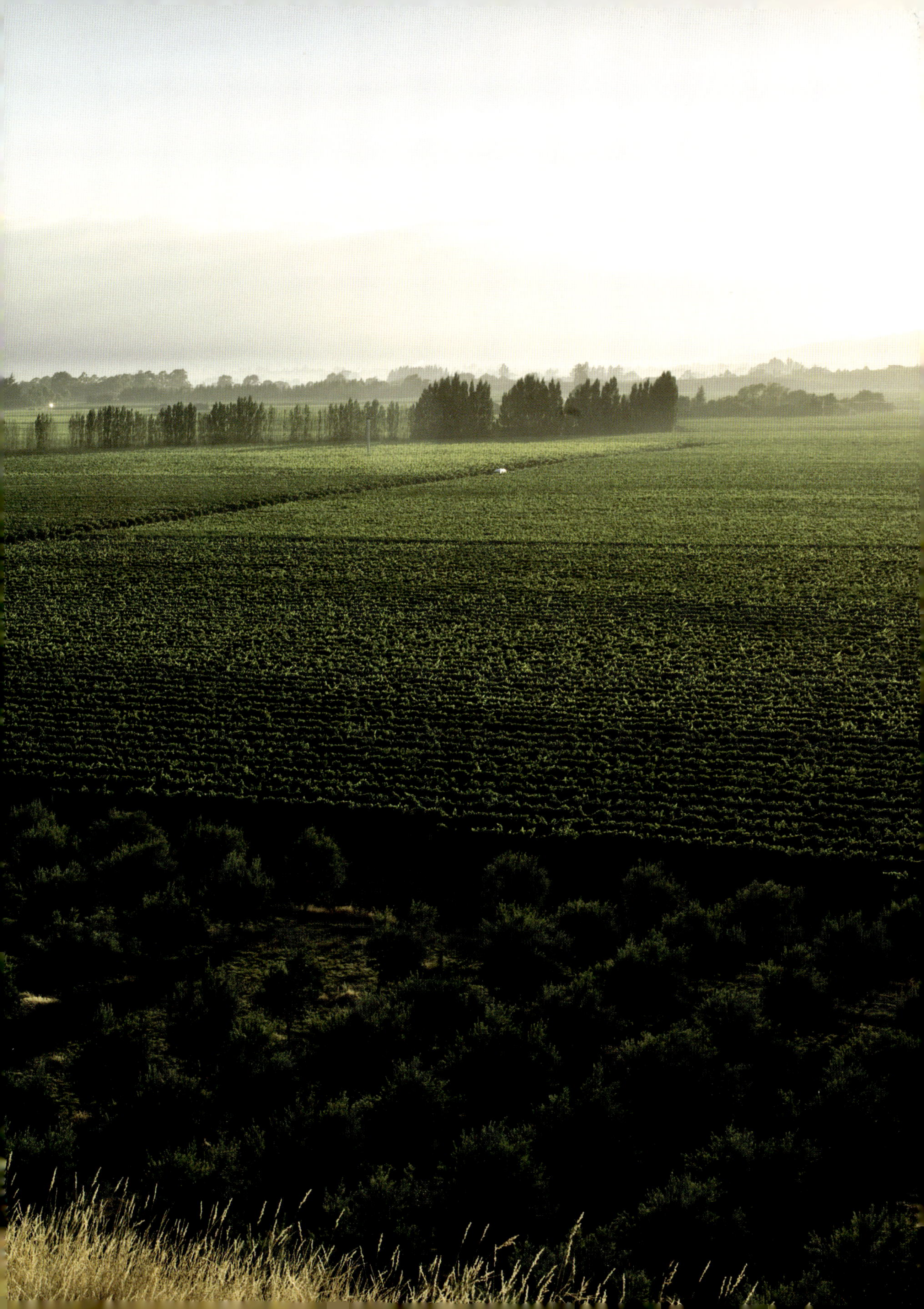

FROMM WINERY

The Fromm Winery is widely regarded as Marlborough's pre-eminent producer of Pinot Noir. It was the first winery in the region to surrender itself wholeheartedly to the quest of making fine Pinot, and was subsequently the first to achieve outstanding results.

Georg Fromm and his wife, Ruth, established the winery in 1992. A fourth-generation winemaker from Grisons, the mountainous easternmost part of Switzerland where Pinot Noir has a strong foothold, Fromm engaged another Swiss (also from Grisons) to make the wine at his Marlborough outpost. Hätsch Kalberer had been working in Gisborne for Matawhero when he met Georg Fromm. Kalberer's name has been synonymous with the Fromm Winery and its Pinot aspirations ever since.

In 2003, the Fromms began relinquishing their shareholding. The company is now owned by a pair of wine-loving Swiss businessmen and a young Marlborough-born New Zealand winemaker, William Hoare. As well as being involved in making the wine, Hoare is general manager of the winery. Before taking up that role, the experienced Hoare made some outstanding Pinot Noir under his own label, William Thomas.

Unusually for Marlborough, the Fromm Winery was established to make only red wine. A raft of varieties were planted initially, but for Kalberer there was always one favourite.

> Of all red wines, Pinot is the one that makes you smile … wow! Great Pinot has a bliss factor. It's also so expressive of site, and it responds to whatever you do or don't do. The efforts you put into viticulture are rewarded with Pinot. You can't compensate with smart winemaking to make up for what you failed to do before. I don't call it a heartbreak grape … it just needs more care and attention.

There was certainly nothing heartbreaking about one of the first Pinots the winery produced, the Fromm La Strada Reserve Pinot Noir 1994, made with fruit from Mike Tiller's vineyard. It received immediate critical acclaim and rearranged everyone's thinking about Marlborough and Pinot Noir.

'We worked the vineyard as if it was ours,' says Kalberer. 'We did nothing else other than show people what the region could achieve with some good viticulture.'

Today Pinot Noir accounts for just over half Fromm's annual production of

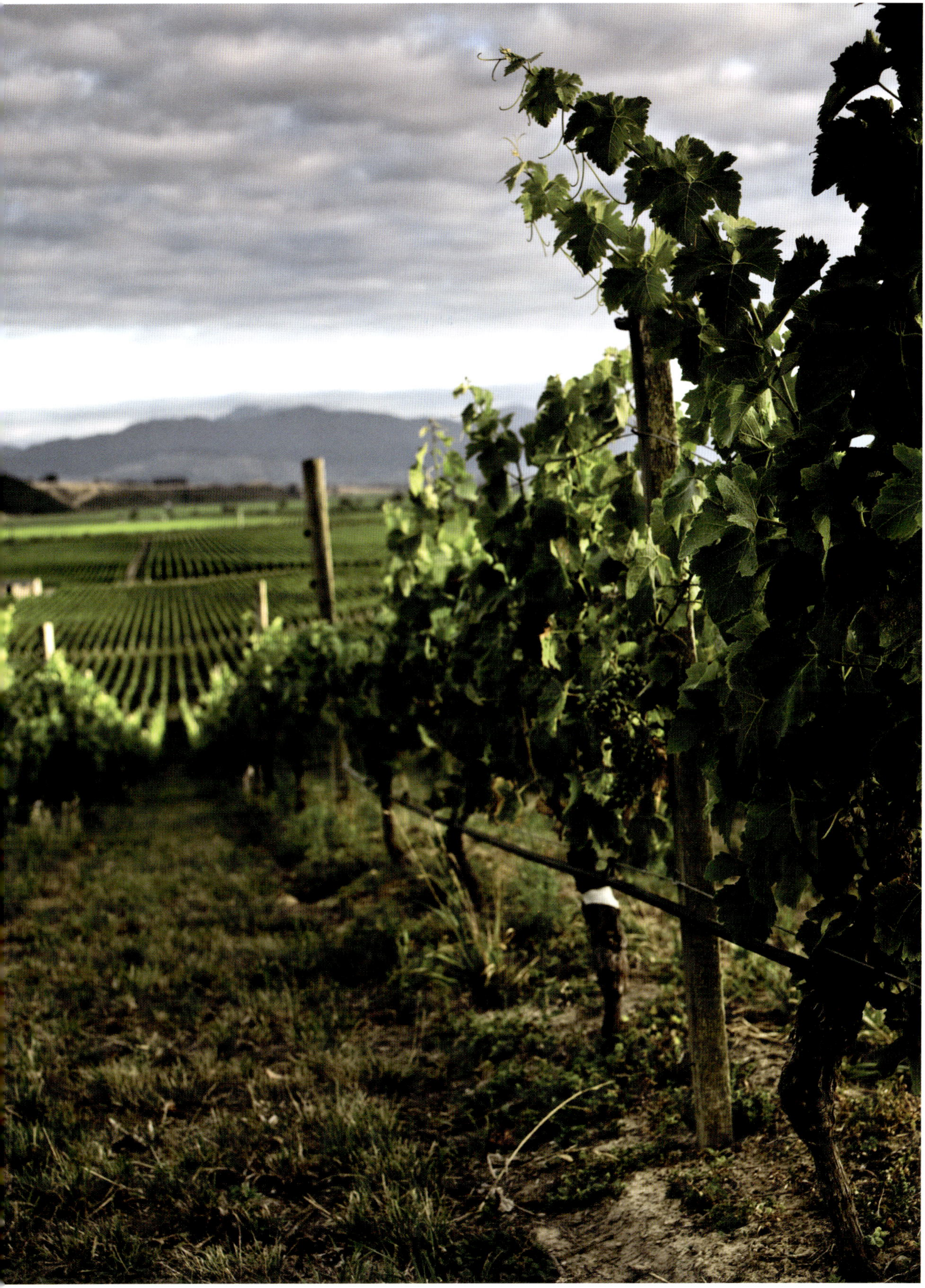

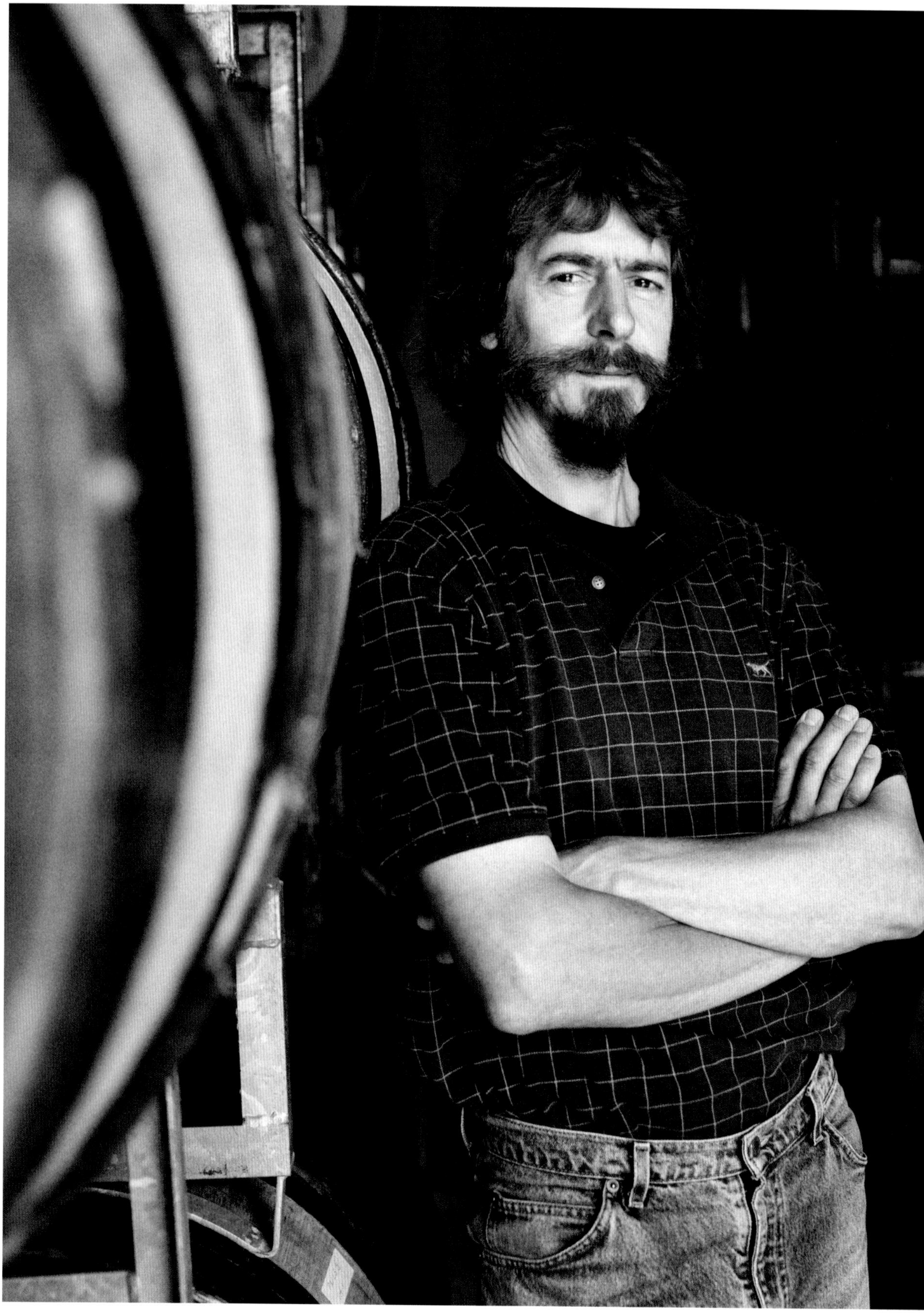

5000 to 6000 cases of wine. Four different Pinots are made. The La Strada Pinot is a blend of fruit from different vineyards in the Brancott Valley as well as the Fromm Vineyard in the central Wairau Plains. It is, says Hoare, 'the most New World style of Pinot we make. It has more forward fruit and is aimed to provide good drinking when released.'

The other three wines all sit under the Fromm banner. The Brancott Valley Pinot, first produced in 2005, is a blend from two Brancott vineyards. The Clayvin Vineyard and Fromm Vineyard Pinots, which both made their débuts in 1996, are single-vineyard wines.

'The Brancott I sometimes call a racehorse for its combination of muscle and elegance,' says Kalberer. 'The Clayvin is more seductive and opulent, and the Fromm is more intellectual, classically European, and as such is outside the square of typical New Zealand Pinots.'

Care in the vineyard is at the heart of the Fromm approach. Kalberer takes such pains with the fruit that Hoare once joked to US scribe Eric Arnold, 'When we crush each bunch, it's "goodbye Doris, so long Fred".' The vines are cropped at 5–7 tonnes per hectare, which Kalberer says is slightly higher than what he considers to be ideal because it gives him 'the freedom to be quite liberal when it comes to eliminating any problem fruit, which we mostly do in the vineyard'. Grapes are hand-picked, destemmed and fermented using wild yeasts. Care is taken not to be too extractive, with two daily plungings or gentle pump-overs during fermentation. But the time on skins is a total of four weeks, giving the wine a more refined tannin structure. The wine stays in traditional Burgundian oak *pièces* (5–15% new) for 16–18 months, and those in the Fromm range are bottled with minimal or no filtering.

Says Kalberer:

The Fromm and Clayvin Vineyard wines are true expressions of terroir. If you adopt the position of coach rather than winemaker and let the fruit express itself, you get a sense of place. The differences shine through. That gives the signature back to the vineyard rather than the winemaker. I think that's what I appreciate the most.

Left: Hätsch Kalberer, Fromm Winery.

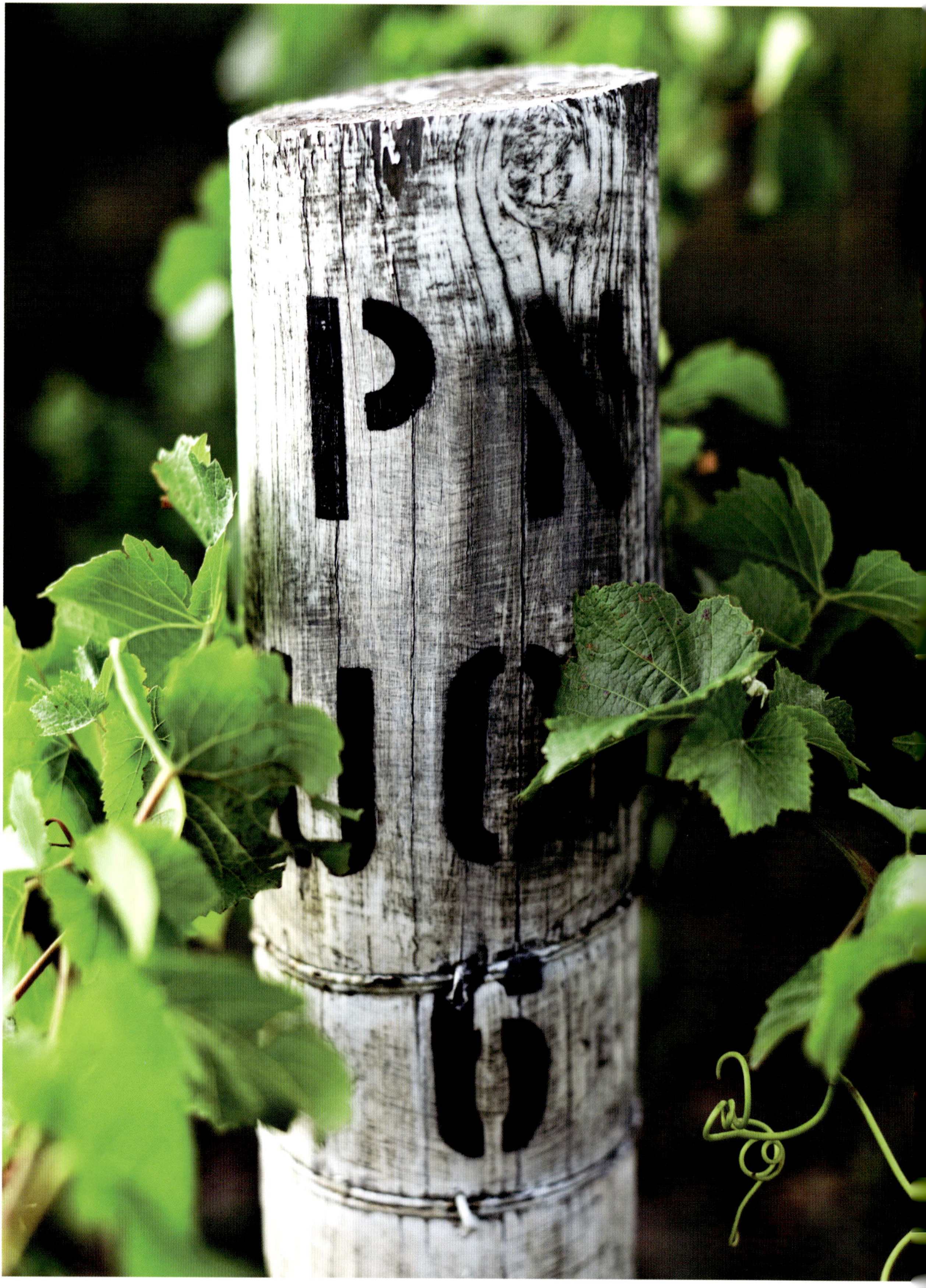
PN

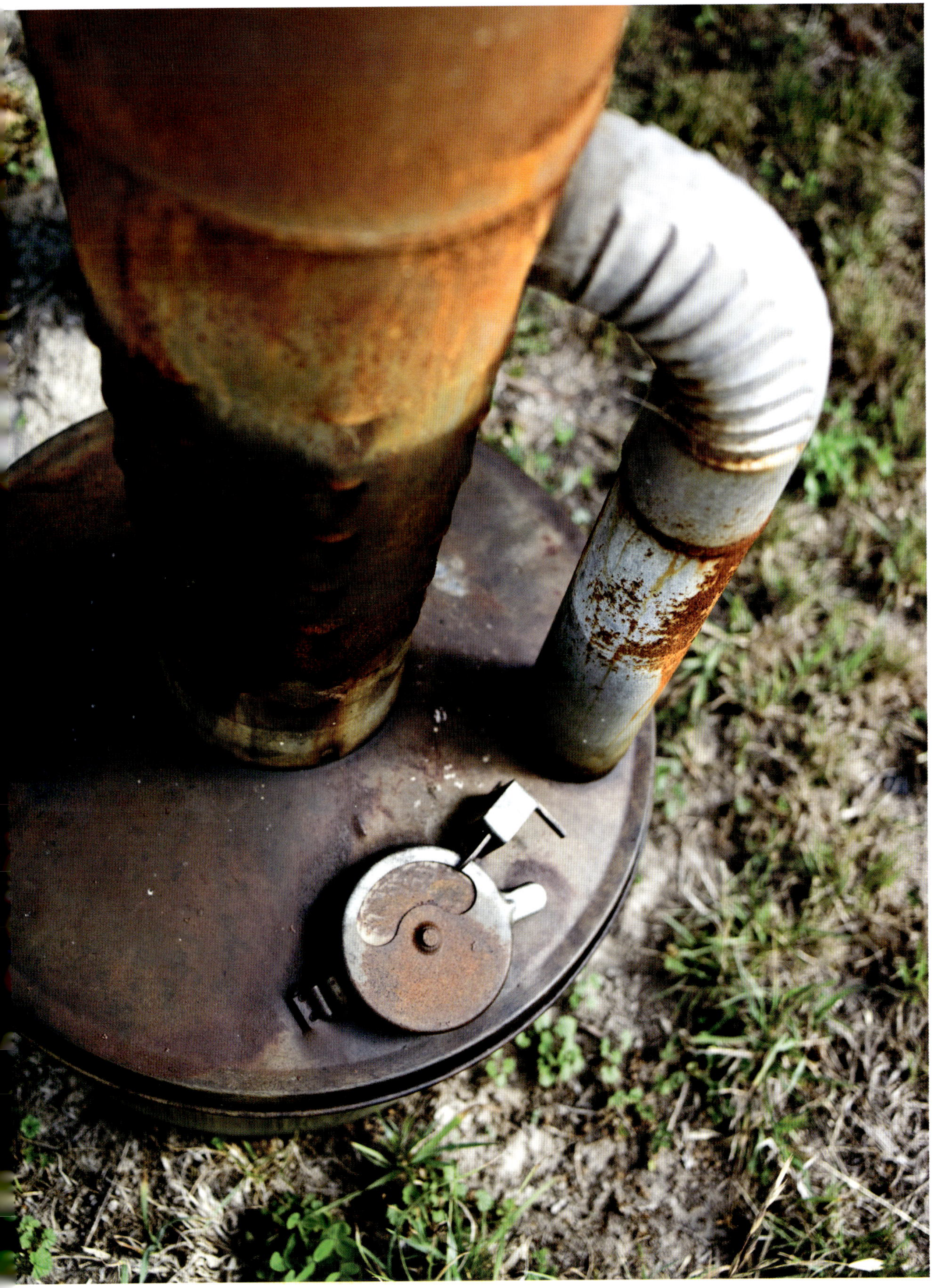

MAHI

It was as the foundation chief winemaker/general manager at Marlborough's Seresin Estate that Brian Bicknell became intimate with Pinot.

> I'll never forget the first table Pinot we made at Seresin, the 1997, which was a cool vintage. We had Dijon clones in, a bit of 10/5. We destemmed it and decided we were going to do a wild ferment. I was out at a movie when I got a call from the winery: 'We've fucked the Pinot Noir!' All this aroma was coming out of the vat and it had this volatile character. We now know this happens when one yeast is involved. From the start that 1997 looked beautiful … and still was eight vintages later. It came down to skin tannins, which were round and full. That's when I moved away from an 'ugly young, beautiful old' mindset. It might be OK for Bordeaux wines, but with Pinot you've got to be good early.

Bicknell left Seresin in 2006 to devote his energies to Mahi, a label he and his wife, Nicola, had been running on a cottage-industry scale for several years. Mahi means 'our work, our craft', which sums up the Mahi ethos: hand-made, enquiring and honest.

Moving into Daniel Le Brun's old winery near Renwick, with its large subterranean barrel halls, the original Mahi concept was to produce wines made from distinct blocks or vineyards. That vision has been modified slightly since then — the Pinot Noir range now includes an estate blend — but the emphasis is still on offering glimpses of the diversity of the region's terroir. Most of the fruit is bought from carefully chosen sites belonging to contract growers; at this stage the home block beside the winery houses the only estate-owned vines are those in the home block beside the winery.

The first Pinot made was the Mahi Byrne Pinot Noir 2001, a wine that immediately established Mahi as an exciting new player. The Byrne vineyard, which recently received organic certification, is located near Renwick on the Wairau Plains and is planted in clone 115 and 10/5. This is a round, energetic wine, with a strong dark fruit presence and a robust structure. The 2004 is still a lively drink with many years to go. The last Byrne Pinot was made in 2006; the fruit is now used for the estate blend.

One other single-vineyard Pinot has been made; the Mahi Rive Pinot Noir made its début in 2005. The Rive vines are planted on silt over gravel at a site near Rapaura. The Mahi Rive is an intense, driving wine — a juicy fruit core framed by firm, upright tannins.

The Mahi Marlborough Pinot Noir is the estate blend of fruit from four vineyards, including a clay-rich site in the Brancott Valley. The wine is typically effusively perfumed with a mix of red and dark fruit notes, with fresh acids and a lithe texture.

Quantities are still small — only 300 cases were made of the Rive 2008, and 1000 cases of the 2009 estate blend. Bicknell knows what he wants:

> We don't chase the market … I don't think you can, anyway. If you wanted a recipe for funky Marlborough Pinot, it's what we do. We hand-pick, use wild yeast and there are no additions. We mostly destem, although we are experimenting more with whole bunch. After ferment we go straight to barrel where it sits on lees and gets stirred once in a while. The fining is minimal and there's no filtration, to try and reflect the site as much as possible. I love the idea that you can make wine in as natural a way as possible … and that it expresses the small piece of ground it comes from.

PERNOD RICARD

The company responsible for producing 23% of New Zealand's wine in 2009, and roughly 10% of its Pinot Noir, officially carries the name of the French drinks superpower in whose orbit it has revolved since 2006: Pernod Ricard. Yet most New Zealanders still refer to the organisation by the name chosen by its Croatian founders — and which remains its central brand — Montana.

The company's Pinot Noir dimensions are, even in a world context, startling. The annual production of just over 250,000 cases of still Pinot Noir puts it among 'the top three or four producers in the world', says Phil Bothwell, the company's national wine ambassador. 'We're bigger than any Burgundian negociant, but we don't want that accolade particularly.' The growth rate that led to that output is equally startling when you consider that, although Montana had started dabbling in Pinot in 1979, only a couple of fermenters were used to make 600 cases of Pinot as late as 1996.

In that year Pernod Ricard's scramble for Pinot Noir took off. The company planted the 250-hectare Kaituna Vineyard on alluvial soils on the northern side of the Wairau Valley mainly in Pinot Noir. It remains one of the largest single Pinot Noir vineyards in the country. Shortly after that the Pinot presence in the established Brancott Estate vineyard was enlarged and the Terraces vineyard was planted in 1998. Acquisitions in the Awatere Valley followed. The company currently has close to 350 hectares in Marlborough planted in Pinot Noir, 285 hectares of which supply fruit for still Pinot. The only other region so far targeted by Pernod Ricard for Pinot plantings has been Waipara, where it has a number of sites, including the sloping Camshorn vineyard. The company also buys in grapes from several growers in Central Otago.

At its Brancott winery base on State Highway 1 southwest of Blenheim, Pernod Ricard produces a large family of Pinots. None would rate among the very top Pinots in the country; the company's strength is the effort it puts into its less-exalted labels, many of which provide consumers with their introduction to Pinot Noir.

That was the role earmarked for the Montana South Island Pinot Noir when it was first released as the Montana Pinot Noir in 2000, and for that it was hailed as an

Right: Patrick Materman, Pernod Ricard.

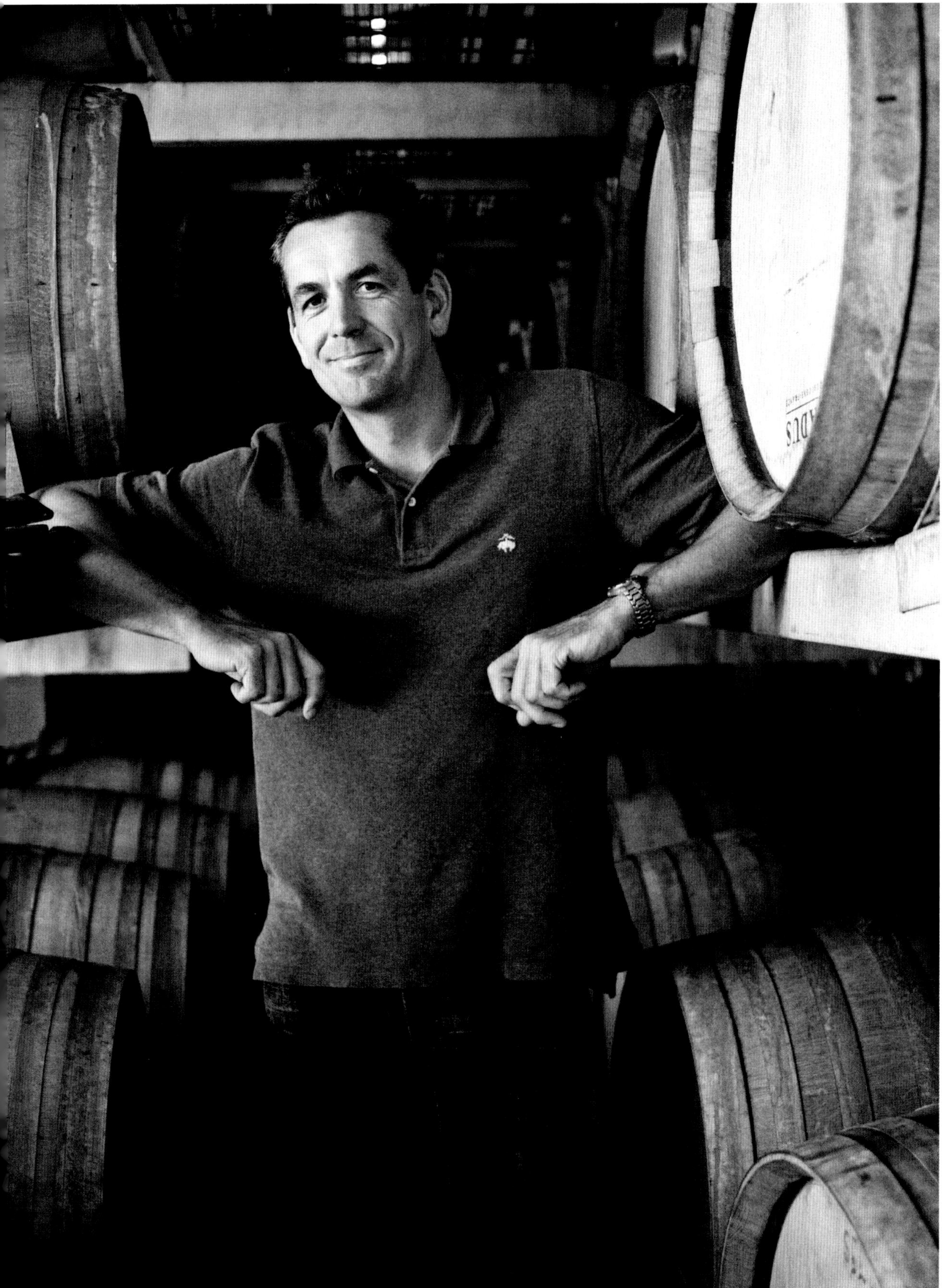

important wine. From 2006 it became a blend of fruit from several South Island regions, hence the slight name change.

Although it now usually sells for just under $20, it certainly isn't all done on the cheap — a portion of it sees new French oak barrels. (As an aside, Pernod Ricard's Marlborough barrel hall contains an astonishing 12,000-plus French oak casks.) It is typically light, soft, cherry-scented and approachable — everything an 'entry level' Pinot should be. 'Delivering all this, at that price point, is actually very difficult,' says Pernod Ricard's chief winemaker Patrick Materman. 'Some years I think we over-deliver.'

In the past, production of this wine has exceeded 100,000 cases; that has now been scaled back to less than 75,000. Says Materman:

> The focus has been to increase the scale of our more premium offerings. We're also mindful of not creating a Pinot surplus. After 12 years or so of looking at each vineyard block, we've identified and started pulling some of the under-performing Marlborough vines. They're being replaced by better clones on more suitable soils.

At the other end of the portfolio is the Montana 'T' Terraces Marlborough Pinot Noir, a wine first made in 2002. Its fruit is mostly taken from a terraced hillside site in Marlborough's Fairhall Valley, with production limited to roughly 4000 cases. From the best vintages a dark, spicy, muscular wine is made.

In between there is a varied quilt of Pinot labels under the Montana, Boundary Vineyards, Camshorn, Stoneleigh and Triplebank brands. Among these are a number of single-vineyard wines, many of them created during an innovative period of expansion in the early noughties. Notable is the Montana Showcase Series (formerly called the Terroir Series), which features a single-vineyard Pinot from Central Otago.

Following pages: pp186-7 Looking north over Pernod Ricard's Triplebank Vineyard in Marlborough's Awatere Valley.

FROM THE BEST VINTAGES A DARK, SPICY, MUSCULAR WINE IS MADE.

MARLBOROUGH

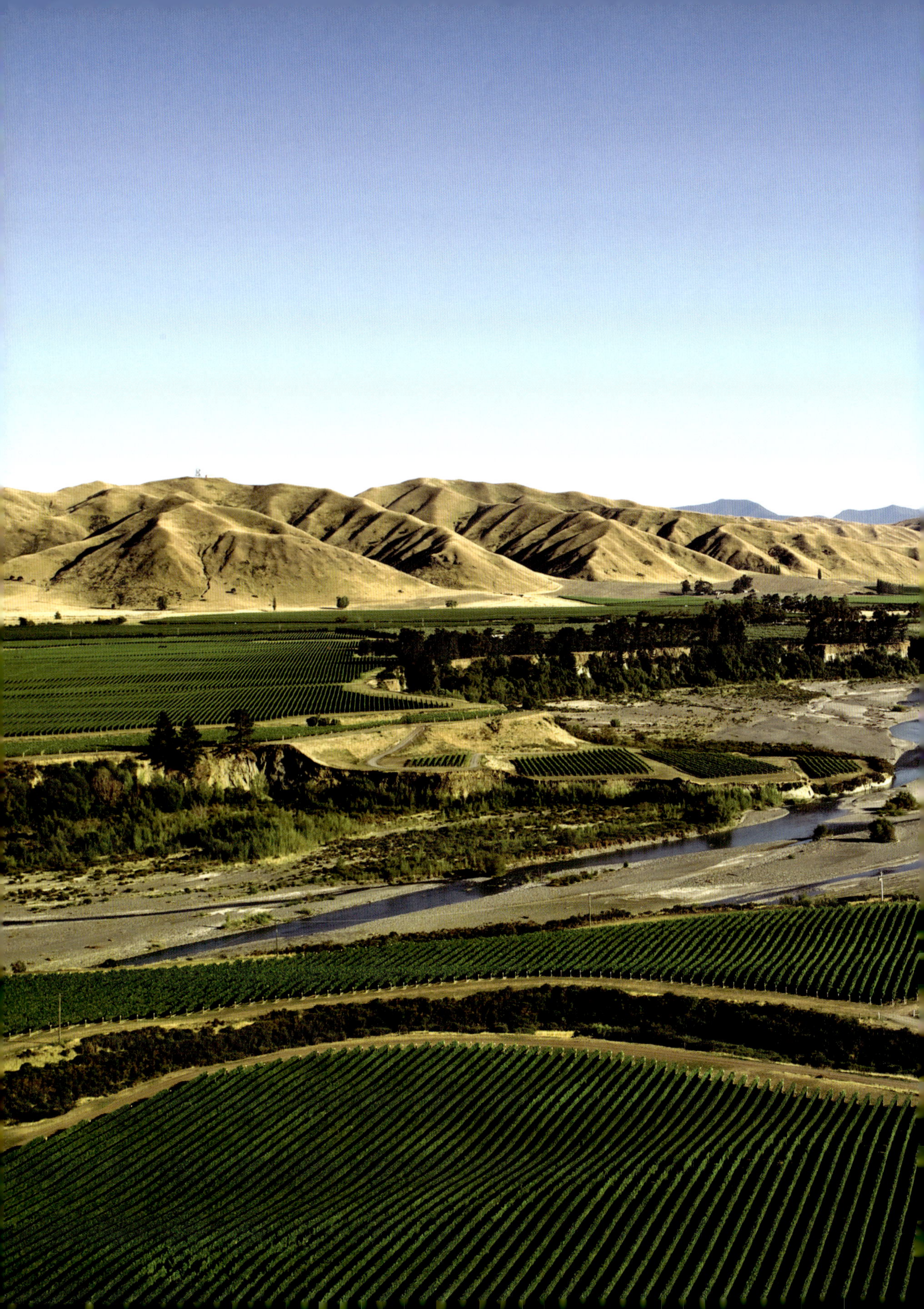

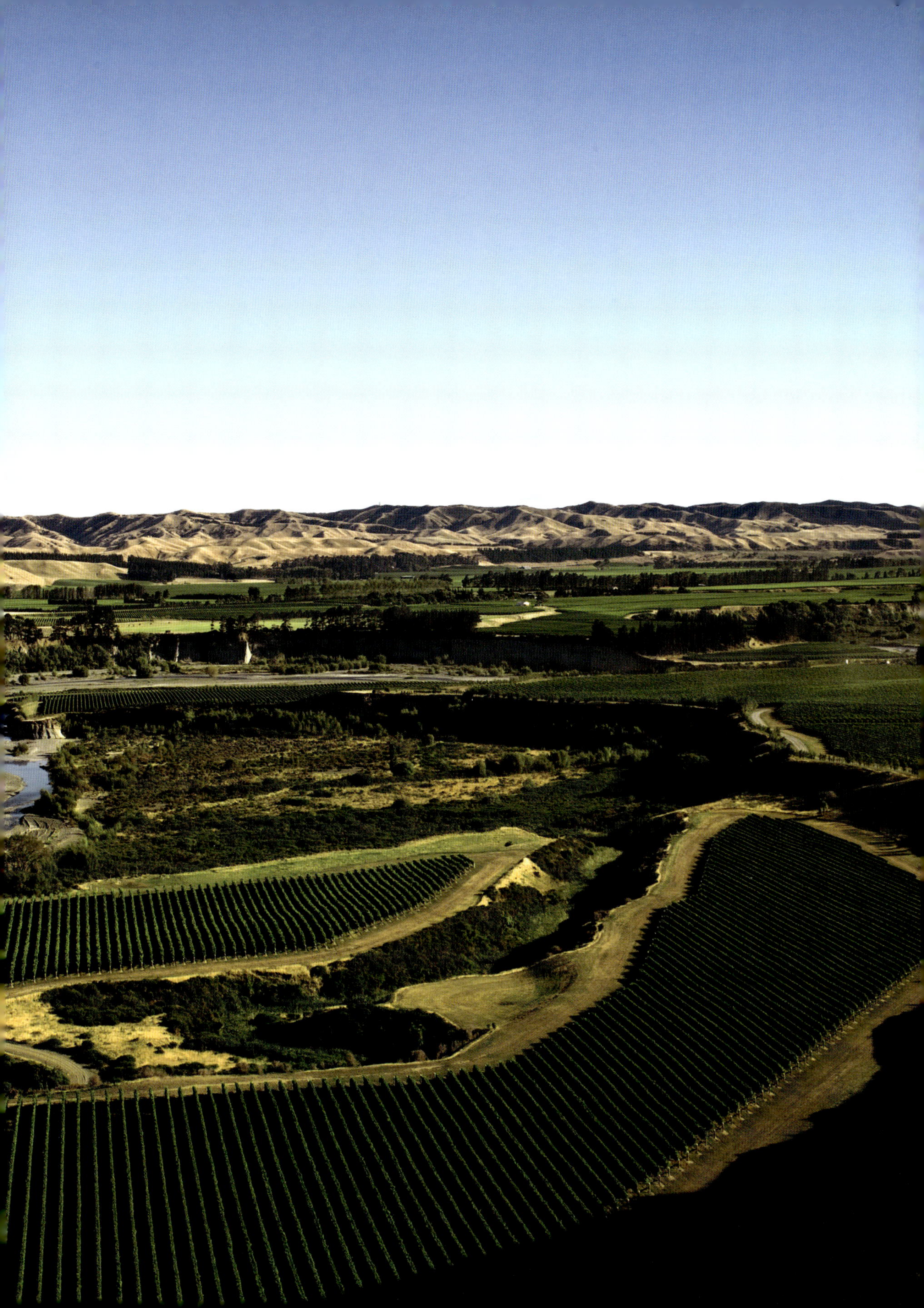

SERESIN ESTATE

The official line from Seresin Estate is that the company's human handprint logo symbolises a number of things, such as 'strength', 'a gateway to the soul' and 'the mark of the artisan'. It could be added that the Seresin approach is less about *la patte du vigneron* and more about the invisible, but telling, hand of Mother Nature.

Founded by Wellington-born Michael Seresin in 1992, Seresin Estate rose swiftly to become a mid-sized winery with a strong export focus — the brand has a high profile in the UK and Australian markets. Seresin is himself a successful local export; he is a cinematographer of world renown whose credits include *Midnight Express* and, more recently, *Harry Potter and the Prisoner of Azkaban*.

Very early on Seresin was seized by the desire to have his wine operation embrace organics without compromise. The decision was made for no other reason than a conviction that it was the right thing to do by the land. Seresin's winemaker and viticulturist at the time, respectively Brian Bicknell and Bart Arnst, set about the task. The company received BioGro registration in 2000, becoming Marlborough's first producer of certified organic wine. From organics, the company plunged headfirst into biodynamics. Full biodynamic certification will be achieved by 2011.

Adding impetus to the move to biodynamics have been a new winemaker, Clive Dougall, and a new vineyard manager, Colin Ross, who both arrived at Seresin in 2006. Ross has an unquenchable enthusiasm for the Steiner 'total farm' approach, which he acquired while managing biodynamic vineyards in Western Australia. Dougall, too, is a believer in biodynamics and, having worked at Pegasus Bay for four years before his move to Marlborough, he is also a believer in Pinot Noir. Like a number of other producers, he sees biodynamics (and hands-off winemaking) as important if Pinot is to express its terroir truthfully and clearly. Around this pair, Seresin has developed an impressive team buy-in and sense of adventure in regard to both biodynamics and Pinot.

Of Seresin's three estate-owned vineyards, Raupo Creek has the largest Pinot planting. The 51-hectare block is situated on clay-rich slopes in the Omaka Valley and

Right: Spraying the vines with biodynamically prepared compost tea at Seresin Estate.

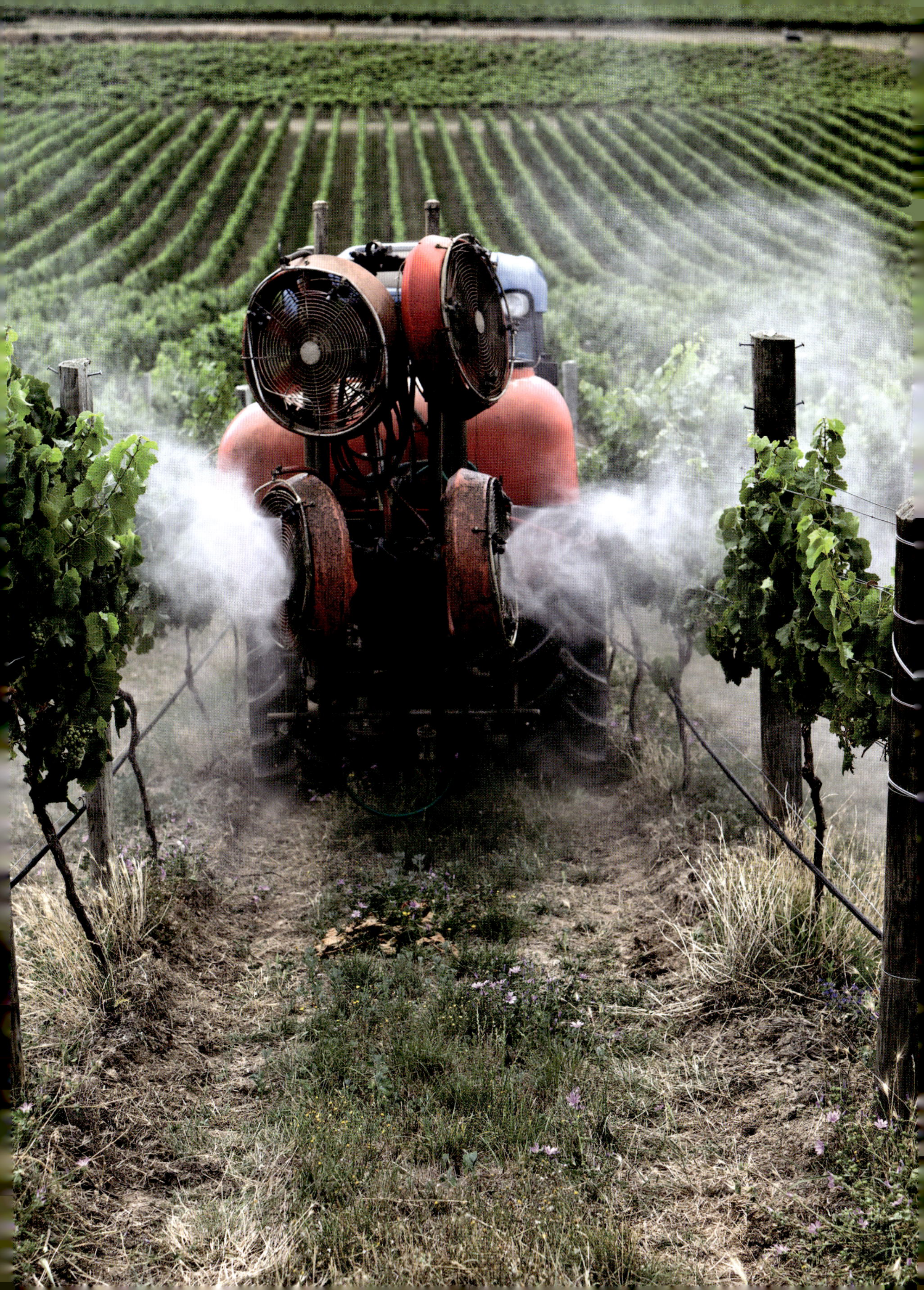

was planted with a mix of clones between 2001 and 2003. The other Pinot plantings are at the Home Vineyard and Tatou Vineyard, both on Wairau Valley alluvial gravels. Tatou is sited close to a bend in the Wairau River and is particularly stony. Some of the Home Vineyard vines date back to 1992.

The exploratory Pinot path was taken further in the 2007 vintage, Dougall and Ross's first at Seresin. In that year seven Pinots were produced. These included two début wines: Seresin Sun & Moon Pinot Noir 2007 and Seresin Home Pinot Noir 2007. Just 70 cases were made of Sun & Moon, which retails for $120 a bottle.

'It's a blend of the best fruit from the Raupo and Home Vineyards, from three barrels that I kept aside,' explains Dougall. 'We won't make it every year. I think it does show New Zealand — and the world — that Marlborough can produce great Pinot Noir.' It is a dense, deep, still expression with a complex flavour profile that leans towards darker fruits. Following its development will be interesting.

The Home Pinot Noir is a single-vineyard wine, as is Tatou Pinot Noir, first produced in 2004, and Raupo Creek Pinot Noir, which made its début in 2005. There are two other Pinots, Rachel and Leah, which are blends of estate-grown fruit. The winery also produces a second-tier label, Momo, that has been very successful. Made using estate-grown and grower fruit, few production short-cuts are taken and the wine offers exceptional value.

Dougall comments:

> I feel humble doing what I'm doing. I want to produce wines that are expressive of vintage, people and, most importantly, place. We try not to manipulate our wines' destiny. We add no enzymes, bacteria, tannin, yeast or nutrients. I'm in a dream job, but I know Pinot is something I'll never master.

Right: The Seresin Estate barrel hall. Following pages: p192 Cow horns, key ingredients for biodynamic preparation; p193 Eggshells from Seresin's free-run hens are kept and eventually join the biodynamic compost pit; pp194-5 Seresin Estate's Raupo Creek Vineyard in the Omaka Valley, source of the company's finest Pinot fruit.

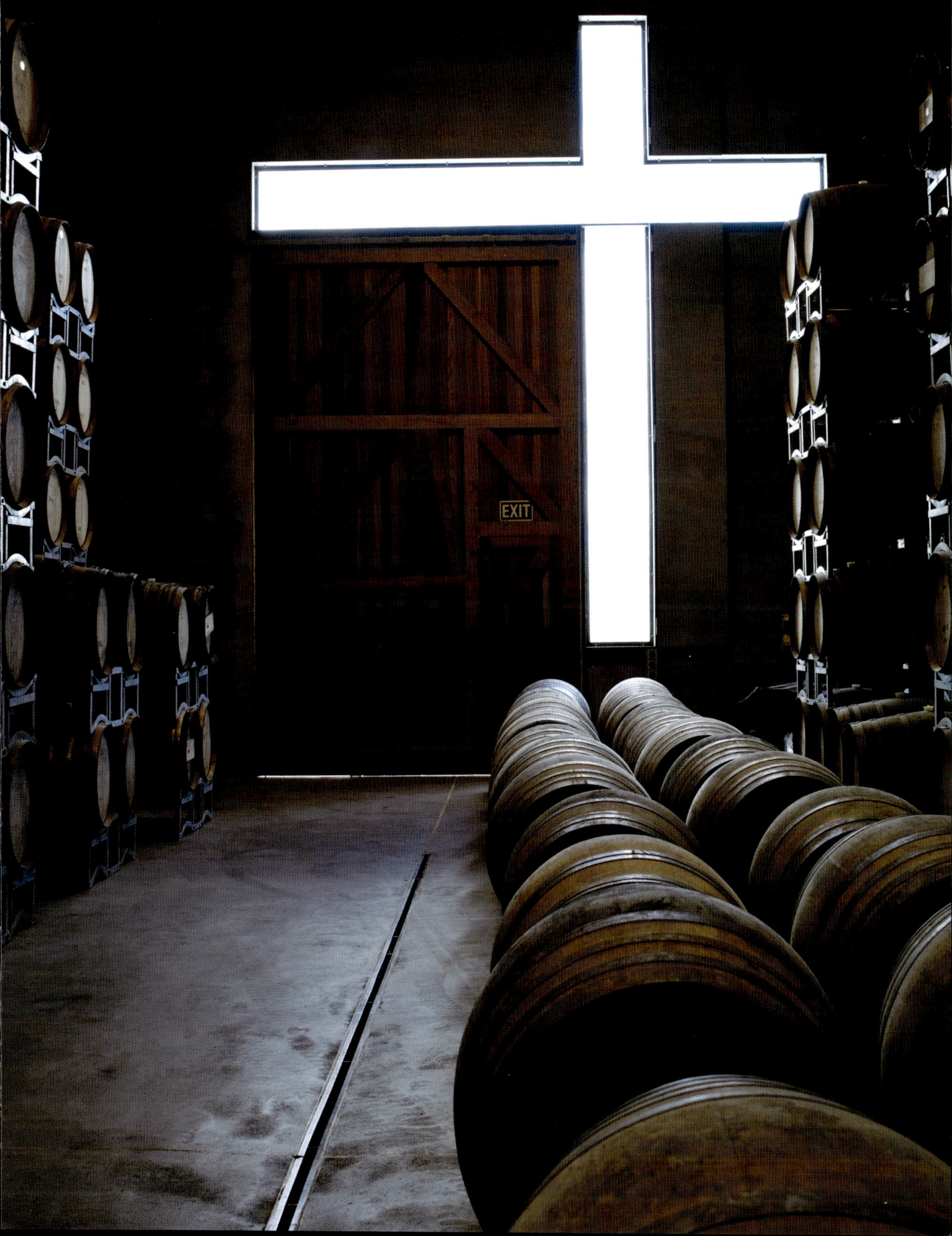
EXIT

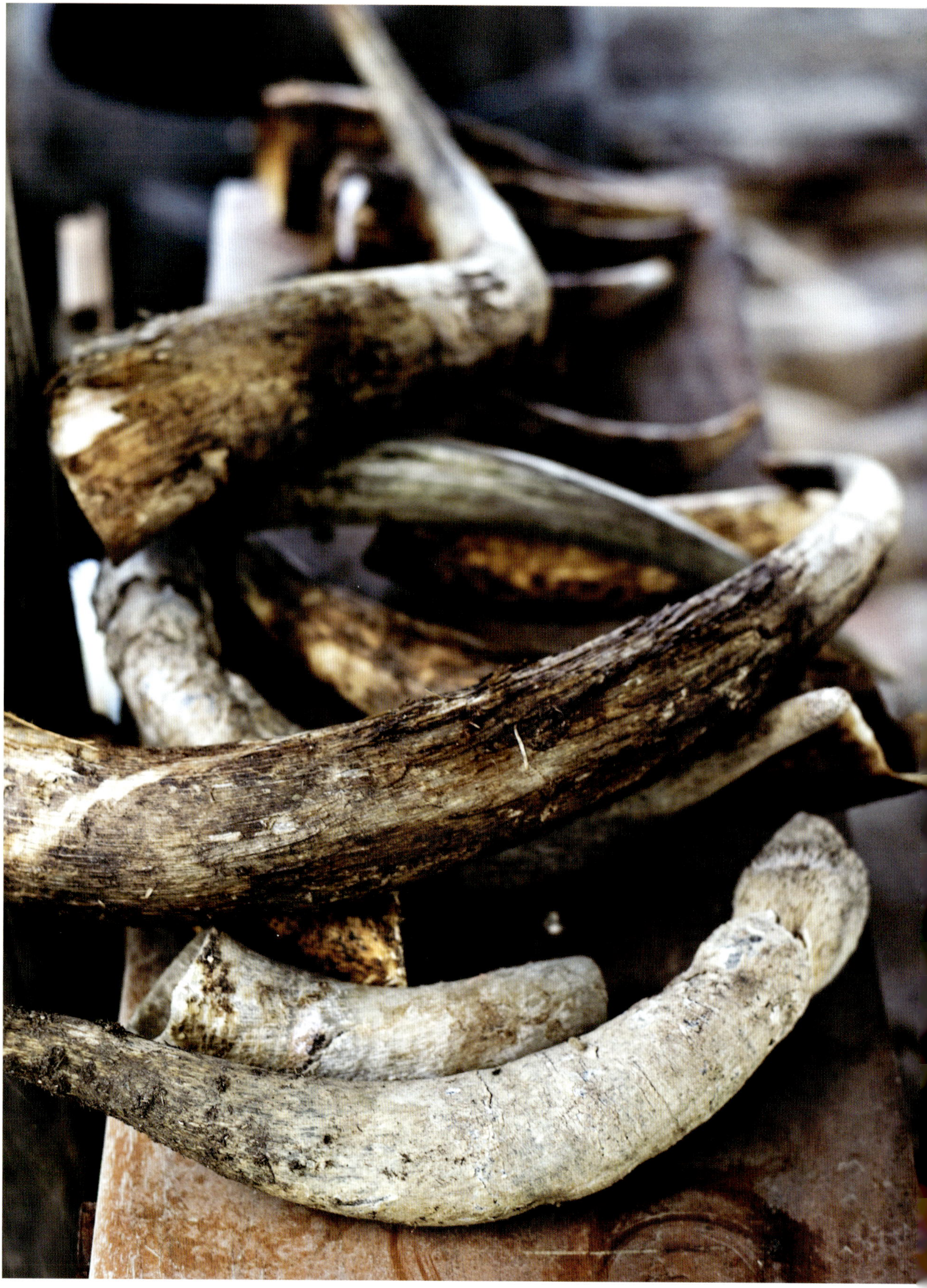

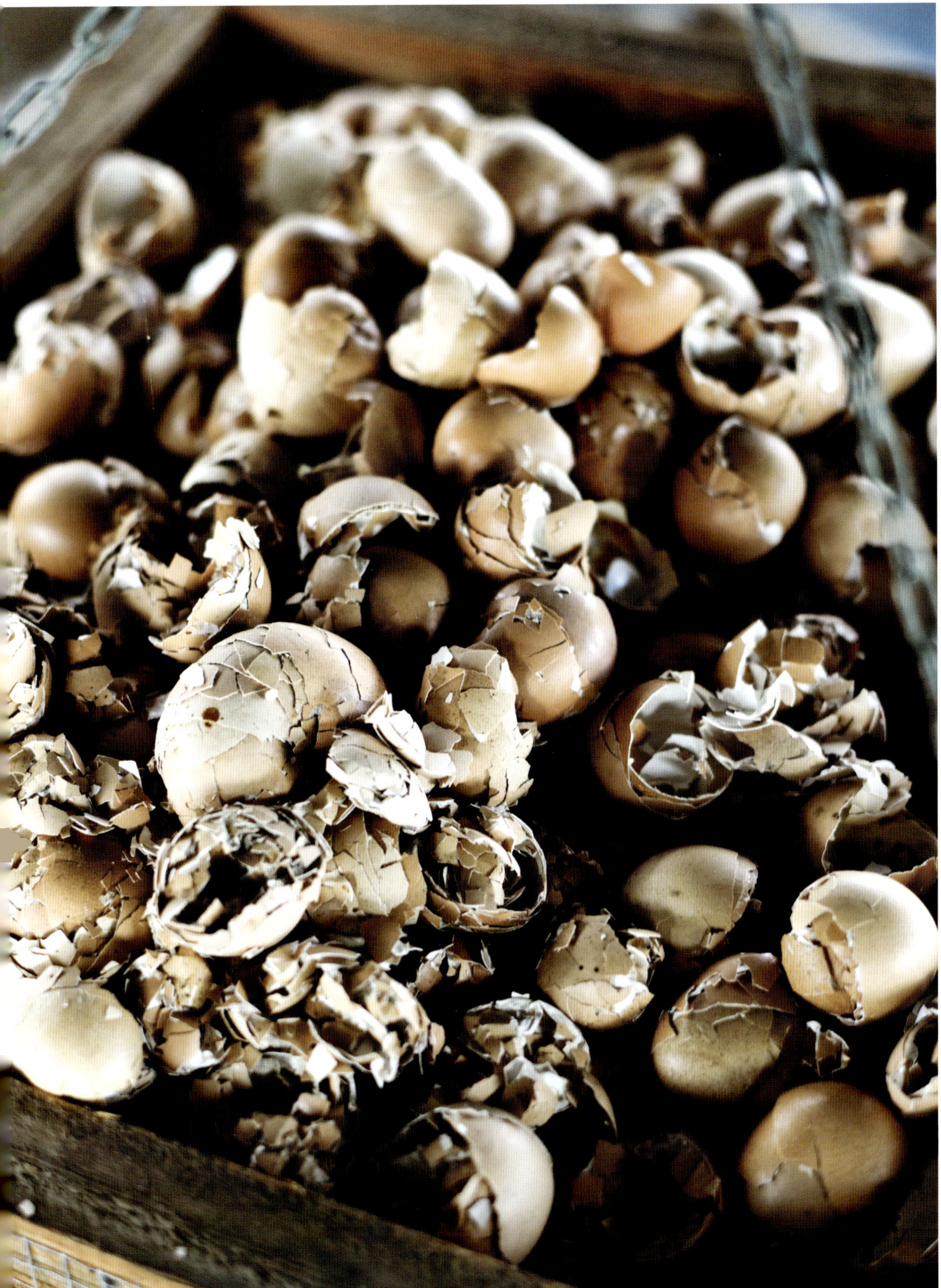

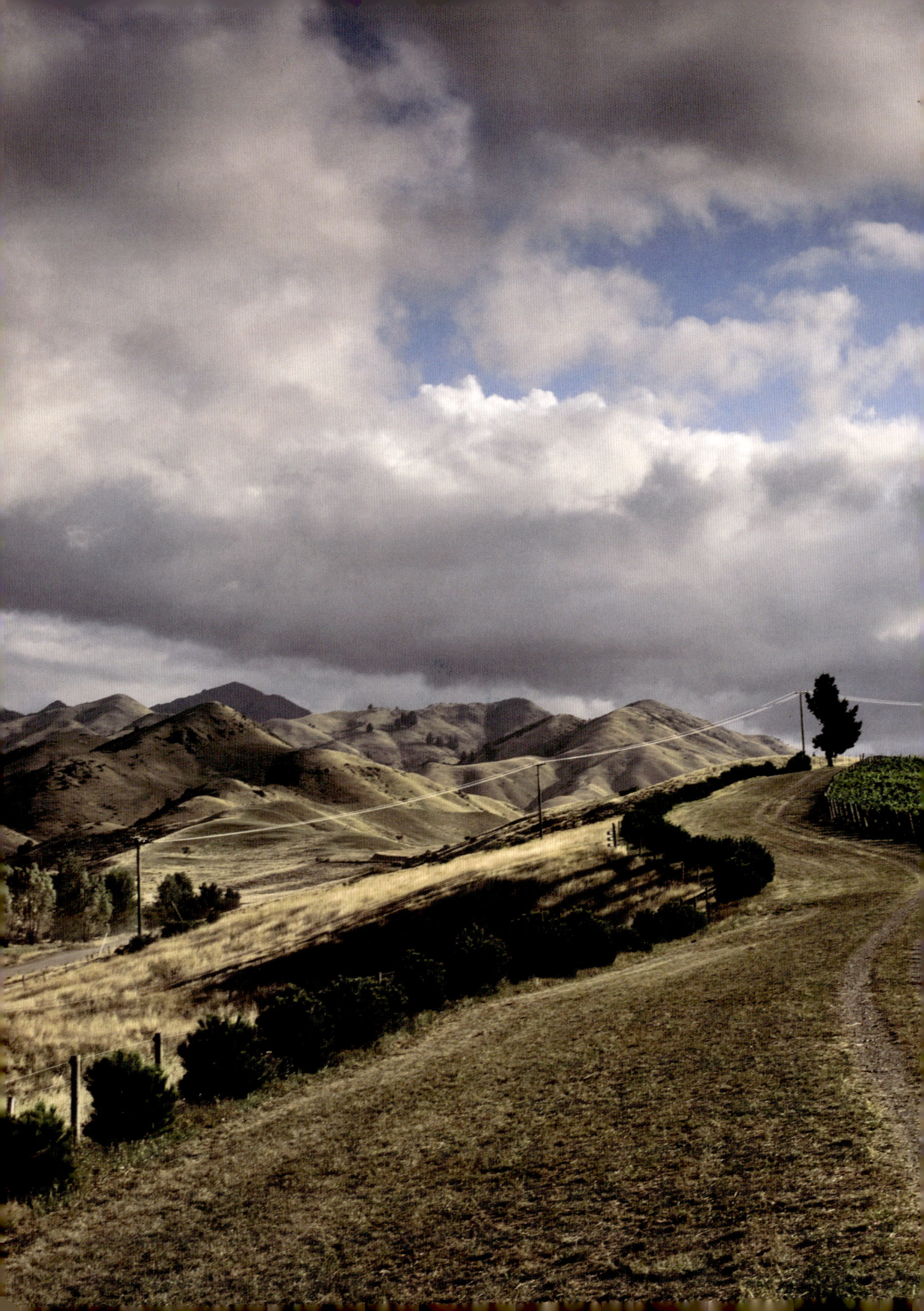

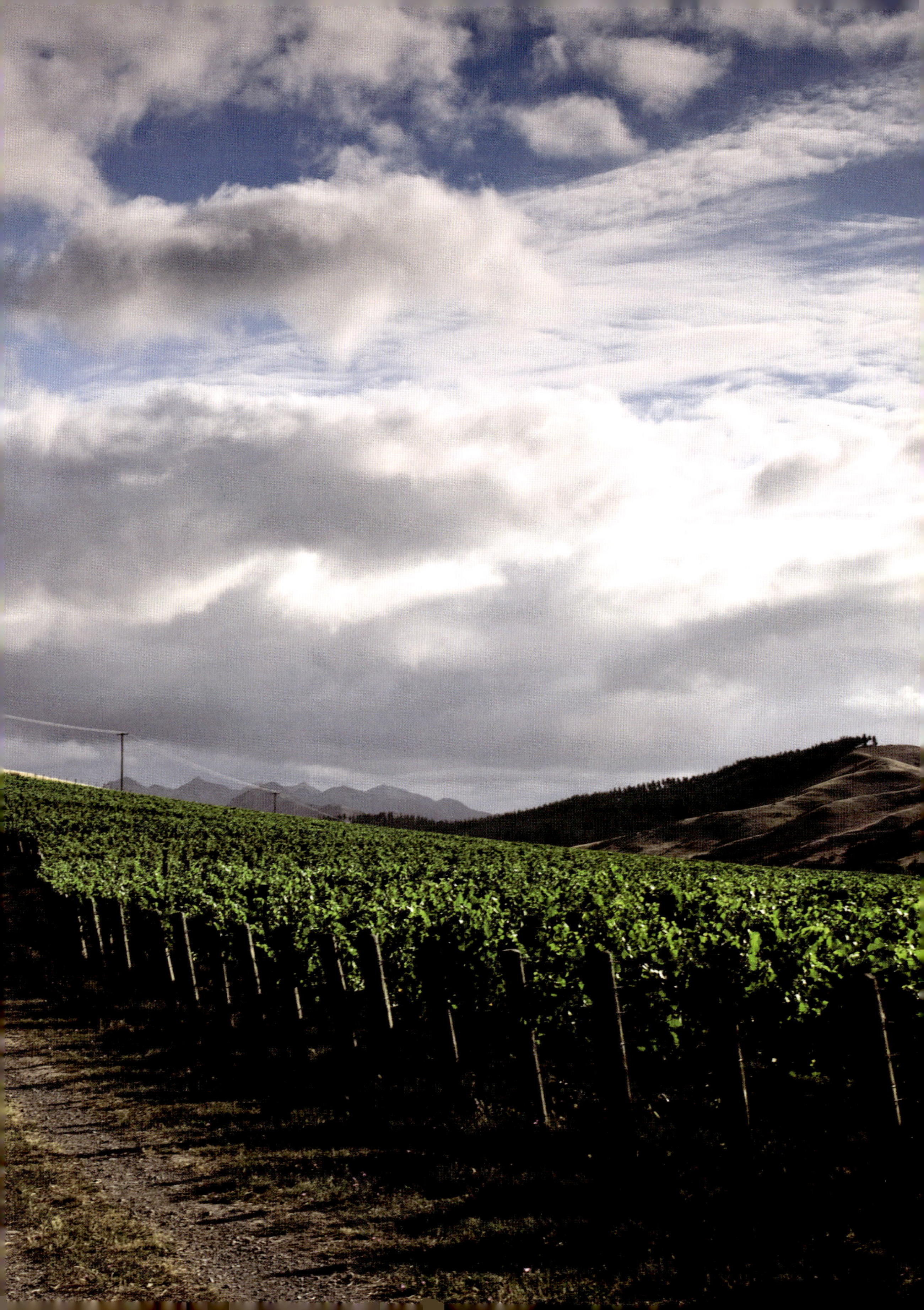

TERRAVIN

Mike Eaton was among the first to begin breaking Marlborough out of its Sauvignon-cosseted comfort zone and get serious with Pinot. That occurred in 1991 when he acquired and planted Clayvin Vineyard, a sloping site in the Brancott Valley.

But the Eaton story in Marlborough begins earlier, close to the beginning of the wine region's modern era. He took a job with Montana as a general labourer in 1981, from there moving into viticulture. Mike Eaton had found his calling. Towards the end of the decade, a working visit to Burgundy, the Loire and the Jura convinced him that Marlborough's southern hillsides were an untapped treasure. He returned and established Clayvin with his wife and business partner, Jo.

At that time Clayvin was Marlborough's largest still-Pinot-designated vineyard. It immediately impressed Georg Fromm, the Swiss proprietor of Marlborough's Fromm Winery, who bought the first and subsequent crops of Clayvin fruit.

By the time he sold the vineyard in 1998, Eaton was Pinot-struck and eager to continue exploring Marlborough's potential with the grape. He found a 12-hectare north-facing slope on Brookby Road in the Omaka Valley. Steeper than Clayvin, it was exactly what he was after. The horror Jo expressed when she was first shown the unruly, pine-studded hillside proved temporary, and TerraVin was born in 1999.

TerraVin's first Pinot was produced in 2001 and the label has grown to become one of Marlborough's standout Pinot producers. In 2007, the Eatons sold some of the vineyard (above which they built a house). The reason for the sale, says Mike, was 'to retire most of our debt and regain our sanity'. They retained 4 hectares.

Eaton sums up his fondness for hillside planting thus:

Each piece of hillside is intrinsically unique in topography, aspect and microclimate, and each person's interpretation of how to handle that slope varies. These things give the fruit from each hillside a unique personality. The same can't be said of plantings on the flat where there are often huge tracts of similar land that can be 'mirror-developed' to produce fruit all of a muchness. For commercial styles of Sauvignon Blanc and more fruit-driven Pinot Noirs that's ideal. But truly great Pinot must have a sense of place — something that reflects site, not just variety or winemaker.

The TerraVin vineyard has a 150- to 200-mm layer of topsoil over clay-bound glacial gravels. It is close-planted (1.6 metres x 1.25 metres). In the portion currently owned by the Eatons the predominant clones are Pommard, Abel, 667 and 777, along with a smaller portion of 10/5. The vineyard is cropped to 6.6 tonnes per hectare.

There are two Pinot Noirs produced: the premium TerraVin Hillside Reserve Pinot Noir and the TerraVin Pinot Noir.

The first of these is made from fruit harvested mainly from what Eaton calls the 'kidney', or centre, of the TerraVin slope. No more than 400 cases, often fewer, are produced of the Hillside Reserve. A barrel selection is made after the wine has had a year on oak. In all, the Hillside Reserve spends 18 to 20 months in French oak, of which 40–55% is new. It is an intensely sensual wine; deep, layered and finely textured.

The TerraVin Pinot Noir is currently made with a blend of fruit from the TerraVin vineyard and two nearby southern valley vineyards that are leased. They are small and sloping, and Eaton has complete control over viticulture. The TerraVin Pinot Noir does not have the finesse of its stablemate — it is a simpler and more robust expression, though still feminine. Eaton likes the expression 'big woman, small dress' to describe it. Between 800 and 1000 cases are made annually.

'With both styles we never force the wine into a mould but allow them to express themselves,' says Eaton. 'The Hillside very much reflects site rather than vintage; the TerraVin Pinot is more about Pinot Noir and the vintage.'

VILLA MARIA

Does being big detract from a producer's ability to make great Pinot Noir? Villa Maria's record suggests not.

Under the canny stewardship of founder and owner, Sir George Fistonich, Villa Maria has achieved a rare synthesis: the resources of a large wine company (it produced roughly 800,000 cases of wine in 2009) are harnessed to a team of gifted viticultural and winemaking personnel who are given much freedom to express themselves and push boundaries. The ultimate test of how well it all works is the obstinate Pinot Noir, and Villa Maria has been remarkably consistent in creating Pinots of verve and personality over the past decade.

After establishing the winery in 1961, Fistonich began growing Pinot in the mid-1970s. The first of the wines to make an impression was the Villa Maria Private Bin Pinot Noir 1983. Made in those days from grapes planted on the Ihumatao Peninsula (locale of the company's current headquarters), the wine carried off a pair of trophies at the 1985 Australian National Wine Show.

'It was earthy, basic Pinot — a quaffing red,' recalls Fistonich. 'It was not until the mid-1990s that we made a conscious decision to get serious with Pinot.'

Villa Maria's viticultural wunderkind at the time, Steve Smith MW, was charged by Fistonich with selecting the region best suited to helping the company achieve its Pinot Noir aspirations. After some deliberation, Smith gave Marlborough the nod. 'He concluded that Marlborough had everything Martinborough had, and also more versatility with its different microclimates and soil types,' says Fistonich. 'We went in there budgeting on not making any money for a while, but were pleased early on with the results.' Contributing to the Marlborough decision was the fact that the company had an existing winery in the region.

Villa Maria's first Marlborough Pinot Noir vines were planted in 1995. The company was quick to seize upon the southern valleys' aptitude for the grape and by 2002 there were 37 hectares planted, most shared between sites in the Awatere, Waihopai and Ben Morven valleys. The growth has continued apace. Villa Maria currently has 97 hectares of Pinot Noir in the ground in Marlborough.

Investment was also being made in people. In 2000, group winemaker Michelle

Richardson made a trip to Burgundy that would have a profound influence.

'I loved the feel of it,' she says. 'I loved the way they put the wine in barrel and didn't touch it until after spring. It [the wine] was precious, but it didn't need its hand held. That experience changed my attitude towards making not just Pinot, but other wines too.'

A strong Pinot-devoted culture quickly developed at Villa Maria's Marlborough winery. Alistair Maling MW succeeded Richardson after her departure in 2002, while George Geris recently assumed the role of Marlborough chief winemaker. The early decision to make reserve and single-vineyard Pinots — and make them without compromise — gave the viticultural and winemaking team the opportunity to make wines of which they can truly be proud. At vintage time there are myriad small fermentation tanks parked around the winery. They hold special parcels destined for these top wines and are all carefully managed and hand-plunged. It is an extraordinary sight in an operation as large as Villa Maria — and 'a logistical nightmare', according to Richardson — but is testimony to the company's Pinot attitude.

The first designated single-vineyard Pinot was the Villa Maria Lloyd Vineyard Pinot Noir 2000. Since then, in years deemed good enough, three other vineyards have authored single-vineyard wines: Seddon, Taylor's Pass and Rutherford. The first two are in the Awatere Valley; the Rutherford occupies a site in the Ben Morven foothills on the southern side of the Wairau Valley. 'A single-vineyard wine doesn't necessarily have to be the biggest or the boldest but it does need to have a level of individuality and uniqueness,' Maling explains.

The 2000 vintage was also the first of the Villa Maria Reserve Pinot Noir, a wine made from fruit grown in a number of Marlborough sites. This is an opulent, heady expression that, like its single-vineyard stablemates, regularly scoops gold at competitions. It has also shown commendable longevity.

Two lower-tier Pinots are also made in Marlborough — the entry-level Villa Maria Private Bin Pinot Noir and the Villa Maria Cellar Selection Pinot Noir. The latter, particularly, offers outstanding value.

....

Following pages: George Geris, Villa Maria.

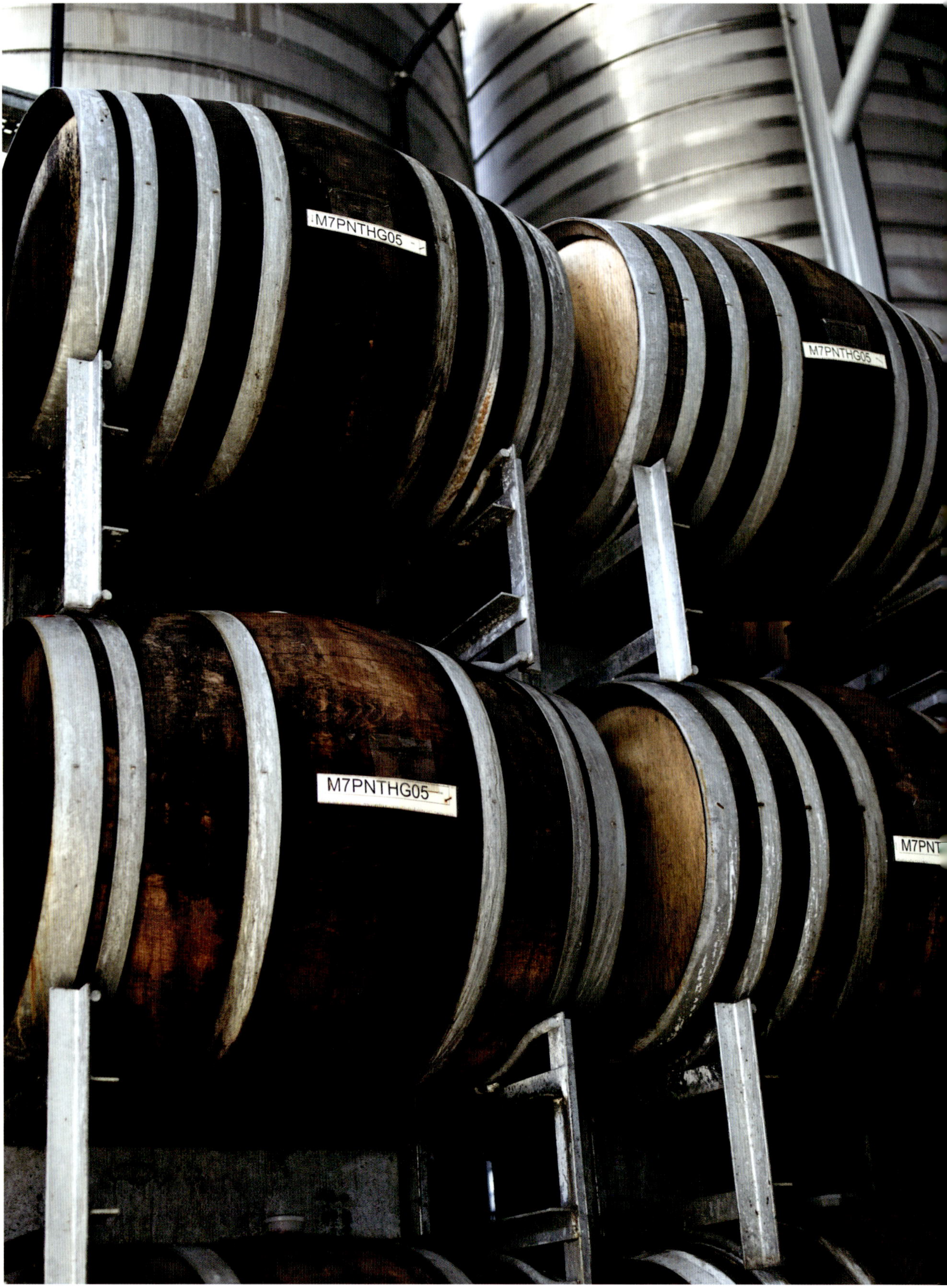
M7PNTHG05
M7PNTHG05
M7PNTHG05

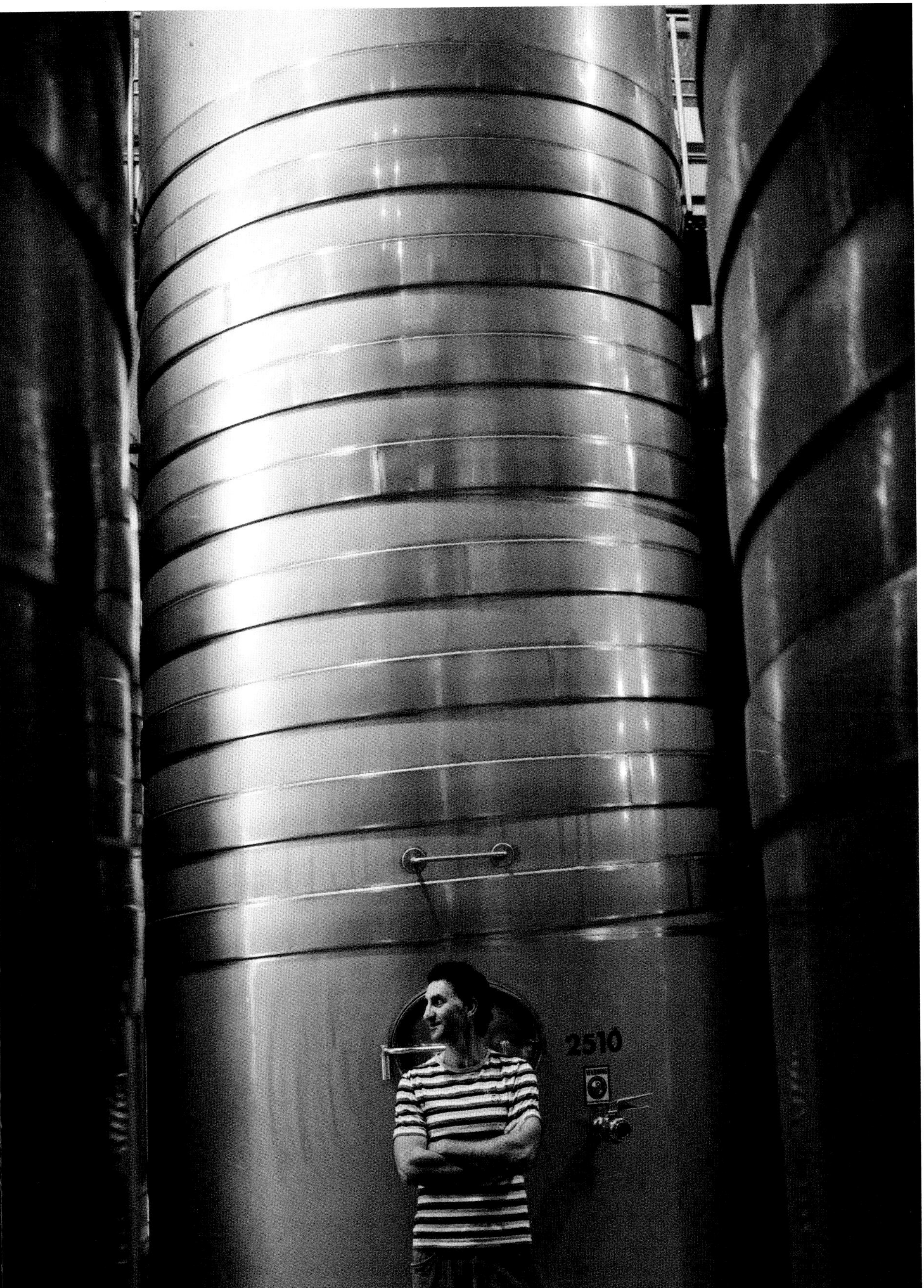
2510

TEN. CANTERBURY

Canterbury's role in the rise of New Zealand Pinot Noir has been seminal. The trials using Pinot and other cool-climate varieties undertaken by Dr David Jackson and Danny Schuster at Lincoln College during the 1970s sent excited ripples through the New Zealand wine industry.

Danny Schuster effectively took the trials out of the lab and into the real world when he joined St Helena, the region's first commercial winery, just north of Christchurch. There he created the gold medal-winning St Helena Pinot Noir 1982, and the excited ripples became seismic rumblings. Schuster's wine was arguably the most influential Pinot in New Zealand history. Doubters were silenced overnight. The variety — and the South Island — set out on what quickly became an ascendant path.

After leaving St Helena, Schuster took to the Omihi Hills in the Waipara Valley in 1986 to establish his eponymous winery. His wasn't the first wine operation in Waipara, but he was among the first serious Pinot producers to set up in the area, which has since become the beating heart of Pinot Noir in Canterbury. 'I needed a warmer place, so I came out to Waipara,' said Schuster. 'I also wanted dense limestone-derived clays and north-facing, gravity-drained grapes. This is what I needed for Pinot Noir.' (Sadly, after more than 20 vintages and some outstanding wines, Daniel Schuster was put in receivership in late 2009).

By the mid-1990s a clutch of Waipara boutique wineries led by quality-focused Pegasus Bay had joined the vanguard of New Zealand Pinot production. Today there is more Pinot Noir planted in Canterbury than any other variety, a distinction the region shares with Central Otago and the Wairarapa. Even so, Pinot Noir accounts for a modest 30% of the region's total plantings, testimony to the interest in, and quality of, Canterbury's aromatic wines, particularly Riesling.

The big growth in Pinot Noir came after 2002 and it was centred on Waipara. In that year Canterbury's total area planted in Pinot Noir was just 152 hectares; by 2009 it had climbed to 432 hectares. Contributing to this were major plantings and big players, including New Zealand's largest wine producer, Pernod Ricard. Its Waipara vineyards, Omihi and Camshorn, share a total of 86 hectares of Pinot between them. Across State Highway 1 from the Camshorn Vineyard, the McKean family established the 325-hectare sprawl of the Waiata Vineyard, which supplies Nobilo and Villa Maria.

Canterbury's is a diffuse terroir, unsurprising given that 300 kilometres separates the vineyards at the extremes of the region, which extends as far south as Timaru. In many ways, Waipara is a region in its own right, and its producers' organisation (Waipara Winegrowers) prefers it to be treated as such. Waipara's area under vine is now five times larger than the rest of Canterbury. Most of the vineyards outside Waipara are clustered on the plains around Christchurch, although there are intriguing Pinot vineyards in other districts. For the purposes of this book, Waipara is treated as a subregion of the large, scattered Canterbury wine region.

The Waipara difference is mainly about warmth. Driving north from Christchurch, you can almost sense the rise in temperature as you arrive in Amberley. This is the effect of the Teviotdale hills on the coastal side, which shelter the area from the cold, persistent northeasterly winds. As a result, Waipara can have 100 or more growing degree days (days between October and April where the temperature exceeds 10°C) than the vineyards located further south on the plains. In common with Martinborough, Waipara has an exposed southern flank (rain-bearing southerlies can hit both regions at flowering yet not affect Marlborough) and is open to rampaging winds from the northwest. Spring frosts pose a regular danger.

Soils in Waipara vary markedly. The meanderings of the Waipara River over the millennia have left stony free-draining terraces high above the existing river course. Several wineries have their plantings on these deep alluvial gravels with coverings of fine loess. A number of others are established on the gently sloping, north-facing

Teviotdale hills where the soils are heavier, with the added allure of limestone.

Inland from Waipara, in the hills to the west of the town of Waikari, are a couple of wineries that are particularly fascinating for the perspective they bring to Pinot Noir production. The hills are rich in limestone, a condition that provided the principal magnet for both Bell Hill and Pyramid Valley Vineyard.

The Canterbury Plains from Amberley south to Timaru have the same low rainfall as Waipara, but also cooler temperatures and an even greater frost risk. It is certainly cool-climate winegrowing country, and the district's Pinot Noir growers are thankful for their normally dry, mild autumn days to ripen the fruit. The soils are typically gravelly silt loams over gravel subsoils.

Canterbury's first vines took root at Akaroa in 1840, planted by the enclave of French colonists. The sheltered valleys and volcanic soils of Banks Peninsula have seen renewed viticultural activity since the 1970s, with mixed success. One vineyard that has been noteworthy, and is interesting from a Pinot point of view, is Kaituna Valley.

Canterbury's older vineyards of the modern era were mostly planted in cuttings from the original St Helena vineyard, many of which in turn were derived from the mass selection used at Lincoln in the 1970s. These clones were predominantly 10/5 and 2/10, which have lent the wines 'a lovely spiciness, developing into truffles over time', according to Matt Donaldson of Pegasus Bay.

Waipara Pinots are denser and more robust than those produced on the plains around Christchurch. The vineyards in the region's other corners provide starkly individual expressions. Canterbury's Pinot Noirs are as varied as their habitats — and their winemakers. Some of New Zealand Pinot's most devoted and enquiring producers are to be found in this region.

SOME OF NEW ZEALAND PINOT'S MOST DEVOTED AND ENQUIRING PRODUCERS ARE TO BE FOUND IN CANTERBURY.

KEY PRODUCERS

BELL HILL

There are few more engrossing Pinot Noir ventures in New Zealand than Bell Hill. Trained vignerons Sherwyn Veldhuizen and Marcel Giesen met in the early 1990s at Giesen, the successful winery Marcel and his brothers had established at Burnham, near Christchurch. A joint fascination with both the Burgundian varieties' interpretive gifts when planted on limestone and limestone-derived soils saw Veldhuizen and Giesen begin a location search. In 1997 (on the day before their wedding) they bought the Bell Hill site.

Limestone there is in abundance. The vineyard is planted on the site of an old quarry from which lime was extracted during the early decades of the twentieth century. It is an old seabed, with active limestone running to a depth of 40 metres. The vines cover the lower northern face of Bell Hill itself, a high knoll with an outline resembling a church bell when viewed from the south. The vineyard is nearly 300 metres above sea level.

Planting began in 1997 and is ongoing. The area under vine is currently 1.5 hectares. Even when the planting is completed over the next year or two, the aim is never to make more than 1000 cases of wine (Pinot and Chardonnay). If that scale seems Burgundian, the vine density most certainly is; the most recent plantings are spaced at 1.1 metres x 0.8 metres. The vineyard is divided into a number of small blocks across which are planted a broad range of clones. Bell Hill has taken the biodynamic path over the last two seasons.

'We didn't know from an elevation point of view if we could make it work,' Veldhuizen says. 'And look at the soil … in parts it's basically chalk. Making sparkling wine was going to be our Plan B.'

Plan B has not been necessary. The first vintage was 1999. Between that year and 2002 the wines were released under the Old Weka Pass Road label. The premium Bell Hill Pinot made its début with the 2003 vintage and both labels have been made annually since then.

'There is not a formula for the ratio of Bell Hill to Old Weka Pass Road each year,' explains Veldhuizen. 'Each block is picked, vinified and put to barrel separately and given a minimum of 12 months in barrel to show us its true potential before we make blending decisions for the two wines.'

The curiosity, ardour, risk-taking and faith of Giesen and Veldhuizen have been answered with wines that have won praise around the world. The Bell Hill Pinot typically treads a path between masculine and feminine, has great savoury depth, a lithe texture and quiet intensity.

> We're looking for structure, not fruit. That was part of the journey. We're not looking to make sweet-fruited Pinot Noir, which is a reason we've planted on limestone. The fruit is there, but it's more savoury. From our experience acidity is very important both for longevity and palate structure. Our wine doesn't show a high total acidity; it's where it presents itself in the mouth … it's very linear and drives things along. Essentially, we're looking for a wine that tells a story of where it's grown.

And yet the vineyard has not been without its problems. The vines in the Limeworks block, planted in 1999, have suffered from chlorosis, a devigorating condition brought on by iron deficiency in the alkaline soil. It is unusual in New Zealand, though not so much in France. It is hoped that replanting the block on different rootstock will rectify the problem.

Veldhuizen sums up the approach thus:

> Bell Hill is a long-term and lifelong project. We started purely to discover the potential of the soil. Now we are beginning to understand our site and are giving it the time to show us its true potential. It is the hardest thing we have ever done. One day the vineyard may tell us it was all worthwhile and live on beyond us. It may also tell us to naff off.

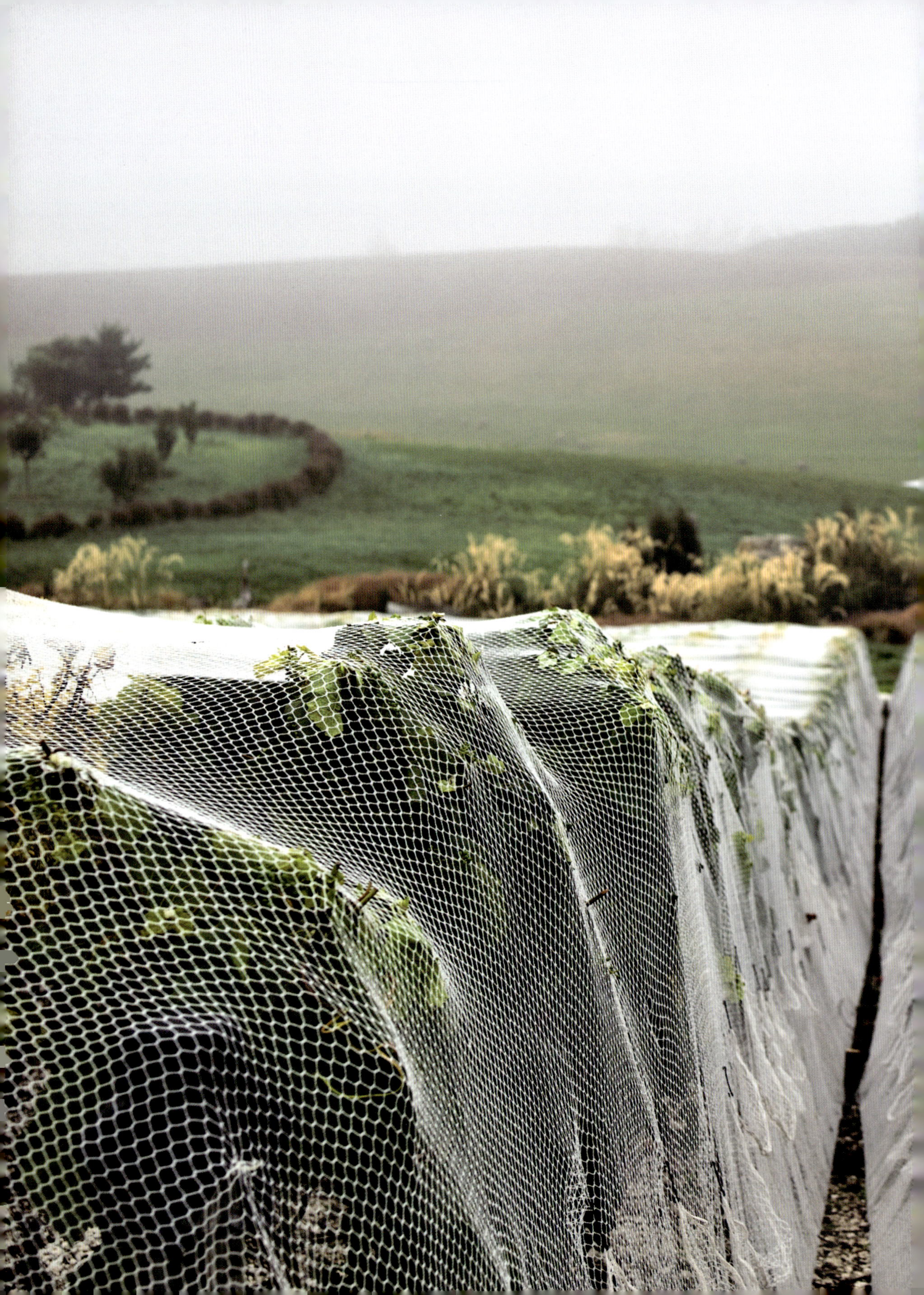

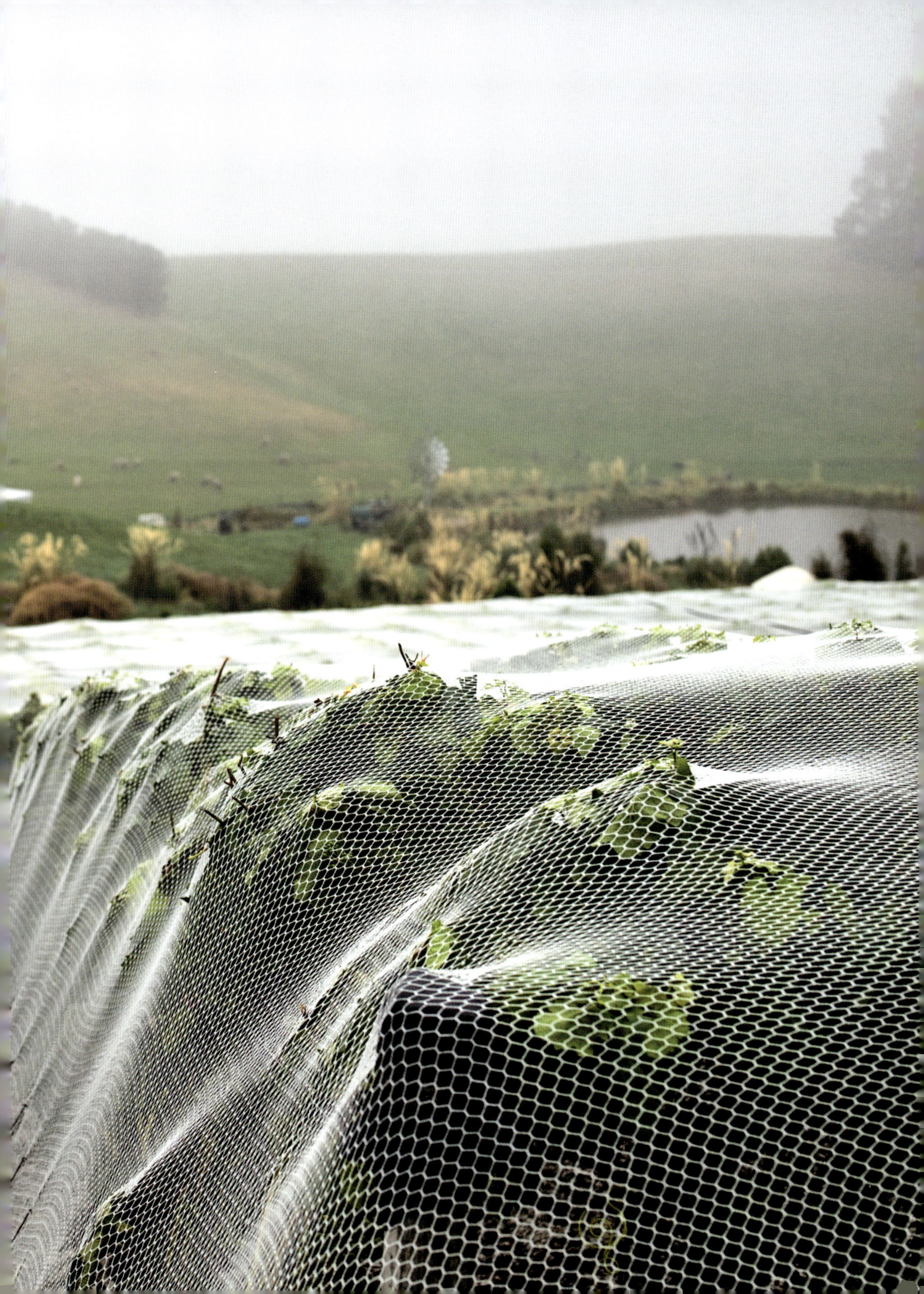

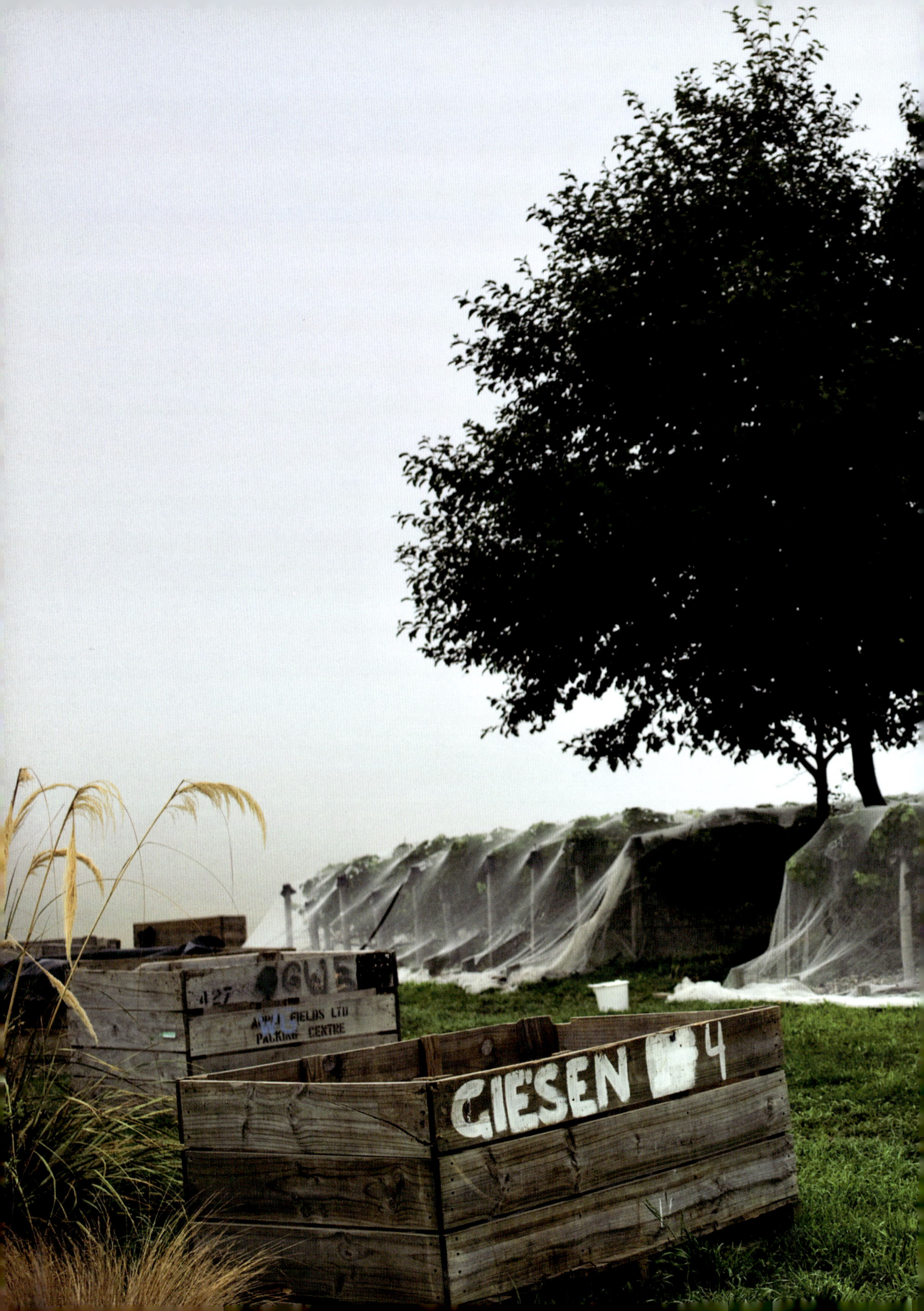
427
FIELDS LTD
CENTRE
GIESEN
4

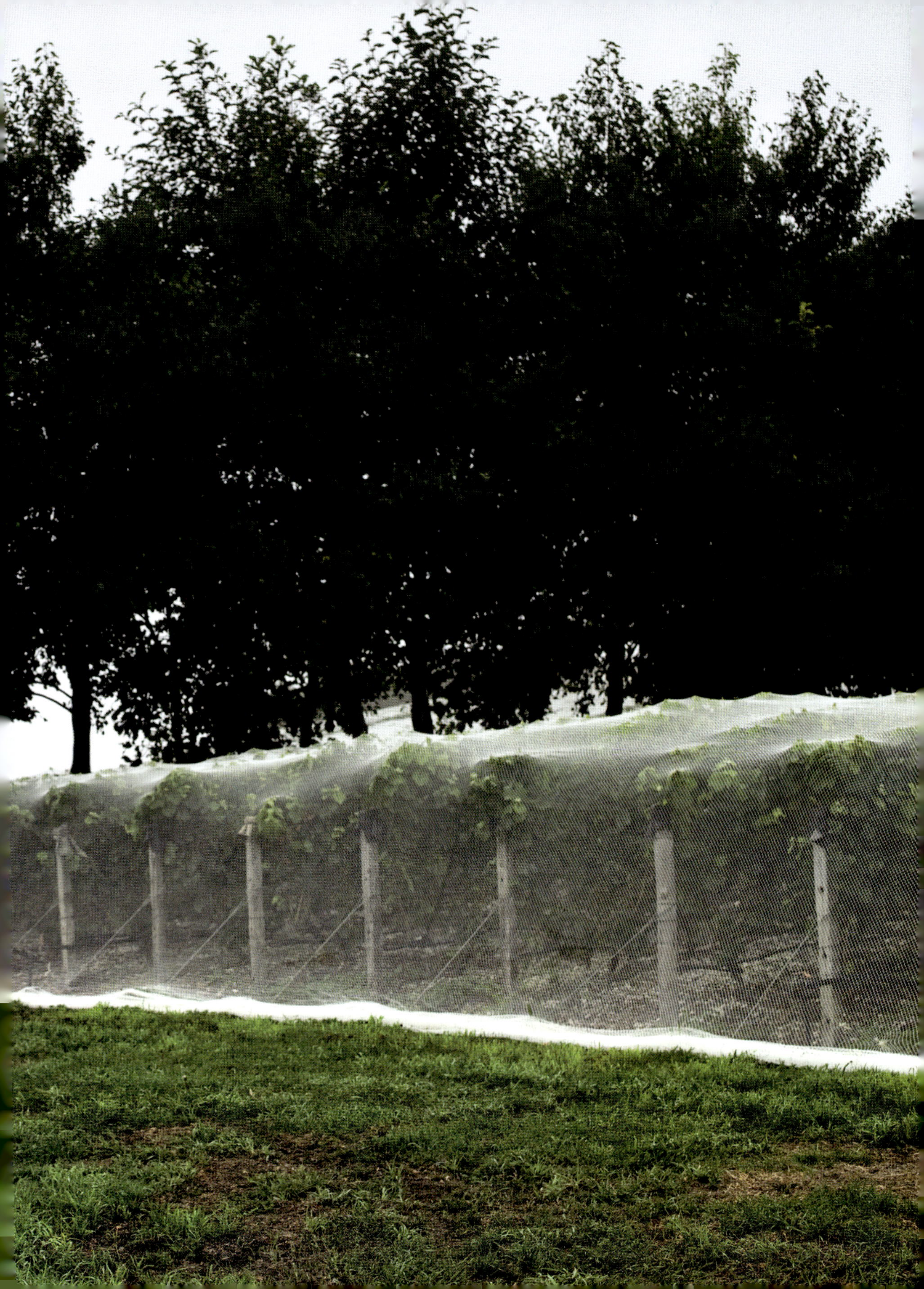

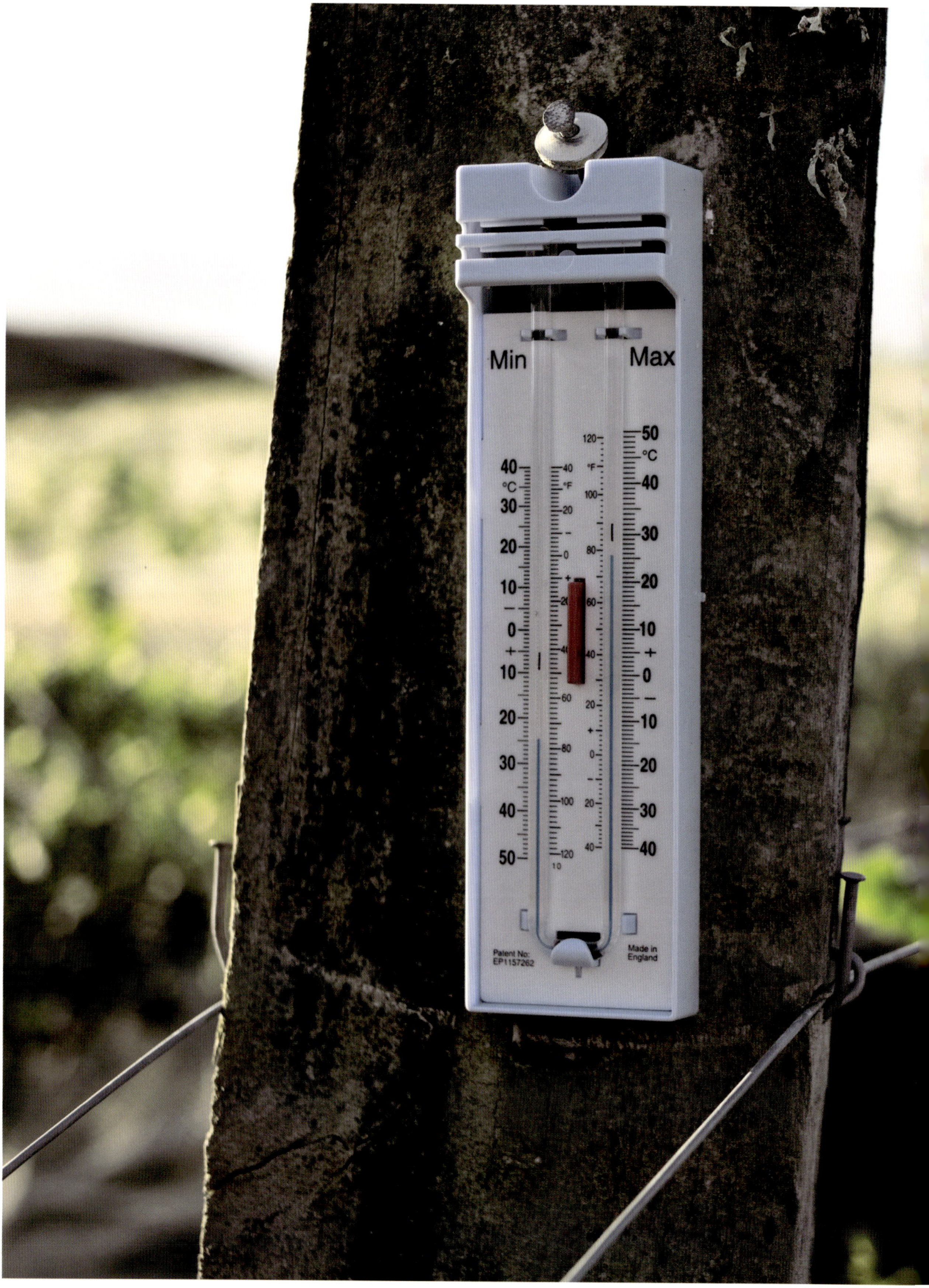
Min
Max
°C
°F
Patent No:
EP1157262
Made in
England

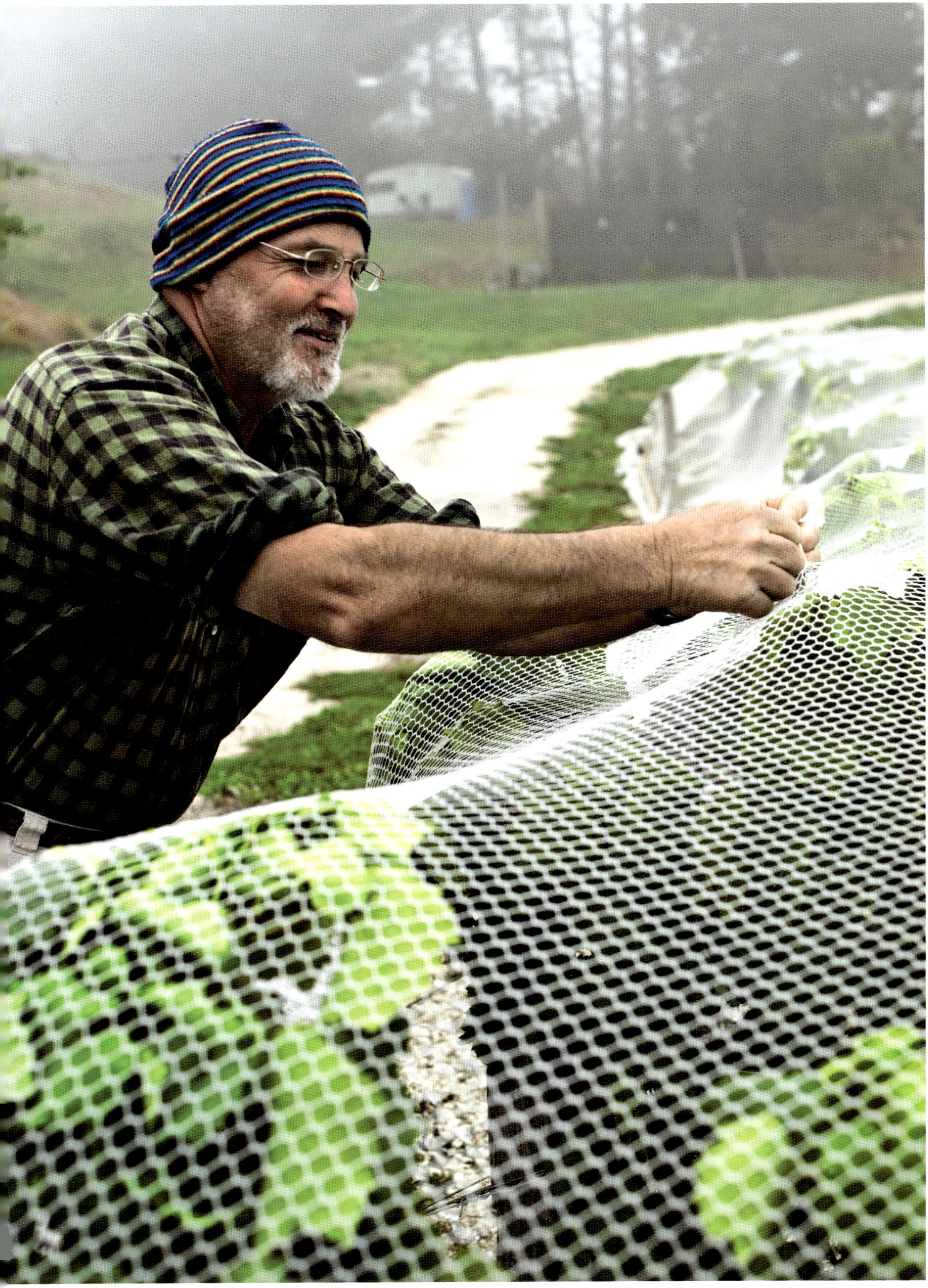

KAITUNA VALLEY

Kaituna Valley's home vineyard is as unlikely as it is storied. It is situated out on a limb, not far from the coast, a few kilometres up the rugged, volcanic Banks Peninsula corridor that is Kaituna Valley. The vineyard's first vines were planted in 1977 by Graeme Steans, then a tutor at Lincoln, assisted by Danny Schuster. At the time it was the first commercial Pinot Noir vineyard in the South Island. Today the original plantings comprise what is regarded as the oldest commercial Pinot plot in the country and have consistently been the source of some excellent wine.

The vineyard has figured in an eye-popping run of success in wine shows. Its first vintages contributed to the historic gold medal-winning St Helena Pinots in the early 1980s. The vineyard's current custodians, Grant and Helen Whelan, say that fruit from the site has been used to make a single-vineyard Pinot that, every year since 1993, has won either silver, gold or a trophy.

The site is blessed in a number of ways. One is the shelter provided by the surrounding hills, which foster a warmth not shared by the vineyards on the exposed plains around Christchurch. The valley experiences low rainfall and high sunshine hours, and is frost-free — the Whelans are also able to grow citrus fruit on the site. The vineyard itself sits on basaltic, high clay-content soil on a north-facing slope.

The Whelans (both of whom are closely involved in the wine production) bought the property in 1997 and extended the vineyard with a planting of Dijon clones. The current Kaituna Valley Kaituna Vineyard Pinot Noir is a 50/50 blend of fruit from these vines and the originals. The wine is a fantastic advertisement for the benefits of vine age. Any shortcomings in the original clonal material (obtained from Lincoln) have been negated by the savoury depth and seamless texture that only older vines can provide. The wine sees up to 60% new oak.

In 1996 the Whelans planted a vineyard in Marlborough's very dry, relatively cool Awatere Valley, shared between Pinot Noir, Pinot Gris and Sauvignon Blanc. The site is a sandy, stony river terrace on the northern bank of the Awatere River. The first Kaituna Valley Awatere Valley Pinot Noir was produced in 1999, the product of mainly Dijon clones, densely planted, with a yield of just 5 tonnes per hectare. 'The nature of the soil and climate restricts vine growth and produces grapes with intense flavour and a distinctive mineral/flinty character,' says Grant Whelan.

Two more Pinots were added to the portfolio with the 2006 vintage, both single-vineyard wines grown in Canterbury. The first was the Kaituna Valley Bone Hill Pinot Noir, made from fruit grown in a new vineyard right beside the original home vineyard in the Kaituna Valley. The other was the Kaituna Valley Canterbury Pinot Noir, which Whelan describes as 'a new "entry level" label using fruit harvested from our favourite Banks Peninsula vineyards'.

> Pinot is our passion and we believe careful viticulture to be the key to achieving top quality,' says Whelan. 'Intensive canopy management — shoot and cluster thinning, lateral removal, extensive leaf plucking — to maximise the exposure and quality of fruit is at the heart of this.

MOUNTFORD

The winemaker at Waipara's Mountford Estate is the multilingual, sight-deprived CP (Chung Pin) Lin.

The CP Lin story is unusual, to say the least. Born in Taiwan, and rendered blind from carcinoma of the retina when he was two, he settled in New Zealand with his family at age 13. He went on to study maths and electrical engineering at the University of Canterbury, became bored with it, and joined a wine club as an antidote. This opened up a compelling new world, made more so by the discovery that he had a particularly perceptive palate. He made up his mind to study viticulture and oenology at Lincoln University and move into the wine industry.

CP Lin was working in a wine-export business when, one day, he lunched at Mountford with friends. Mountford's then owner, Michael Eaton, struck up a conversation and asked what CP thought of Mountford's wines. CP told him he thought they were 'crap' and proceeded to offer some viticultural advice. Eaton wasn't sure what to make of the blunt-talking, blind Asian visitor. He lit up a cigar. CP Lin sniffed the air and asked, 'Who is smoking a Montecristo Cuban cigar?' Indeed, the cigar was a Montecristo. Eaton could not help but be impressed at this display of olfactory cognisance, and the session finished with CP Lin engaged as winemaker at Mountford. That was 1997 and he has been in the role ever since.

Eaton and his wife Buffy began planting their clay-over-limestone site in the Teviotdale foothills in 1991. The original vineyard was 4 hectares, evenly divided between Pinot Noir and Chardonnay. The Pinot clones were the earlier arrivals — 10/5, Abel and the UCD clones — all planted on their own roots. Ten years later, Dijon clones were planted on two small sloping blocks. Dubbed the Gradient and the Rise, these north-facing blocks have high levels of limestone and were planted with a spacing of 2 metres x 1 metre. The site has one of the steepest gradients in the region.

In 2007, the Eatons sold the vineyard and winery to Kathryn Ryan and Kees Zeestraten, a couple who share a working background in agriculture. Since then, CP Lin has continued to make the wine while the operation has expanded around him, Ryan and Zeestraten having decided the winery's tiny 2000-case output could be increased. The winery has been extended and in 2008 another 4 hectares of vines (a third of which is Pinot Noir) were planted.

Mountford Estate currently makes four Pinots. Fruit from the original plantings is used for the Mountford Estate Pinot Noir. Among CP Lin's stated objectives are: low cropping levels (3–4 tonnes per hectare), hand-managing the vineyard, taking a traditional approach in the winery and allowing the site to have the starring role. Over recent vintages this has been a dark, succulent, harmonious wine, firm but fine in texture, with a singular, savoury edge.

The 2006 vintage was the first time the Mountford The Gradient Pinot Noir was produced, made with single-vineyard fruit from the site of the same name. It is produced only in exceptional years, the decision coming when the barrels are blended. Only three barrels contributed to the 2006.

From the 2000 vintage the Mountford Village Pinot Noir has been made with fruit sourced from both the older estate plantings and Waipara contract growers. From 2007 the Mountford Liaison Pinot Noir has been produced with contract-grower fruit only.

MUDDY WATER

Pinot Noir lured Jane East and her husband, Michael, to Waipara. The Christchurch pair — she trained in viticulture and oenology and he is an obstetrician — tasted a 1990 Pinot made by Mark Rattray at Waipara Springs. 'We were blown away by it,' East remembers. Soon after, in 1993, they were planting vines of their own on the clay and limestone lower slopes of the Teviotdale hills. The name Muddy Water is a direct translation of Waipara; its closeness to the name of a famous Chicago bluesman was an added attraction for music-mad Michael.

The label's first three vintages were made at the Waipara Springs winery. Since 2000, Muddy Water has had its own winery and its own winemaker, the accomplished Belinda Gould.

Gould has become synonymous with Muddy Water. Her own roots in the district run deep. She was christened at the Glenmark Church, something of a Waipara focal point, and trained as a viticulturist at Lincoln. Years were spent working in Germany, Italy and France before she returned and made the move from vineyard to winery.

For a decade of vintages at Muddy Water, Gould has sought to produce Pinot Noir without affectation — vineyard-reflective wines. 'She's her own harshest critic,' Jane East says of her winemaker. 'If she says she's happy with a wine, it's going to be good.'

Muddy Water has gone down the organic path over recent years, with full organic accreditation for the vineyards and the winery through AgriQuality.

Close to half of Muddy Water's vineyards are planted in Pinot, and future plantings will be devoted to the variety. There are usually three Pinots produced annually, all from estate-grown fruit: Slow Hand and Hare's Breath, which are both single-vineyard wines, and the Muddy Water Waipara Pinot Noir. Gould needs to be happy with the fruit for the single-vineyard wines to be made, hence no Slow Hand from the frost-affected 2008 vintage.

Slow Hand used to go by the name Mojo until 'trademark issues' forced the change. Its fruit is drawn from a block of 16-year-old 10/5 vines, originally obtained from St Helena, planted on dense calcareous clay. It is a fine example of what 10/5 with some vine age can produce — concentrated, well-structured, finely textured wine, quite savoury in flavour.

Hare's Breath fruit is sourced from a sloping, limestone-rich plot. A broad mix of close-planted clones are co-fermented. 'The components are always more complete than the other vineyard of the same age,' says Gould. 'Some of us think it will be our best vineyard in the long term.' It is typically a dense, sensual wine oozing red and dark fruit charm, again with an underlying savoury seam.

The Muddy Water Waipara Pinot Noir does not have the individual flair of the other two, but is always an interesting, often funky, rendition.

All the wines are fermented with indigenous yeasts, hand-plunged, spend 15–16 months in French oak (30–35% new) and are bottled without filtration.

'I'm not into big chunky tannins,' says Gould. 'I like the fine-grained tannins you get from long maceration. Total time on skins will vary from year to year — anywhere between 21 and 32 days. We're looking for wine that ages up to a point … we don't want people to have to hang around for ages waiting for them.'

PEGASUS BAY

The Waipara winery named after the large nearby bay, named after the sailing ship that surveyed part of the South Island, named after the winged horse of ancient myth, has become the Cathedral Square of Canterbury wine — a much-visited icon.

It was established by Ivan Donaldson and his wife, Christine, in 1986. Outside his career as a neurologist, Donaldson had been active in the Canterbury wine scene since the 1960s — as a judge, as a wine writer for *The Press*, and as a hobby vineyard dabbler. The creation of Pegasus Bay was undertaken with the whole family in mind. Ivan and Christine's son Matthew Donaldson studied oenology at Roseworthy in Australia and — along with his wife, Lynnette Hudson, also a trained winemaker — has been making the wine at Pegasus Bay since the 1992 vintage. Two other sons, Edward and Paul, are involved on the administrative side of the business.

Pinot has always figured higher than any other variety on Matt and Lynnette's preference list. Their personal Pinot voyage at Pegasus Bay is not unlike that of many other winemakers of their generation in New Zealand. As Matt says:

> When I started, I had the Australian mentality that bigger is better. I still do enjoy full-bodied reds. But back then I was trying to get maximum tannins … cold soaks, a lot of plunging. The result was a big, extracted style. In 1998 we were tasting some of the earlier wines and wondering if they would ever come into balance. So in 1999 we changed completely … a much shorter time on skins, one plunge a day. But it was a bad vintage to try it; we needed more seed tannin that year. So we went from one extreme to the other — from over-extracted wine to one lacking in structure, which also hasn't lasted.

'From 2001 on, we've become more flexible,' adds Hudson. 'We've been assessing vintage conditions and modifying our approach as to how we see the fruit, swinging back and forth with tannin extraction. The key is long skin contact — 28 days — which, mixed with controlled oxidation, reduces up-front fruitiness in the wine. New Zealand has enough of that anyway, so we sacrifice some of it for palate structure and complexity.'

The Pegasus Bay Pinot Noir has been made since the 1991 vintage. It is produced with fruit from the estate vineyard, planted on a free-draining old river terrace to the south of the Waipara River. There is a varied clone population, from 22-year-old 10/5 and 2/10 plantings (what Matt Donaldson calls the 'Lincoln mix') on their own roots through to a spread of Dijon clones.

The aim is for a maximum crop level of 2.5 tonnes per hectare, often achieved naturally. Grapes are harvested when the seeds are ripe, usually in mid-April. The fruit is typically destemmed (and not crushed) and given a one-week cold soak. The ferment starts naturally; plunging is carried out twice daily. The wine is matured for 18 months in French oak (30% new) and goes through malolactic fermentation naturally in the spring.

Recent vintages have been outstanding — a sinewy, sexy expression with beguiling sweet/savoury balance and firmness in structure.

Care is taken during fermentation and after to keep older clones and parts of the vineyard separate. In hot, excellent vintages, a selection is made of the best batches to make Prima Donna, a reserve-style Pinot that was first produced in 1996. The oldest vines on their own roots, unsurprisingly, always perform best and provide the fruit for Prima Donna. It is a well-structured, robust, stylish, balanced Pinot that sees 40% new oak and ages gracefully.

The Main Divide Pinot Noir is a second-tier wine first made in 1994. Fruit is sourced from growers in Canterbury and Marlborough, and the wine spends a year in old barrels. It was joined in 2004 by the Main Divide Tehau (a selection of Waipara contract-grown fruit) and in 2006 by Main Divide Tipinui (a blend of two Marlborough southern valley sites). The Main Divide wines regularly offer Pinots of exceptional quality and interest for their price.

Following pages: As grapes ripen, the threat posed by hungry birds also rises. The protective nets are put on at Pegasus Bay.

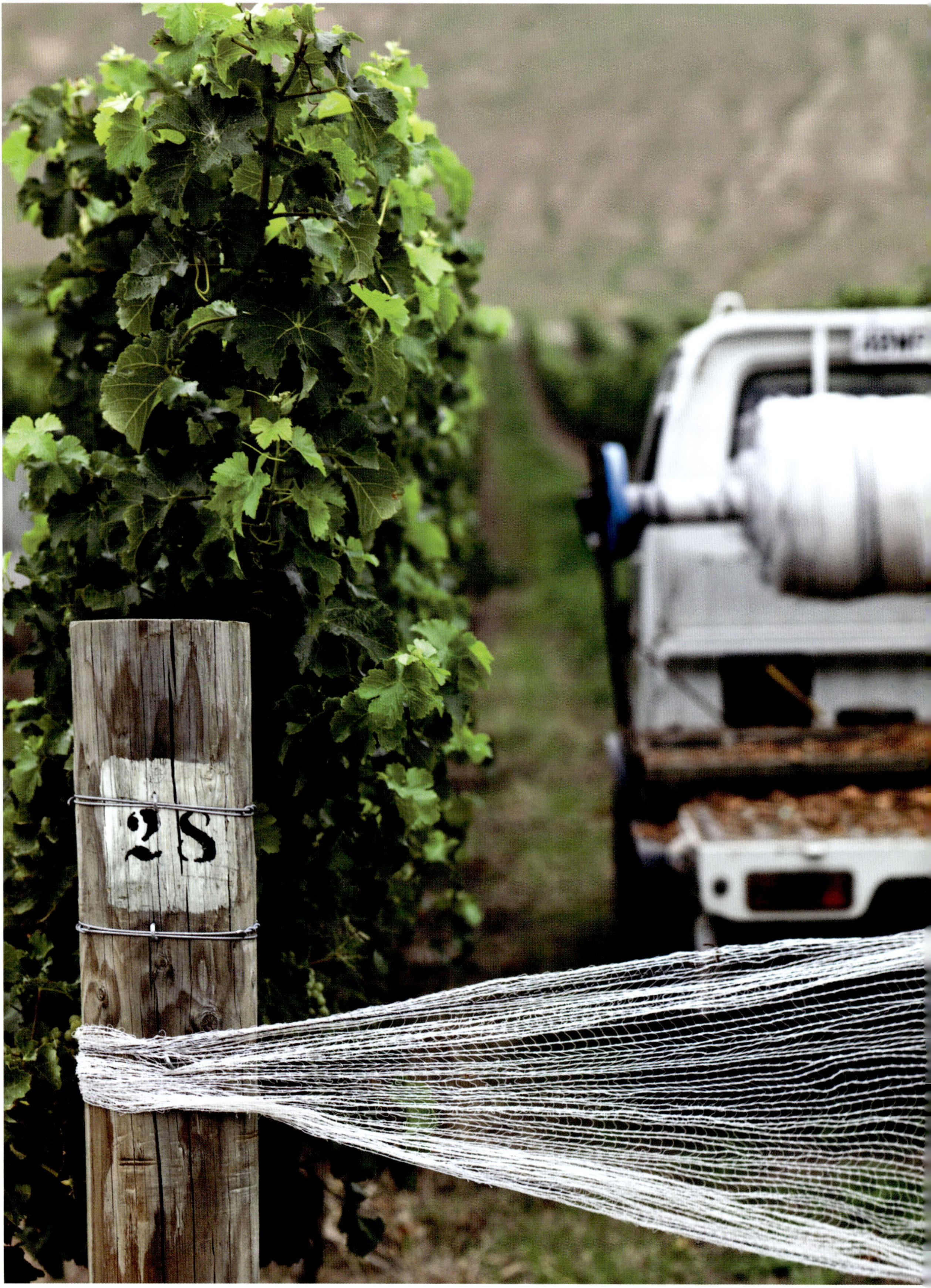
2S

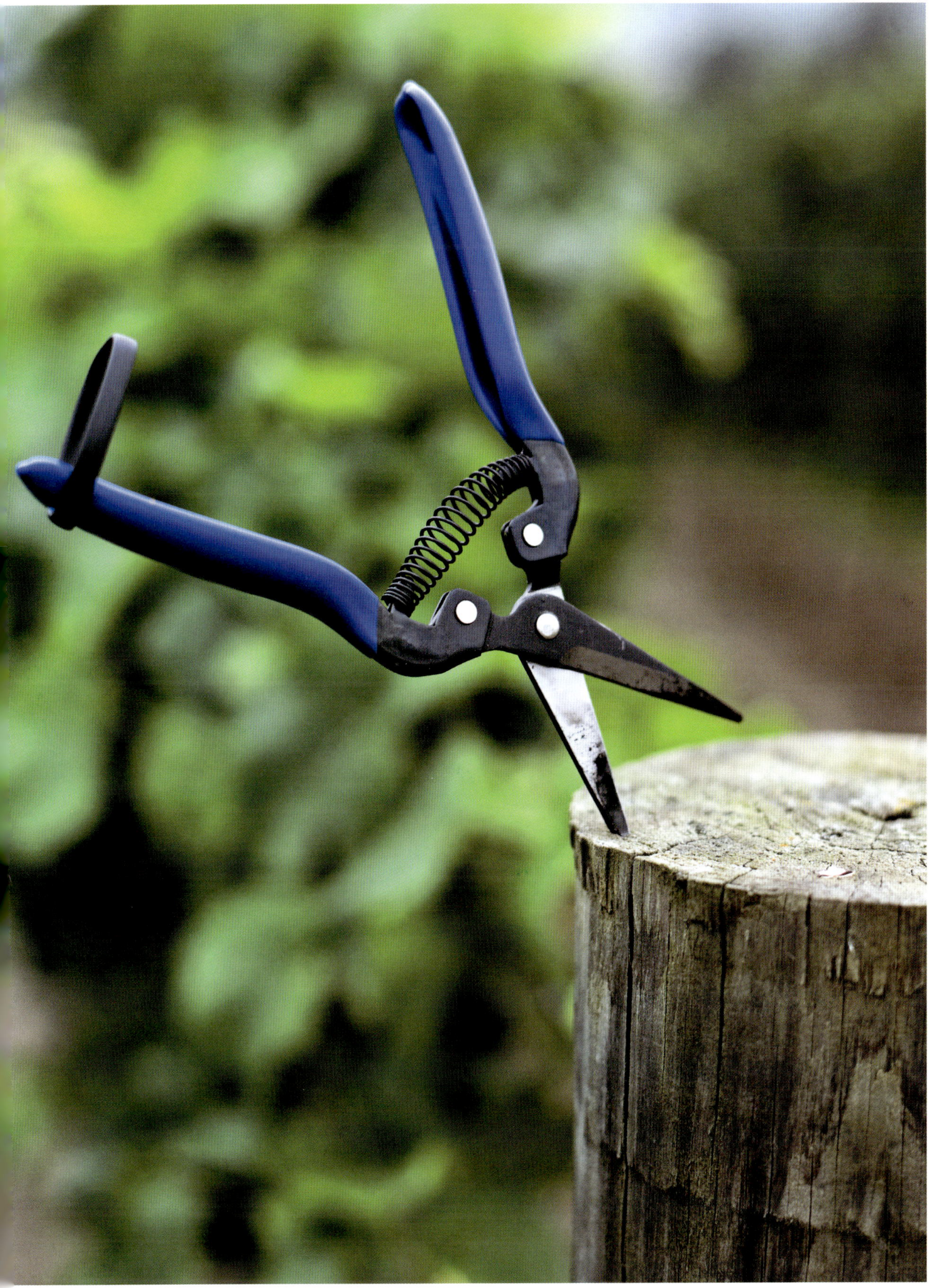

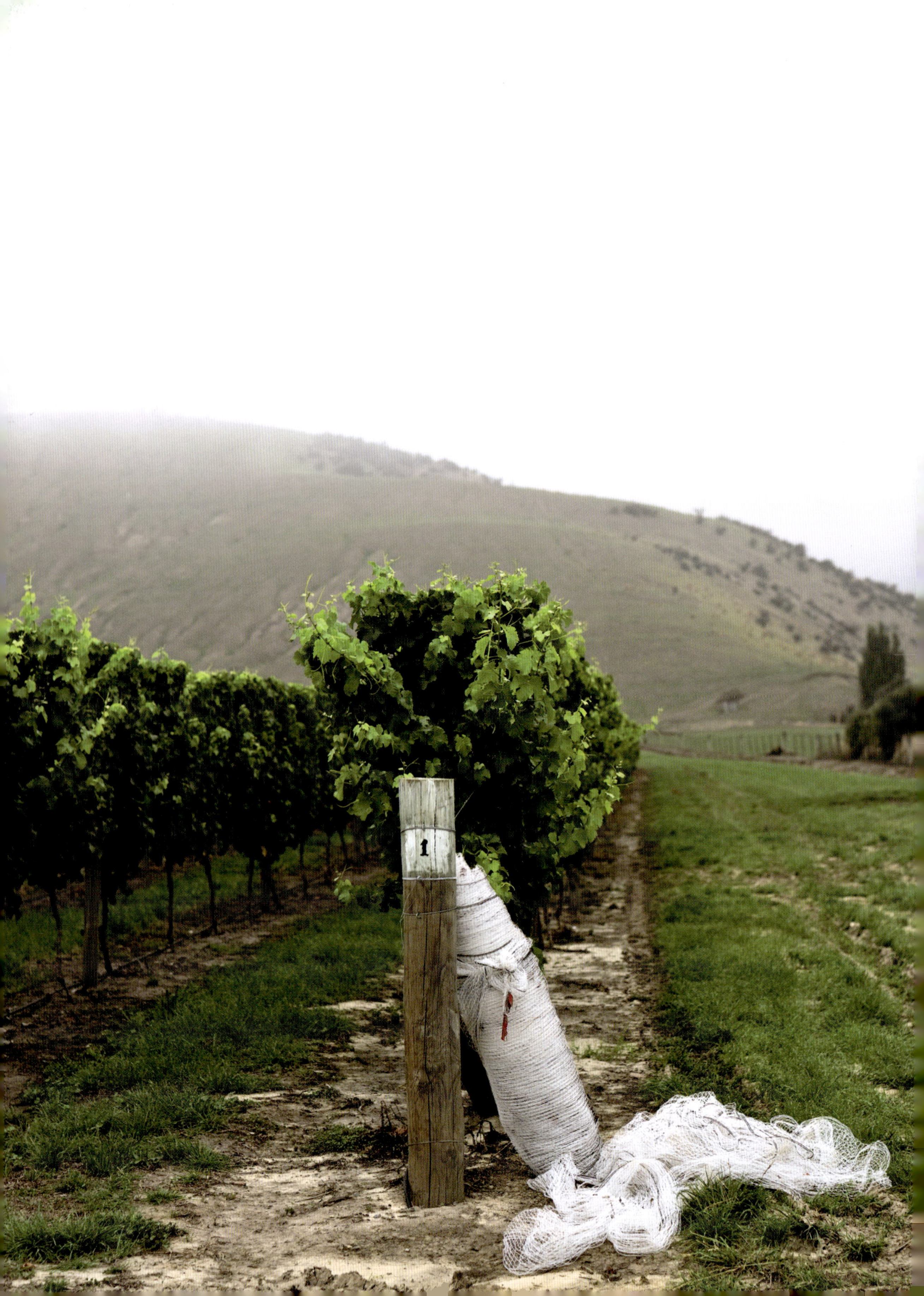
1

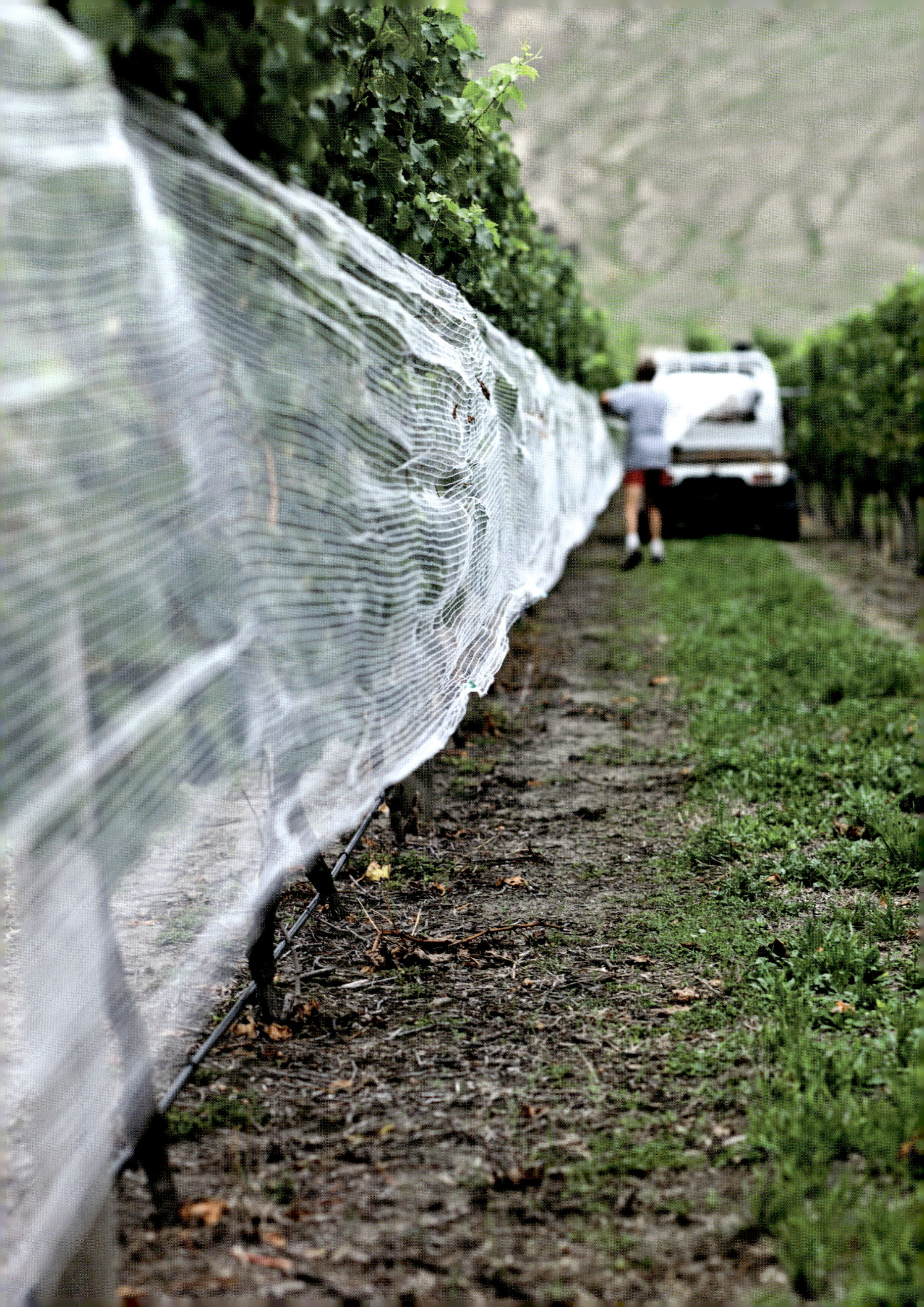

PYRAMID VALLEY VINEYARDS

Although winegrowing has taken Californian Mike Weersing to many countries and regions around the world, his vinous spirit is sheeted to Burgundy. He studied viticulture and oenology at the Université de Bourgogne (Dijon) and has worked for a number of the region's great domaines.

It was after consultation with friends in Burgundy that Weersing and his wife, Claudia, acquired their sloping vineyard site in Pyramid Valley (north Canterbury). The Weersings sent several soil samples to Burgundy for analysis, and the verdict on the soil taken from the lower part of the slope was positive: 'Here you can make a *vin de terroir*.' The land was bought in 2000, ending a worldwide, five-year search for the right location to grow Pinot Noir and Chardonnay.

What they had been seeking was a combination of clay and limestone. Weersing explains:

> Not because we wanted to ape Burgundy. Simply because the European experience is that different soil types have different stylistic effects. For instance, schist augments bright fruit flavours, clay gives you flesh and voluptuousness, limestone gives you structure and ageability, and so on. Once you accept that, and that every variety has its strengths and weaknesses, you match the soil to the kind of wine you want to make. Pinot is thin-skinned, aromatic and low in tannin and tends not to age as well as other reds. So it follows that if you plant Pinot on clay and limestone, the clay will provide body and the limestone will give you structure. You'll get a completeness and longevity.

Pyramid Valley's site, like that of Bell Hill, is in the hills directly west of Waipara, beyond Weka Pass. It is a beautiful place. One part of the slope faces north before curving around so that most of the vineyard has an eastern aspect. At the foot of the slope is a small lake, created by the Weersings.

Frost is not an issue on the slope. The vines are trellised with low fruiting wires and the canopies are trimmed very low. The vine spacing is a dense 1 metre x 0.83 metres. From the start, biodynamic philosophy has governed the management of the vineyard.

Irrigation has been eschewed since the first year.

There are four blocks planted — two in Pinot Noir and two in Chardonnay. The Pinot blocks are the north-facing Angel Flower and the east-facing Earth Smoke, the source of the names being a weed species common in each block. A varied palette of clones has been planted and Weersing harvests all at the same time. 'In Burgundy they pride themselves that some are ahead and some behind … that way you get this spectrum of flavours rather than the similarity that comes with uniform ripeness.'

The first vintage, 2006, produced 30 cases of Angel Flower and 60 cases of Earth Smoke. They are quite different. The former's floral aromatics usher in a savoury but bright, intense palate; Earth Smoke is fleshier and less linear — a low, still, funky expression. Each is made with minimal additions (for example, little or no sulphur) and fermented using indigenous yeasts. The wines are matured in 30% new oak and bottled unfined and unfiltered.

At full capacity, the current plantings at Pyramid Valley will only produce 600 cases of Pinot and 400 of Chardonnay.

Two other Pinots are produced as part of the label's Growers Collection. These use fruit drawn from the leased areas of vineyards belonging to a small band of like-minded growers. One is the Eaton Family Vineyard Pinot Noir, with fruit from Marlborough, last made in 2007. The other is the Calvert Vineyard Pinot Noir, sourced from the biodynamically managed Bannockburn vineyard from which Felton Road, Craggy Range and Pyramid Valley each make a Pinot. It is typically dark, floral, alive and sensual. Says Weersing:

> One of the joys of the Growers Collection is it allows us to make wine and charge less. We never want to be people who make wine for rock stars and millionaires. At $55 a bottle, we've tried to make the Growers Collection Pinots more accessible. Pinot is, of course, the one red grape that is least suited to being an inexpensive wine.

ELEVEN. CENTRAL OTAGO

To visit Central Otago, even for a few hours, is to fall under its spell. It is an unrestful, demanding landscape, overpoweringly beautiful, a place that ignites dreams and seems to enlarge the spirit. These qualities have made it our most photographed wine region. Its upstaging presence in so many scenes of *Lord of the Rings* took its snow-crested, rock-strewn mountains and plunging river canyons to a global audience of hundreds of millions.

Many a New Zealand Pinot Noir producer from regions further north has expressed the view that the seductive physical beauty of Central Otago has played a key role in drawing consumers to the region's wine, the alleged thought sequence being that such god-given beauty must surely provide god-given wine.

Central Otago winegrowers would be the last to deny their landscape is a marketing asset. But what they have achieved with Pinot Noir in a short time has been as remarkable as any of their mountain ranges, and speaks more than anything about their dedication and their wine's quality. The region has gone from zero to 1200 hectares planted in Pinot Noir inside 22 years. It is now our largest producer of varietal Pinot Noir. More significantly, it has achieved a region-grape synonymity in several key international markets.

'Central Otago Pinot has become New Zealand's red wine equivalent of Marlborough Sauvignon,' US wine writer Matt Kramer noted in the April 2005 issue of *Wine Spectator*: 'I cannot recall a winegrowing region, let alone one committed to a variety as demanding as Pinot Noir, that has vaulted to such a level of accomplishment in so short a time.'

Central Otago's growth in the modern era can be divided into three basic phases. The first belongs to the pioneers of the mid-1980s — Rippon, Taramea, Gibbston Valley and Chard Farm, who were later joined by Black Ridge, William Hill and Leaning Rock. In 1992 there were only six producers, who collectively harvested just 71 tonnes.

What is known as the second wave started in the early 1990s with the establishment of the likes of Olssens, Mt Difficulty and Felton Road. Some of these later arrivals came for lifestyle reasons, but many others brought a new layer of expertise to the region, evident in the superior sites in the Cromwell Basin chosen by many. In the case of viticulturist Robin Dicey, a founding partner at Mt Difficulty, his influence would be transformative. South African born and trained, Dicey became an important force in the expansion of winegrowing in Central Otago through his vineyard development and management company Grape Vision.

It was from 2000 that the floodgates opened or, in the words of Gibbston Valley Wines founder Alan Brady, 'all hell broke loose'. The area under vines expanded 625% between 1998 and 2009. Central Otago had by now shed its 'marginal' tag, its success with Pinot Noir ushering in serious recognition and the subsequent tide of investment. However, the country's larger wine companies remained circumspect and have not been a part of this land grab, although most of them do now buy fruit to create their own Central Otago Pinots. The region's producers are almost all small, the exceptions being medium-sized wineries Mt Difficulty and Chard Farm.

What happened with Pinot Noir was the early emergence of a distinctive regional style that became an instant hit with consumers. It was Pinot Noir expressed in a most

Following pages: The Central Otago Wine Company (CowCo), one of the region's two large contract winemaking operations.

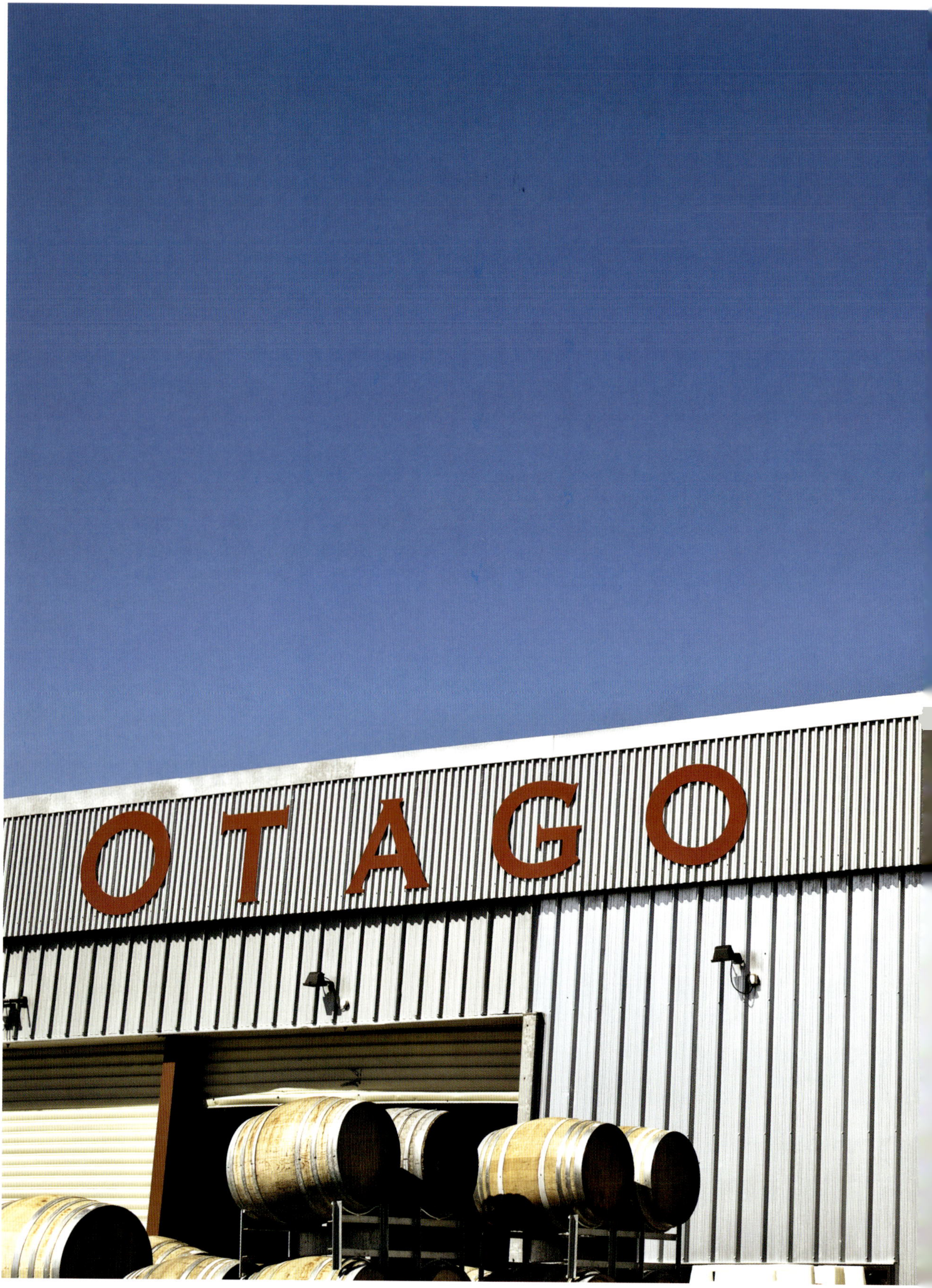
OTAGO

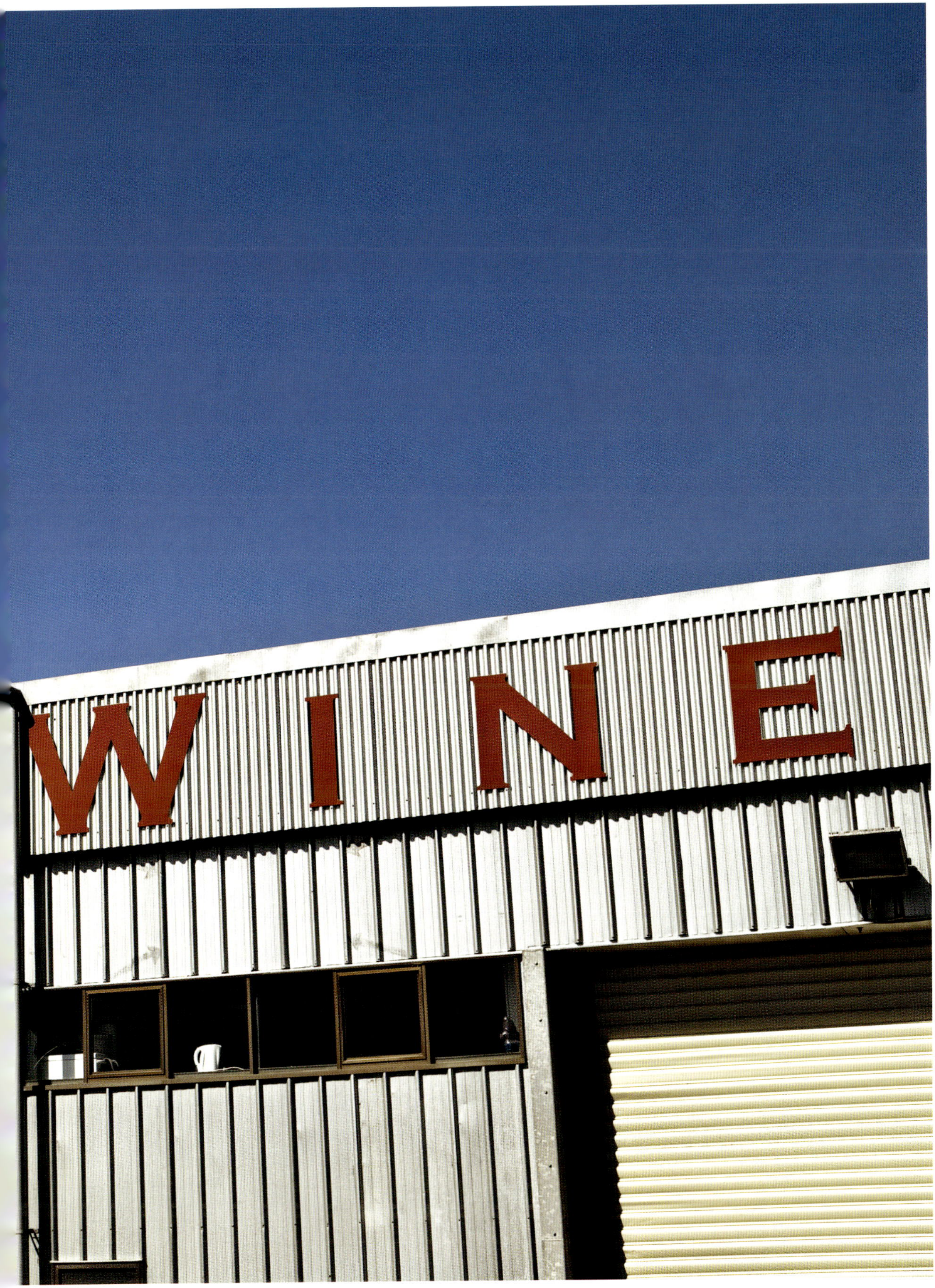
WINE

seductive, approachable way — richly perfumed, plump wines bursting with enticing dark fruit and spice flavours. Central Otago's was the first New Zealand Pinot to declare its origins so boldly and sassily in the glass. Some critics have derided it for being a fruit-driven, 'Pinot for beginners' style. There is some validity in that, but the kind of fruit expression Central Otago is achieving with young vines and one generation's winemaking experience can only bode well for the future.

A side-effect of Central Otago's rapid rate of growth, and the region's large number of small players, has been the rise of contract wineries. These are in effect commercialised wine cooperatives where growers can bring their fruit and have it made into wine by an experienced winemaker. Tank space is leased, while the growers buy their own barrels which are stored in the facility for them. The economies of scale of such a system are not hard to grasp. It opens the door for smaller operators to have their own label and sidestep the considerable costs of setting up a winery.

The twin towers of the Central Otago contract winemaking scene are VinPro and the Central Otago Wine Company (CowCo). They are neighbours, both occupying old warehouses in Cromwell's industrial zone. It all seems agin the artisan Pinot spirit until you talk to and observe the talented winemakers at the centre of each establishment (Dean Shaw at CowCo and, until recently, Carol Bunn at VinPro). They are not commercially driven people, they don't lack Pinot passion and they do get out into their clients' vineyards. While it is not a perfect situation, Bunn and Shaw manage it extremely well and their winemaker thumbprints are becoming less marked as the vineyards age. It's worth noting that the large negociants in Burgundy vinify even larger volumes of Pinot from different sources.

There are now 95 wineries and 75 grapegrowers in the region. Alan Brady laments he'd be 'lucky to know half of them', but one of the region's strengths is its unity. From the beginning, when their isolation brought them together to share equipment and ideas, Central Otago growers have had a strong collective spirit. Today the Central Otago Winegrowers Association (COWA) has 141 members and 31 associate members. From COWA in 2003 sprang a marketing arm which tightly ties Pinot to the region's identity. Central Otago Pinot Noir Limited (COPNL) is a company wholly owned by COWA that is charged with spreading the Central Otago Pinot gospel. It does this very effectively through events such as the Central Otago Pinot Noir Celebration and international media visits.

The event that many Central Otago wineries point to as the great leap forward internationally was COPNL's first foray to London. The 2002 vintage was an excellent one, giving most of the region's producers dark, floral, voluptuous, instantly lovable Pinots. Armed with their wares, a group of Central Otago winemakers descended on London. They held a tasting for trade and media, and in the evening hosted key representatives of those groups at Peter Gordon's Providores restaurant. It mirrored what happened with Marlborough Sauvignon Blanc two decades before; the Brits loved the wine, enjoyed the company of the winemakers, and told the world.

New Zealand's (and the world's) southernmost wine region, Central Otago lies on latitude 45° South. It is our most landlocked region, hemmed in by mountains on all sides, and no New Zealand town is further from the coast than its hub, Cromwell.

The continental feel of the alpine landscape is echoed by the climate, which is as continental as it gets in New Zealand. Freezing, snowswept winters with temperatures as low as -20°C are backed up by scorching dry summers where the mercury regularly tops 30°C. The diurnal shifts are similarly marked. During the growing season nightfall can bring a 20°C drop in temperature.

Dryness is a feature of this climate. Alexandra holds the New Zealand record for the lowest rainfall over a 12-month period (167mm). The region's normal annual rainfall is between 300mm and 750mm.

The dominant feature below ground is readily visible on much of the surface: schist. The region's soils have all been formed by glaciers, with schist being the parent rock. Mixed schist-greywacke alluvium, windblown loess, loamy sands and river gravels can also be present in differing degrees above the schist bedrock. The free-draining nature of the soil is so pronounced that irrigation is seen as a necessity. Generally low in organic matter, the soils result in relatively low vine vigour.

The Pinot Noir growing season in Central Otago is comparatively compressed. Budburst typically occurs much later than it does in Martinborough, which also counts many more growing degree days than the southern region. The winegrowers in Central

Otago point to their summer heat and a long, benign autumn as being crucial to their ability to ripen fruit. The Pinot Noir harvest is a drawn-out affair in Central; it can end in Gibbston six weeks after it usually begins in Bendigo.

A result of Central's short, intense growing season is a berry size typically larger than that achieved by regions further north. This leads to a lower skin-to-juice ratio, a lower tannin presence and a heavier reliance on acids as a structural component.

Threats to the vines can come in the form of spring frosts and the fierce northwest wind. Overall, however, low disease pressure, abetted by breeze and low humidity, is a feature of the region.

A number of winegrowers in the region are convinced the amount of light that shines on Central Otago during its growing season is an important contributor to its distinctive Pinots. The summer days in Central Otago are naturally longer than they are in regions further north in New Zealand. But they are also longer than those in regions such as Oregon's Willamette Valley, which shares the same latitude in the northern hemisphere. The reason for that comes down to the tilt of the earth.

Increasingly, the Central Otago story has become a number of smaller, different stories — those of its subregions. The differences between them are becoming more and more apparent.

There are four main subregions. They are, moving from west to east, Gibbston, Wanaka, Cromwell Basin and Alexandra. The west to east progression is interesting on a number of levels: the soils become lighter and less dense as you move east, the temperatures rise, the rainfall drops.

The largest subregion by far is the Cromwell Basin, from which 75% of the region's Pinot Noir is sourced and which has its own stratum of sub-subregions, or microclimates. It is the Central Otago wine hub, in every sense. Looking at a map, the other subregions surround it like satellites, while Cromwell itself has become the industry's working heart.

The first area of the basin to be planted was Bannockburn, a north-facing, crescent-shaped area at the southern end of Lake Dunstan. While there is variation between the wines from Cairnmuir at the eastern extreme of Bannockburn and those of the Felton Road district in the west, Bannockburn Pinots are typically finely structured, with darker fruit and dried herb notes. The marginality, though, has reduced the number of wines made solely from Gibbston fruit to just a handful.

The vines have extended up either side of Lake Dunstan. On the western side are Lowburn and Pisa Flats, which produce effortlessly ripe, intensely aromatic Pinots with an abundance of dark fruit charm. To the north on the lake's eastern side is Bendigo, a north-facing heat trap. Bendigo Pinots are rich and deep, sometimes kirschy in their dark cherry intensity.

Gibbston is Central Otago's shop window, a beautiful steep-walled valley with a number of well-patronised cellar doors. It is, though, the most marginal subregion, high in altitude, cooler and wetter than the others. When ripened, Gibbston Pinots are typically lively with good acid structure and red fruit notes.

Wanaka is out on a limb and home to a very small number of wineries. It is the closest subregion to the Main Divide but is warmer than Gibbston. Rippon's is the best known of its Pinots, subtly aromatic with a distinctive slatey, savoury character.

At the region's southern end is Alexandra. The driest subregion is also the least glamorous. Yet its sandy soils and wide diurnal shifts are behind some ebulliently aromatic, spicy Pinots.

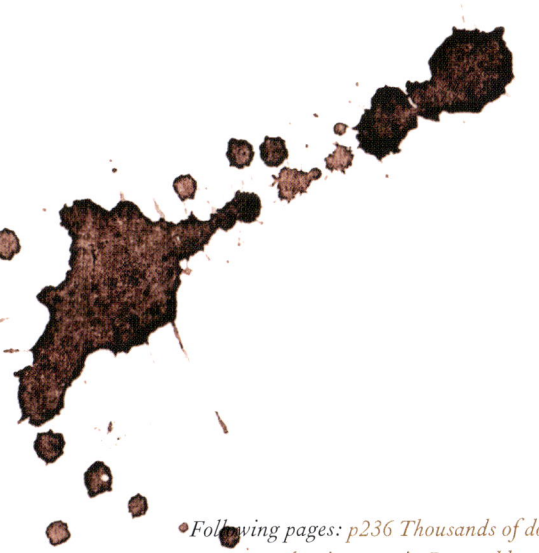

Following pages: p236 Thousands of dollars' worth of French oak; p237 Lombardy poplars punctuate the vine rows in Bannockburn; pp238-9 The Bannockburn ridgeline was sculpted by the goldminers sluicing in the nineteenth century.

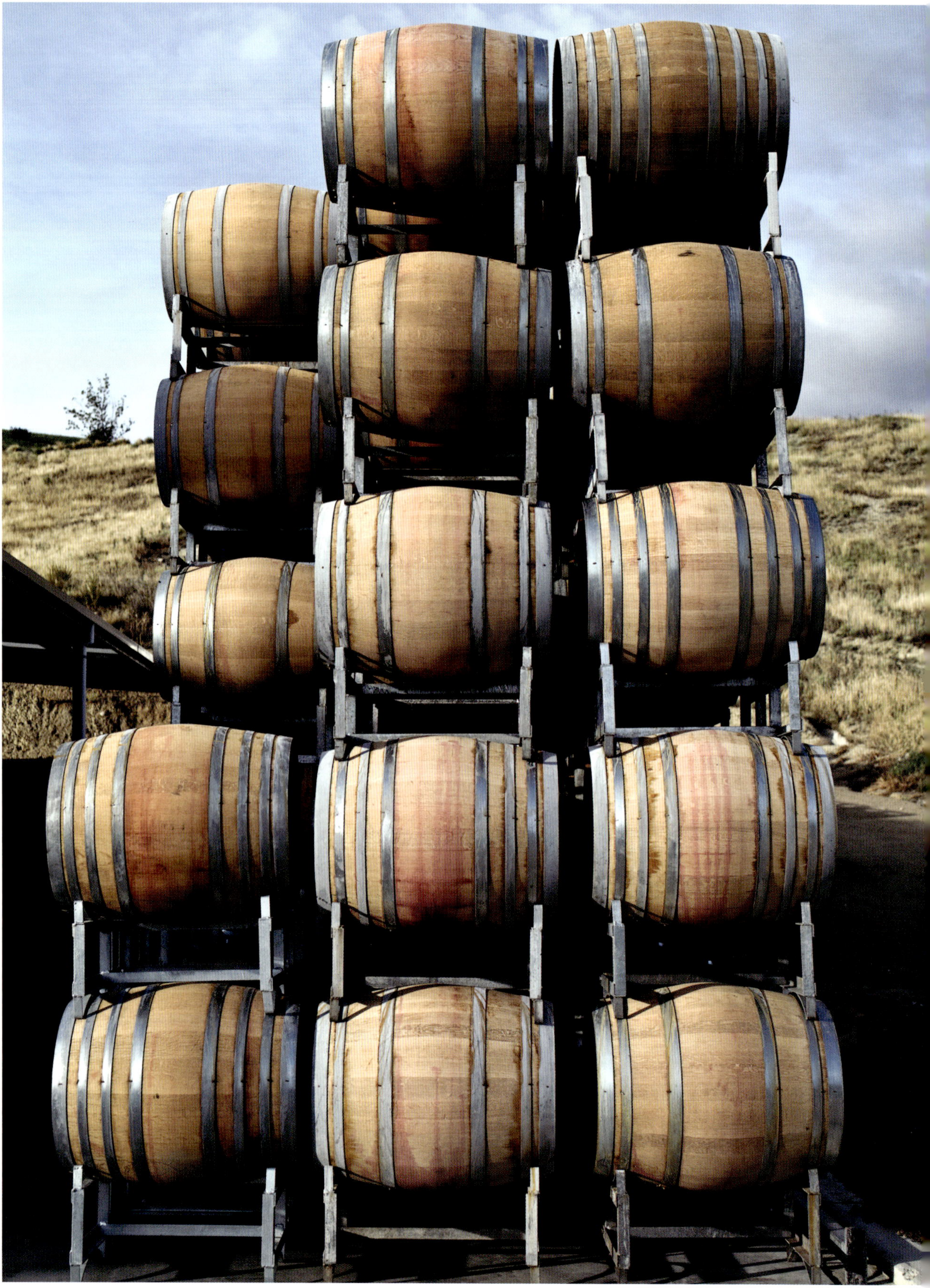

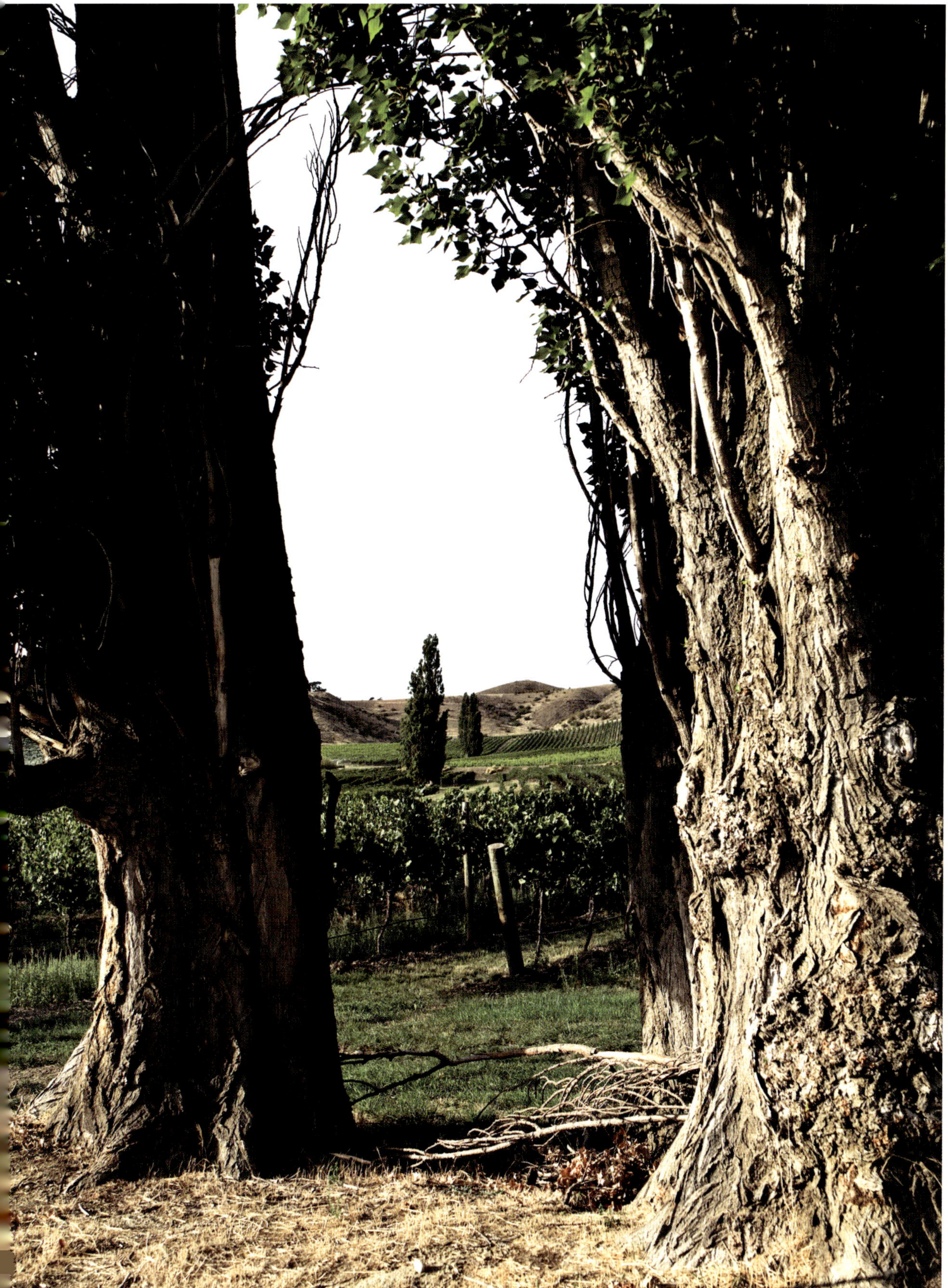

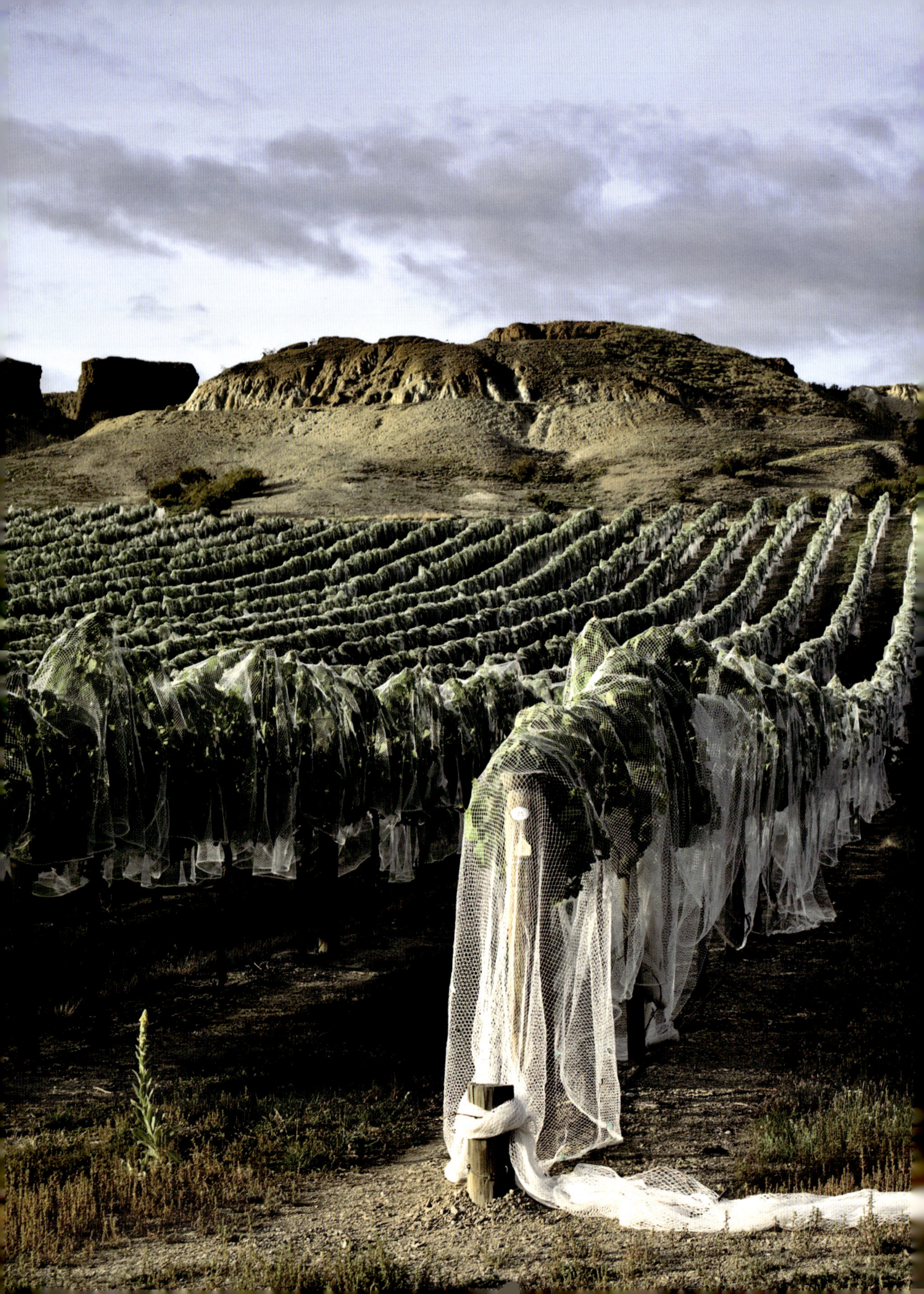

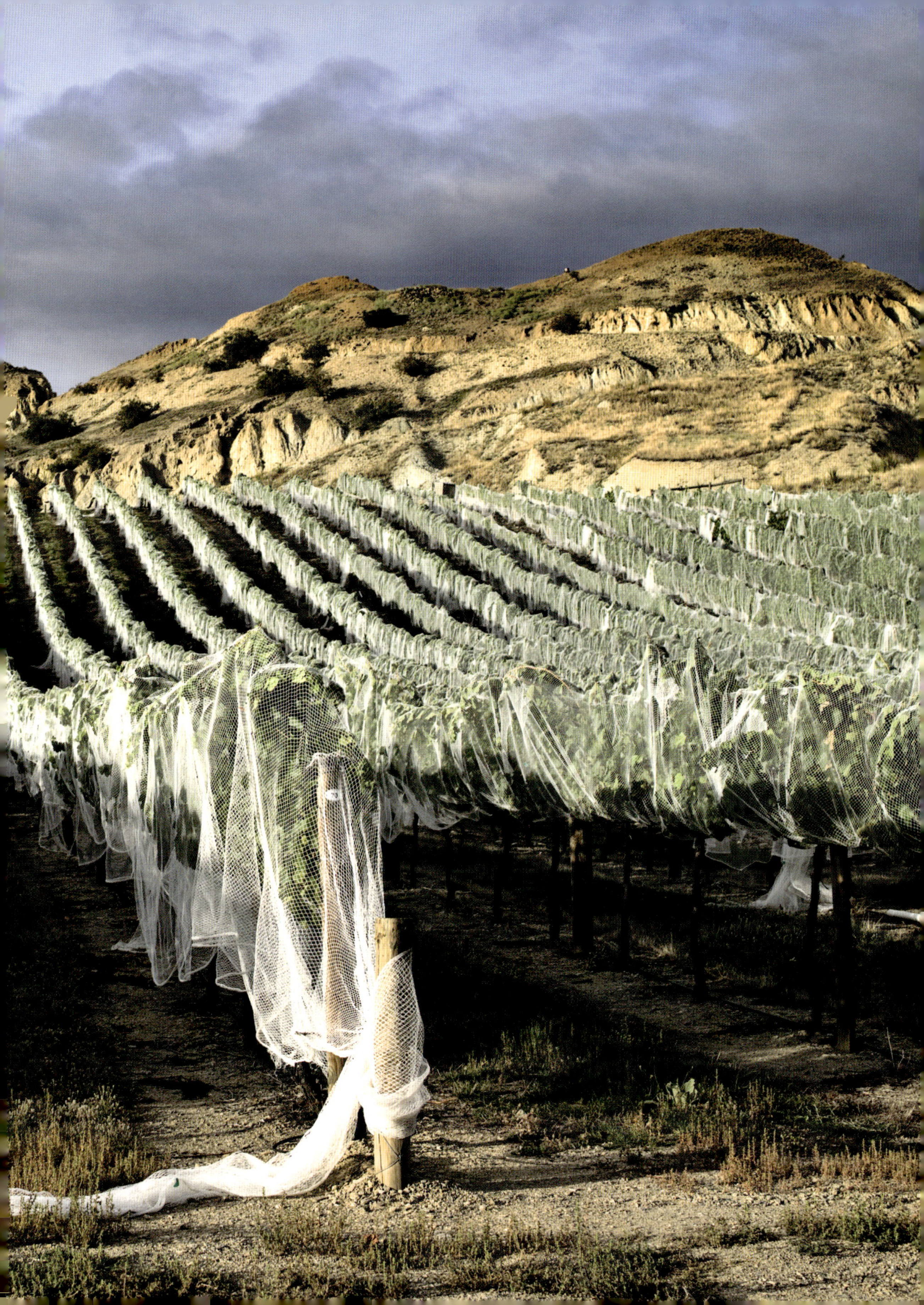

KEY PRODUCERS

CARRICK

Carrick's début Pinot in 2000 made a modest ripple that subsequent vintages turned into a sizeable wave. In a relatively short time the Bannockburn label achieved recognition as a serious Pinot Noir player. With the added lure of an attractive and excellent restaurant, the winery has become a star attraction on the Central Otago wine trail.

Carrick was established in 1999 by Steve Green and his wife, Barbara, along with two other business partners. All were previously part of Mt Difficulty before it changed from being a joint venture partnership to a private company.

The winemaker for the first couple of vintages was Grant Taylor. The talented Steve Davies took the reins from 2002 to 2007, during which time the label's reputation grew strongly. Following his departure another product of the local winemaker nursery that is Felton Road, Jane Docherty, took over. Like Davies before her, Docherty has experience in Oregon and Europe.

Carrick takes fruit from three vineyards, each owned by one of the three shareholders and all leased on a long-term basis to the company. The viticulturist is Blair Deaker.

On Bannockburn's Cairnmuir Terraces, sloping down northwards and westwards to the Kawarau arm of Lake Dunstan and the still waters of Bannockburn Inlet, is the 18-hectare Cairnmuir Vineyard which incorporates another of the vineyard trinity, Le Chat. The namesake Carrick Range looms to the south. Soils here are notably free-draining and lacking in fertility — a layer of glacial loess over broken schist.

The oldest vines on the vineyard were planted in 1994 and are still on their own roots. This presents Carrick with a challenge. The interesting effects of vine age are coming through (in that regard they are particularly pleased with the results obtained from the UCD5 and 10/5 clones), but the threat of phylloxera (first discovered in Central Otago in 2002) is ever-present. The rest of the vineyard's plantings are of varying age, with a broad spectrum of clones. The most recent plantings took place in 2008, when the Abel clone was introduced. Vine density has increased with every new planting, from 1700 vines per hectare in 1994 to 3420 vines per hectare in 2008.

The much smaller, northwest-facing Lot 8 vineyard is only 200 metres from the Cairnmuir Vineyard, but has heavier soils with a higher clay content. It was first planted in 1997 with clones UCD5, UCD6 and 10/5, all on their own roots. Further plantings occurred in 1998 and 2000 using Dijon clones.

Over recent years Carrick has been embracing organics with increasing ardour, driven by what principal Steve Green describes as 'a desire to improve quality and flavours'. The arrival of winemaker Jane Docherty from biodynamic-centred Felton Road has increased momentum in this direction and Carrick now incorporates a number of biodynamic methods in its overall organic philosophy. The winery has just embarked on the three-year BioGro organic certification process. Viticulturist Deaker is also a believer, having once managed the organic Joseph Soler block for Villa Maria.

Carrick aims for a maximum yield of 6 tonnes of fruit per hectare, although yields are often naturally lower due to poor flowering in December or other climatic issues.

Carrick produces three Pinots. The Carrick Central Otago Pinot Noir is the 'old original' and consistently offers beguiling fruit, tension and interest. I recall a tasting of noted 2003 vintage Pinots from Burgundy, Oregon, Australia and New Zealand where the Carrick was a standout performer. It is produced from fruit drawn from all three of the winery's Bannockburn vineyards. Taste and ripeness determine the timing of harvest, and different vineyard parcels are fermented separately in open fermenters using wild yeasts. There is a 12-month maturation period in French oak barrels, of which typically 35% are new.

From 2003 the Carrick Unravelled Pinot Noir has been produced as an approachable, easy-drinking, value-for-money Pinot. Earlier bottling and a lower proportion of new oak are parts of the Unravelled recipe.

The Carrick Excelsior Pinot Noir was introduced with the 2005 vintage. Named after the old Excelsior coalmine which runs under the Cairnmuir Vineyard, this wine is drawn from a selection of barrels from a specific block of mostly older vines on the vineyard. Fruit depth and balance are its hallmarks.

CHARD FARM

The audacity and beauty of the Chard Farm home vineyard, brazenly perched above the Kawarau Gorge at the entrance to the Gibbston Valley, seem to shout out 'Only in Central Otago!' to the passing traffic on the Queenstown–Cromwell highway opposite.

It embodies the vision of Rob Hay, who arrived in Central Otago in 1986 following three years' winemaking study and work in Germany. Convinced the accepted temperature tables presented a false picture of the region's winegrowing potential, Hay fell in with the region's tiny winemaking colony that year, supplying it with a new layer of expertise. He served as winemaker at Gibbston Valley Wines for three vintages as he waited for his vines at Chard Farm to begin providing fruit.

The site's growing potential had been tapped previously, first by Richard Chard with a market garden in the 1870s, in later years with the establishment of a stonefruit orchard. Hay, helped by his brother Greg, planted much of its first few hectares in Pinot, although without a lot of conviction. There was an awareness of Pinot's success in Canterbury several years before, but Rob Hay had had little experience of the variety. It was, he said, planted 'as a default red grape'.

The Hay boys lived rough on the vineyard during the first years, at war with the teeming local rabbit population. The inaugural 1989 vintage yielded a little over 100 cases of Pinot Noir.

From those small beginnings, Chard Farm has grown into one of Central Otago's largest producers of Pinot Noir. After making a single Pinot Noir from the home block for the first decade, the winery spread its wings, acquiring four vineyard sites in the Cromwell Basin. These are all in the Lowburn/Parkburn area along the western side of Lake Dunstan. It now has over 33 hectares planted in Pinot Noir. It is also the 'production partner' with a number of growers for the inexpensive Rabbit Ranch label. Making its début with the 2003 vintage, the light, easy-down-the-hatch Rabbit Ranch Pinot Noir was instantly successful.

A distance is kept between the Chard Farm and Rabbit Ranch brands. You won't see any Rabbit Ranch at the Chard Farm cellar door, for instance. There are currently

Right: Chard Farm's home vineyard in the Kawarau Gorge.

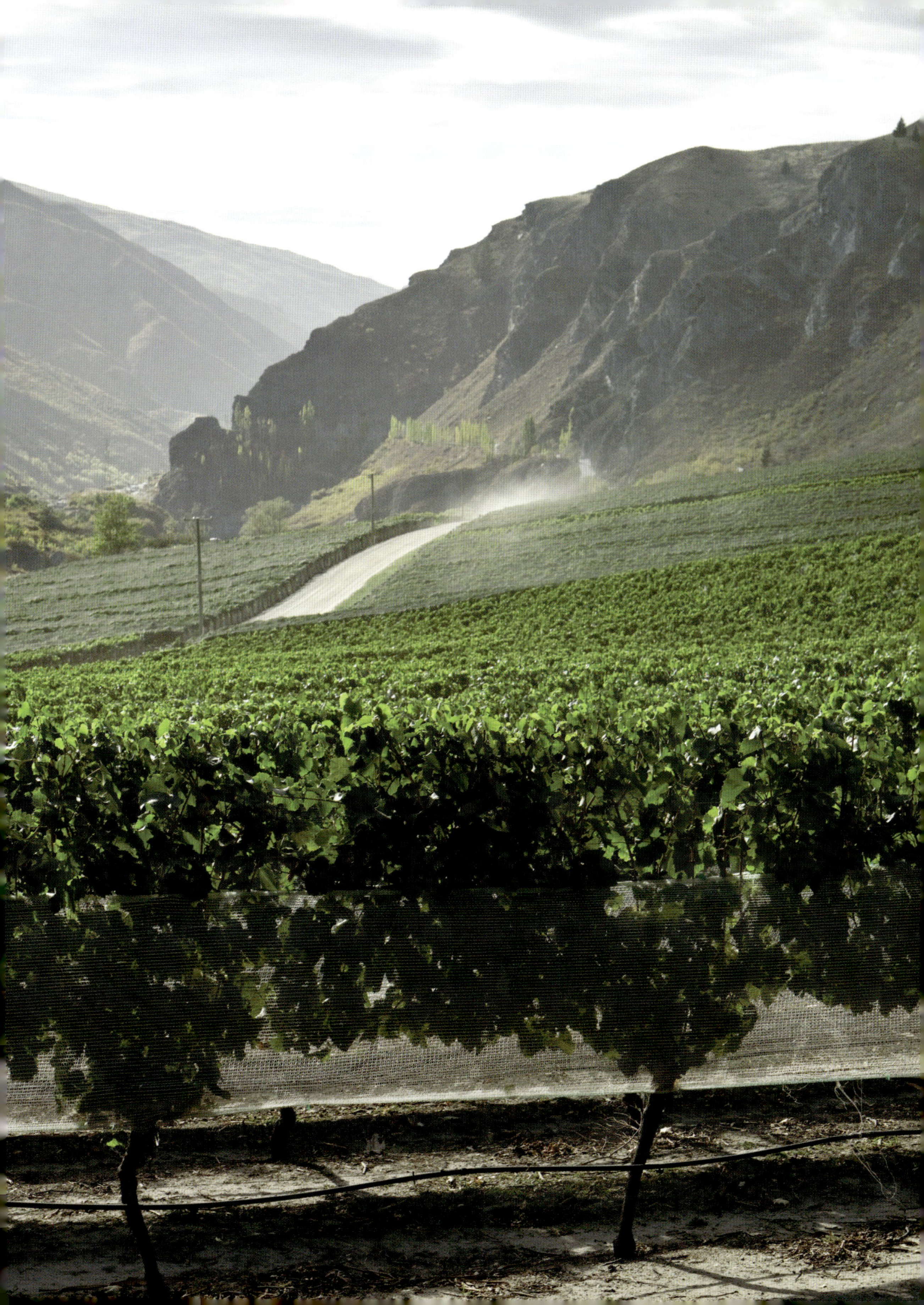

four Pinots released under the Chard Farm label. The Chard Farm River Run Pinot was launched with the 1998 vintage and is an approachable, fresh wine, a blend of estate-grown fruit from the four Cromwell vineyards and the 6.2-hectare Chard Farm home block. A couple of hectares of the original 1986 10/5 and 2/10 vines remain on the home block. The remainder has
been replanted with a spread of newer clones, the most recent being the inclusion of 1.5 hectares of Abel in 2008.

The silky, dark, generous Chard Farm Finla Mor Pinot Noir has come to be regarded as the winery's standard Pinot, though it is made entirely of fruit grown outside the Gibbston Valley. It's a blend of estate fruit sourced from all four of Chard Farm's Cromwell vineyards — the Tiger Vineyard, the Viper Vineyard, the Sinclair Vineyard and the Cook Block.

Two single-vineyard Pinots are only made when the vintage is deemed good enough. The Tiger Vineyard is planted in a range of clones, but the single-vineyard bottling is usually based around three vine parcels — one each of UCD5, 667 and 777, grown on different parts of the 9.4-hectare site. The result is a red berryfruit-infused, beautifully structured wine.

The Viper Vineyard was planted in 2001 to 2002 entirely in clone 777. It is a cooler site than the Tiger, producing a darker, more tannic Pinot.

'Generally, we don't want to make jammy or even fruity wine, so we pick earlyish — as soon as the skins have lost their green tannin flavours,' says John Wallace, Chard Farm's senior winemaker, who has been with the winery since 1998. 'To some extent we would like to get the fruit out of the way to expose a longer-term prospect that shows finesse, elegance, balance of fruit and acid, with fine silky tannins that give movement to the palate rather than hurdles to jump over.'

....

Following pages: p247. Rob Hay, Chard Farm; pp248-9 Autumn creeps over the Chard Farm home vineyard on its 'brazen perch'.

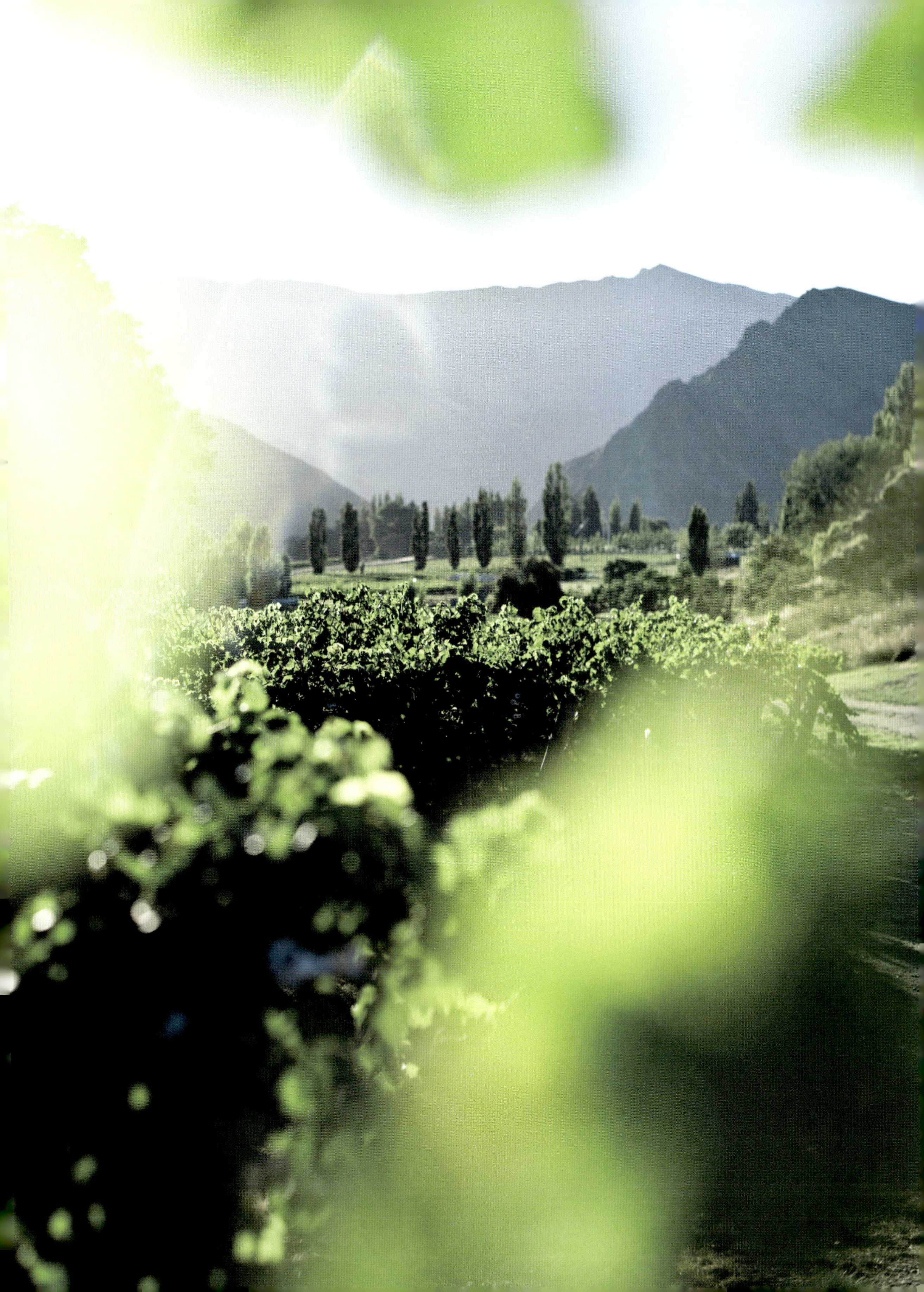

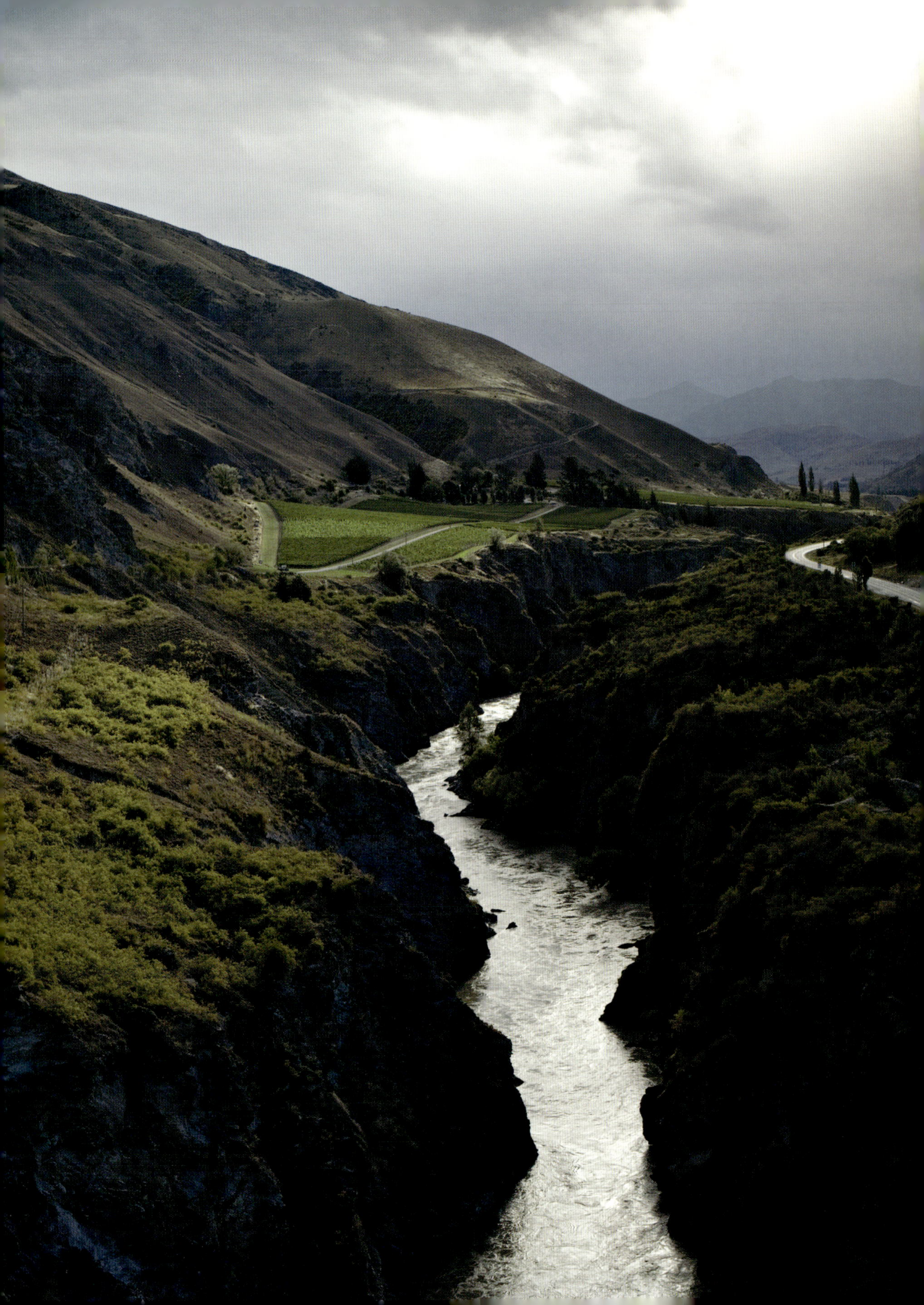

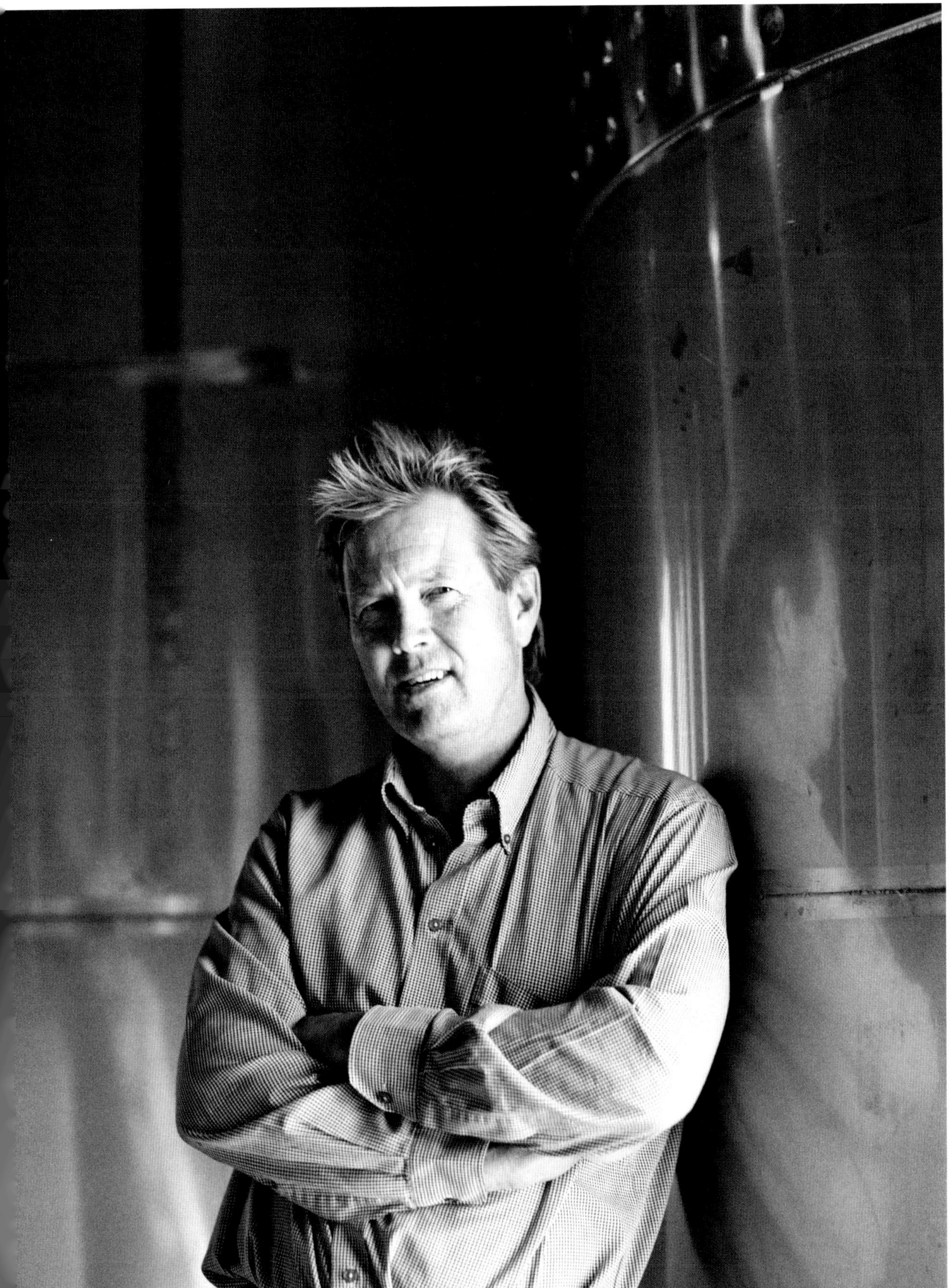

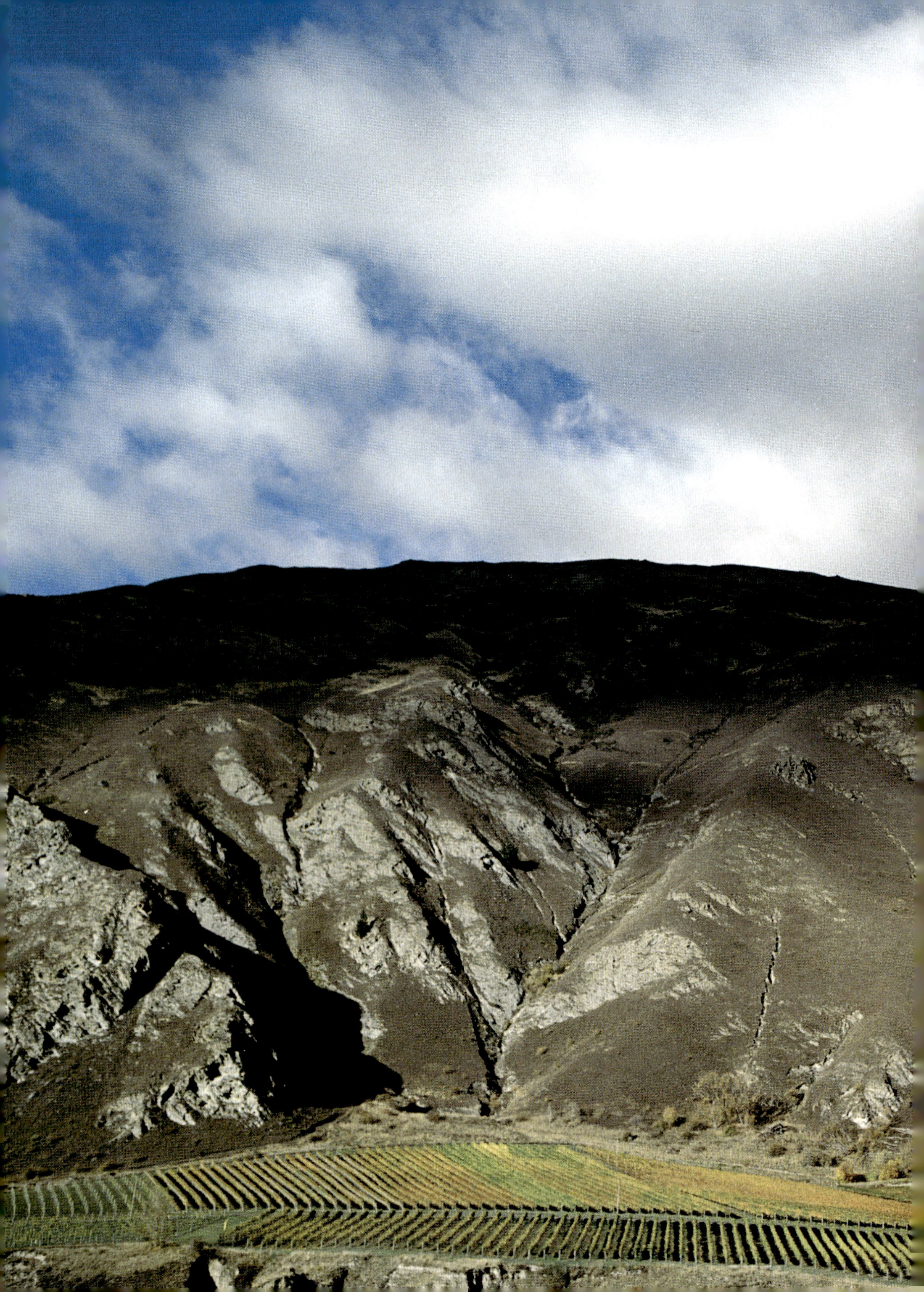

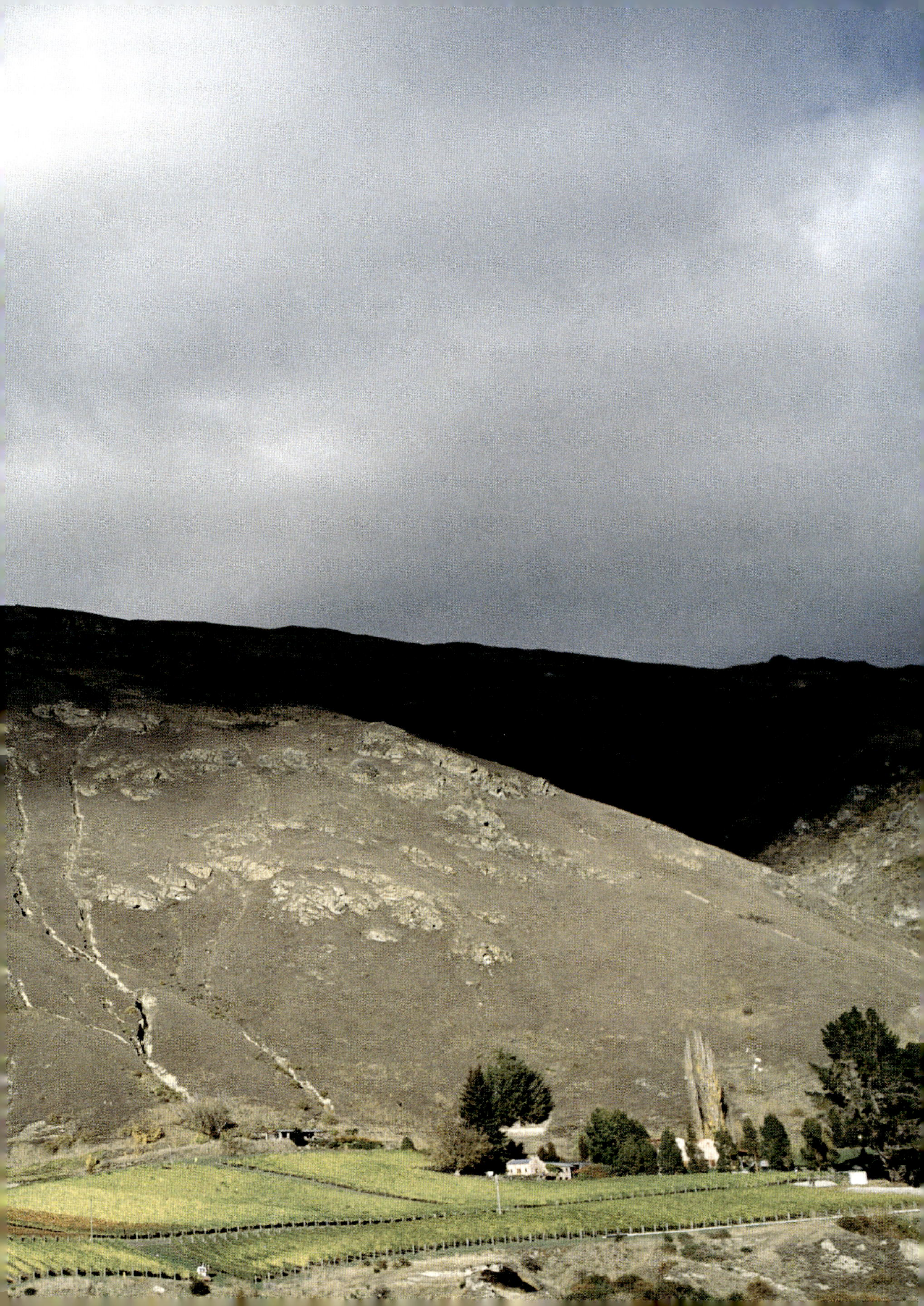

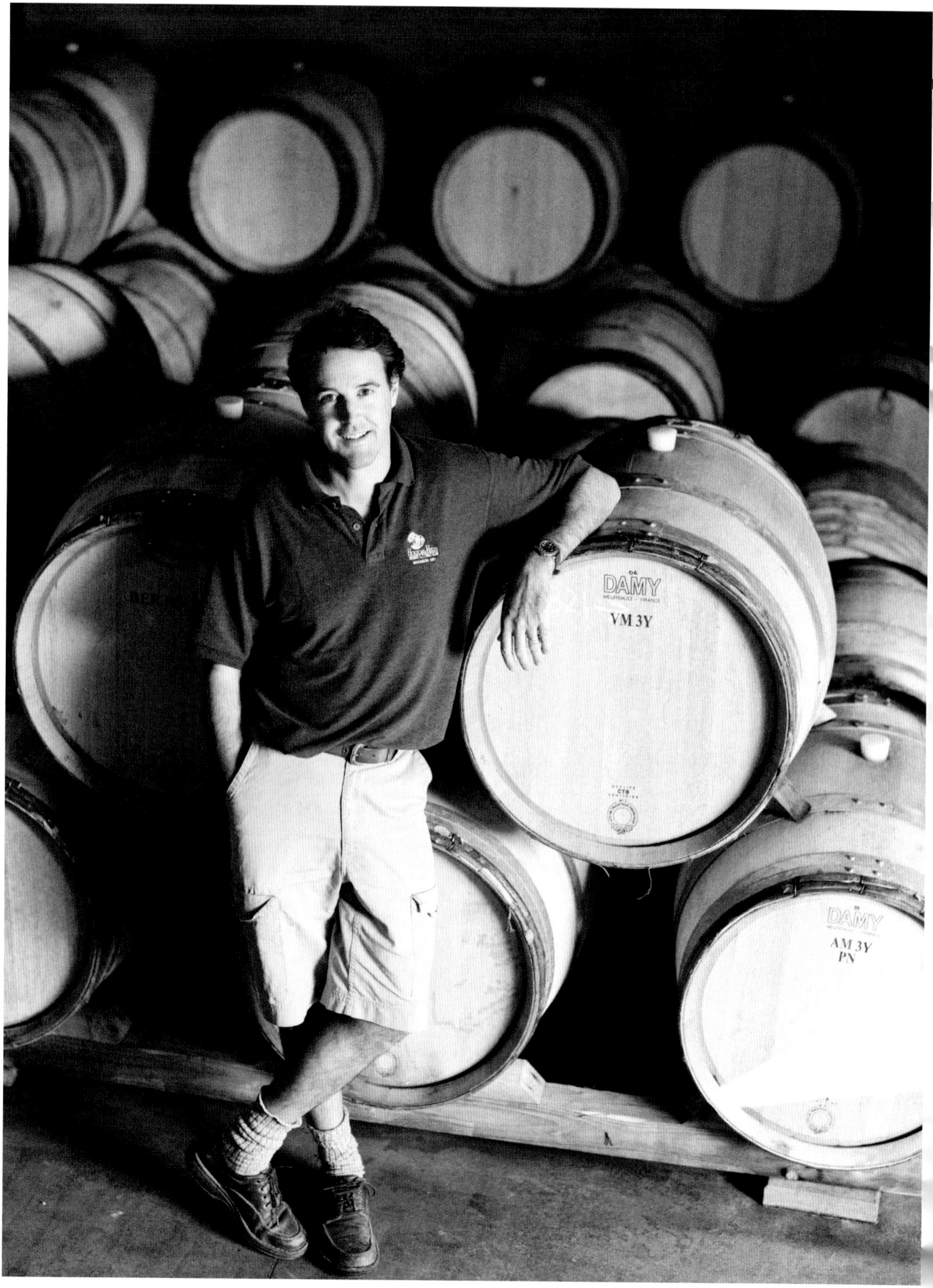
DAMY
VM 3Y
DAMY
AM 3Y
PN

FELTON ROAD

It seems barely credible that Felton Road's first Pinot Noir was a product of the 1997 vintage. So rapid and frictionless has been the rise of this winery, and so quietly authoritative is it in its presentation, that it seems to have been a part of the New Zealand Pinot Noir scene for a couple of decades at least.

Its vines were the first to appear on the spaghetti western-like terrain above Felton Road in Bannockburn, the initial plantings carried out in 1992 by founder Stewart Elms. Elms sold the business to exuberant Englishman and Pinot fanatic Nigel Greening in 2000, but his work is still appreciated by the young winemaker he hired to make his wine, Blair Walter. '[Elms] went to great effort to protect the drainage and aspect,' says Walter. 'And back then he figured that Pinot Noir would do well on the heavy soils, while he planted Riesling and Chardonnay on the lighter, gravelly soils.'

Walter's interest in Central Otago was piqued when he tasted Rudi Bauer's Rippon Pinots in the early 1990s. 'I was working in Oregon at the time and I still remember them vividly … the 1990 was incredible.' Walter now has a shareholding at Felton Road. In tandem with Greening, he is the driving force behind what has become one of New Zealand's iconic Pinot brands.

Both men are strong advocates of biodynamics. Felton Road began to follow the methods of Rudolf Steiner in 2003, and in 2009 became the first New Zealand winery to achieve full organic and biodynamic certification through Demeter. At Felton Road there is much pride in the well-being of the 'whole' — the compost, the cover crops between the rows, even the small colony of hens that live among the vines and the Boer goats grazing on the briar on the steep hills behind the vineyard. Among the benefits biodynamics has brought, says Walter, is 'greater harmony, consistency and balance in our wines'.

Felton Road produced two Pinots in 1997 — Felton Road Pinot Noir and Felton Road Pinot Noir Block 3. Over the intervening years three others have joined the club: Felton Road Pinot Noir Block 5 (first vintage 1999); Felton Road Pinot Noir

Left: Blair Walter, Felton Road.

Calvert (first vintage 2006); and Felton Road Pinot Noir Cornish Point (first vintage 2000 and from 2007 under the Felton Road label).

The original 14-hectare vineyard that slopes down from the winery building is called The Elms. Half of it is planted in Pinot Noir in a variety of clones: 10/5, UCD5, UCD6 and a small portion of UCD13 made up the early plantings; Dijon clones were added later. There are now 11 clones in all and 'not one of them I want to get rid of', declares Walter. The Pinot vines are concentrated on the deep, fine, sandy loam (of which the parent rock is schist) that forms a bench across the lower portion of the vineyard.

Fruit from The Elms finds its way into three Felton Road Pinots. The first is the Felton Road Central Otago Pinot Noir, which accommodates roughly equal parts of fruit from each of the winery's three vineyards. This wine has evolved in style to become deeper, more robust and high-toned than its predecessors of several years ago. The Block 3 comes from a part of the vineyard that Walter says 'generates particular finesse and complexity'. Predominantly one clone, 10/5 (though in several subtypes), contributes to Block 3, which is notable for an enticing dried herb character. The weightier Block 5, just 50 metres away, is influenced by greater clay content in the soil.

The Cornish Point vineyard occupies an extraordinary site at the end of a finger of land almost entirely surrounded by the waters of the Kawarau River and Lake Dunstan. All 8 hectares of the vineyard are planted in Pinot, with 18 different combinations of clones and rootstock matched to different soil profiles. It is described by Walter as a 'geeky vineyard — a laboratory of Pinot Noir and its possibilities'.

Finally, there is the Calvert, a vineyard across the road from The Elms over which Felton Road has taken out a long-term lease. The interesting aspect of this vineyard is that while Felton Road handles all the viticulture, it shares the fruit with two other 'friends in Pinot' — Pyramid Valley and Craggy Range. These two wineries also make single-vineyard wines from Calvert fruit, each receiving the same mix of the seven clones planted. And yet the wines are noticeably different. This points to winemaking decisions having a strong say in the outcome of the finished wine. It will be fascinating to watch the expected effect vine age will have on strengthening the influence of the site.

In the winery, Felton Road goes with the flow — wild yeasts, gravity feeding, spontaneous malolactic fermentation. Typically there is a 20–30% whole bunch component, and a similar percentage of new oak used. At the Pinot Nioir 2010 conference in Wellington, Felton Road was an inaugural recipient (along with Ata Rangi) of a Tipuranga Teitei o Aotearoa award. The award was designed to recognise the top tier of New Zealand Pinot producers.

Following pages: pp254-5. Felton Road's The Elms Vineyard, Bannockburn; p256 Harvest is imminent in the Gibbston; p257. Autumn in the Gibbston Valley.

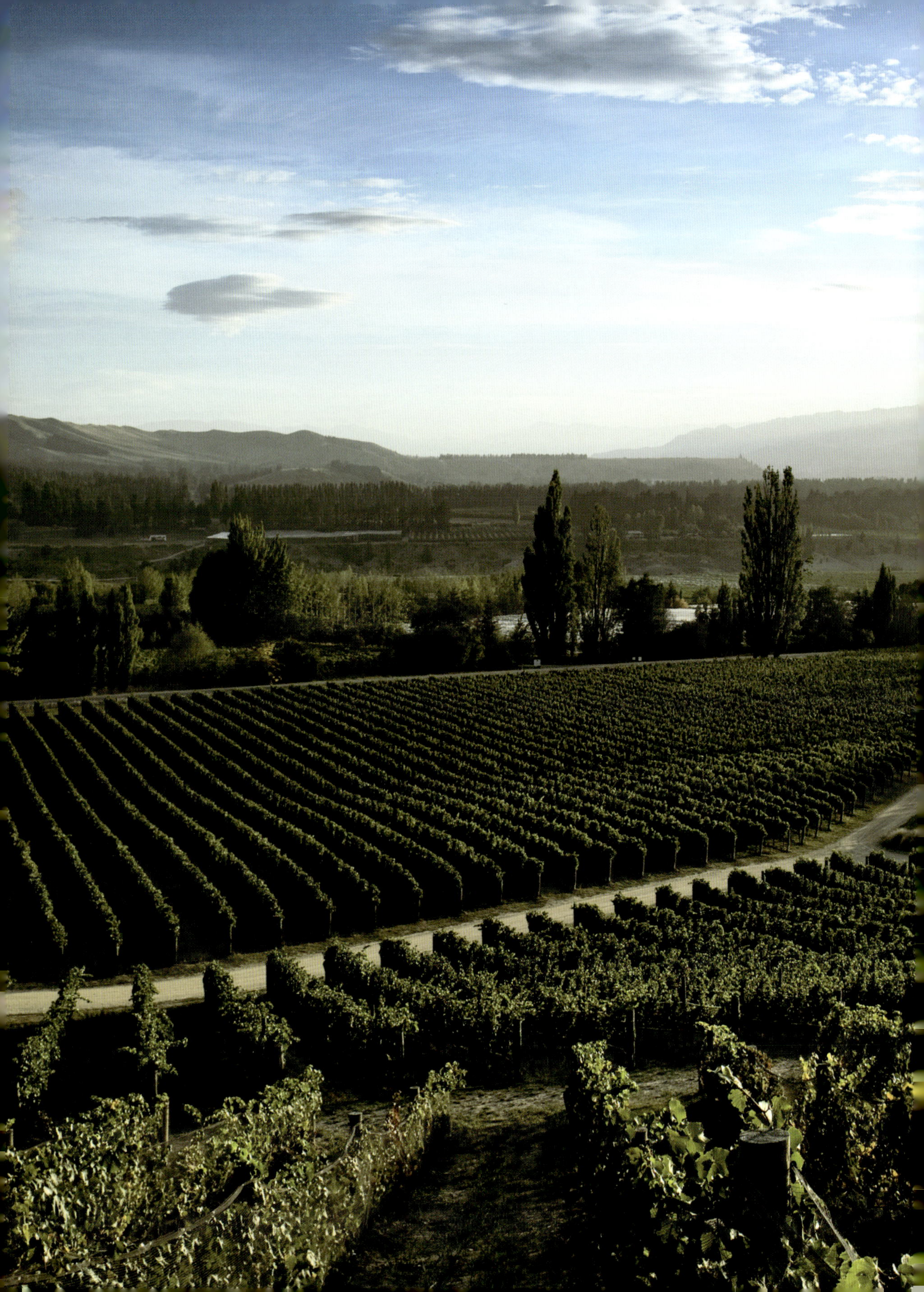

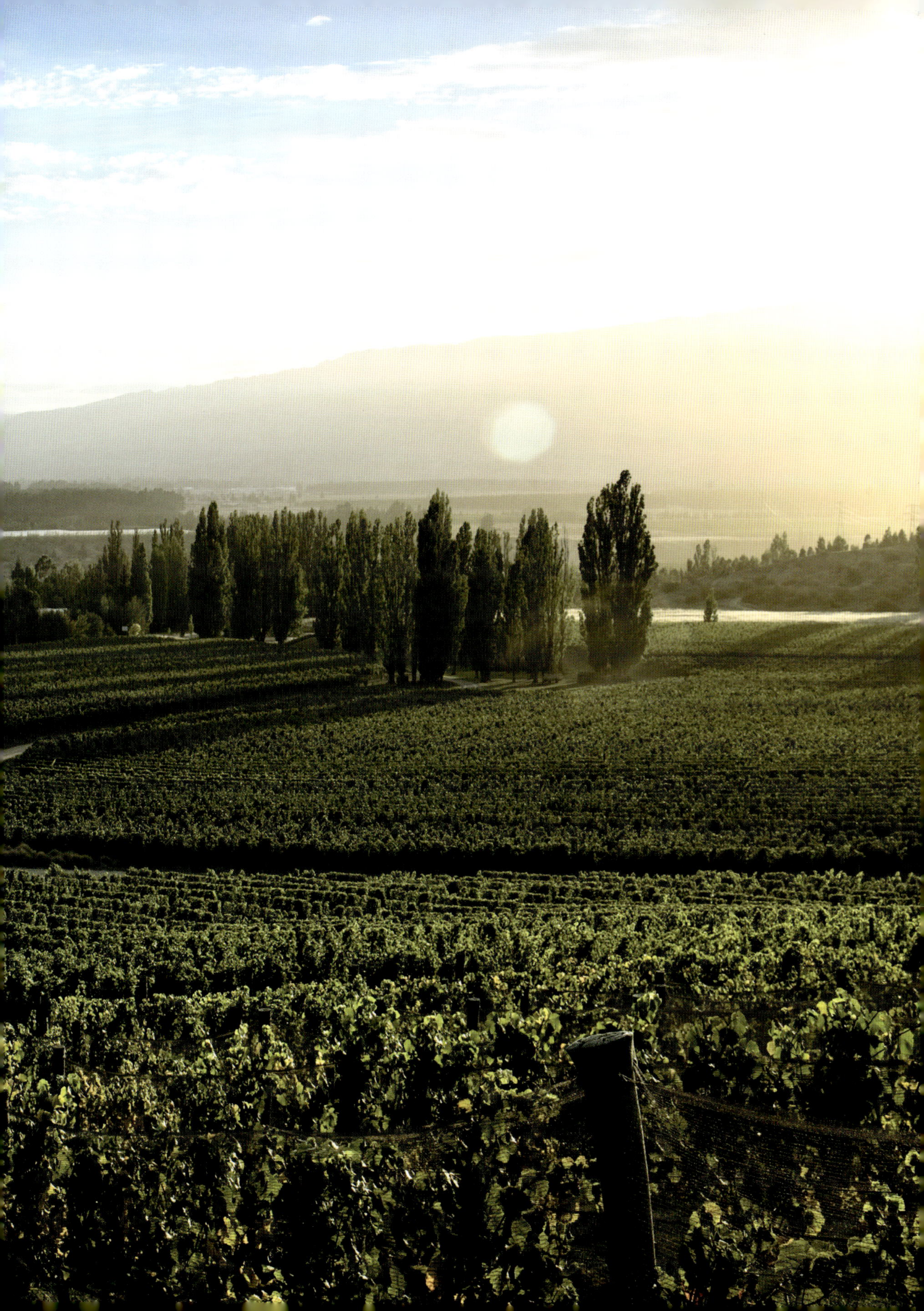

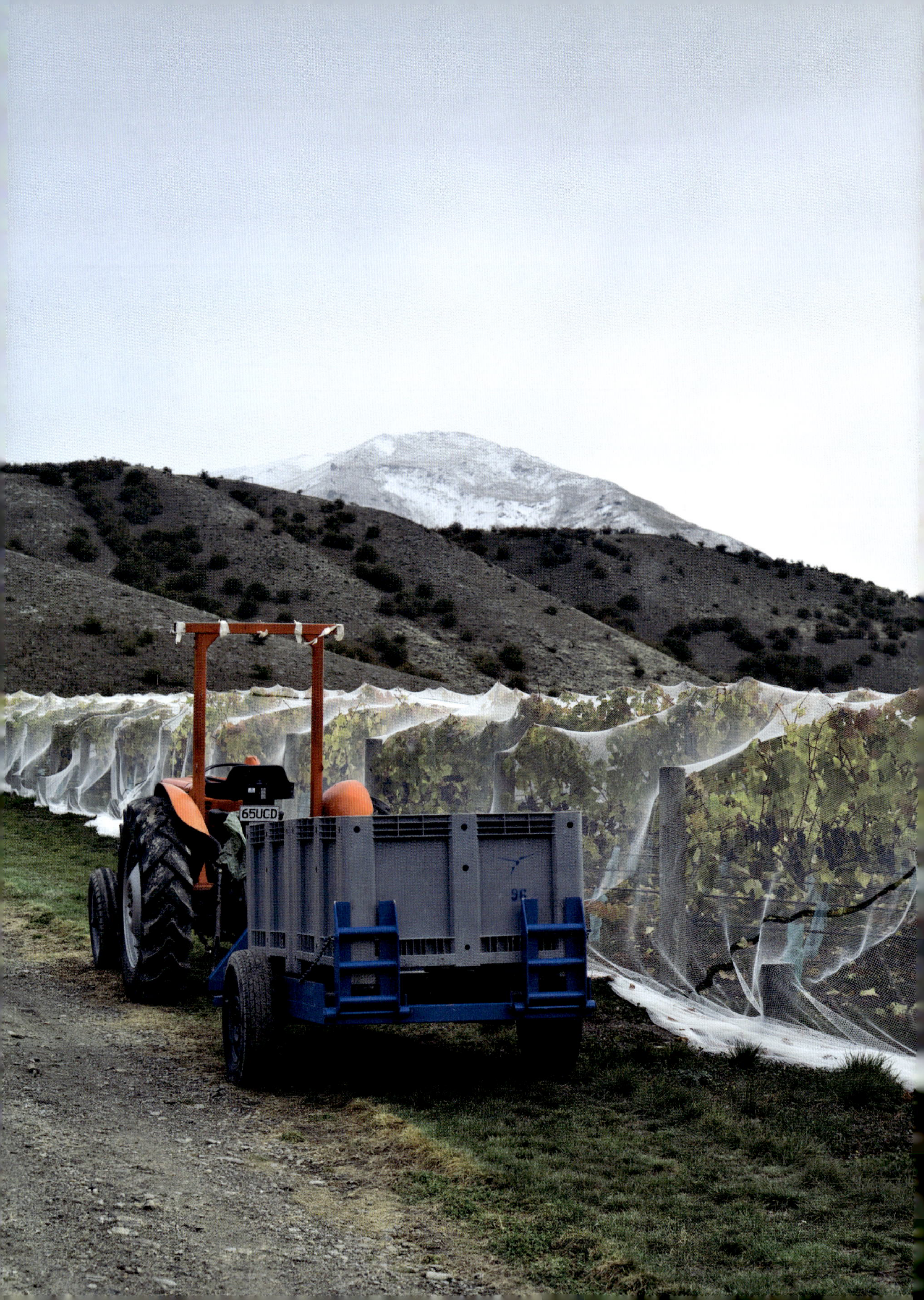
65UCD

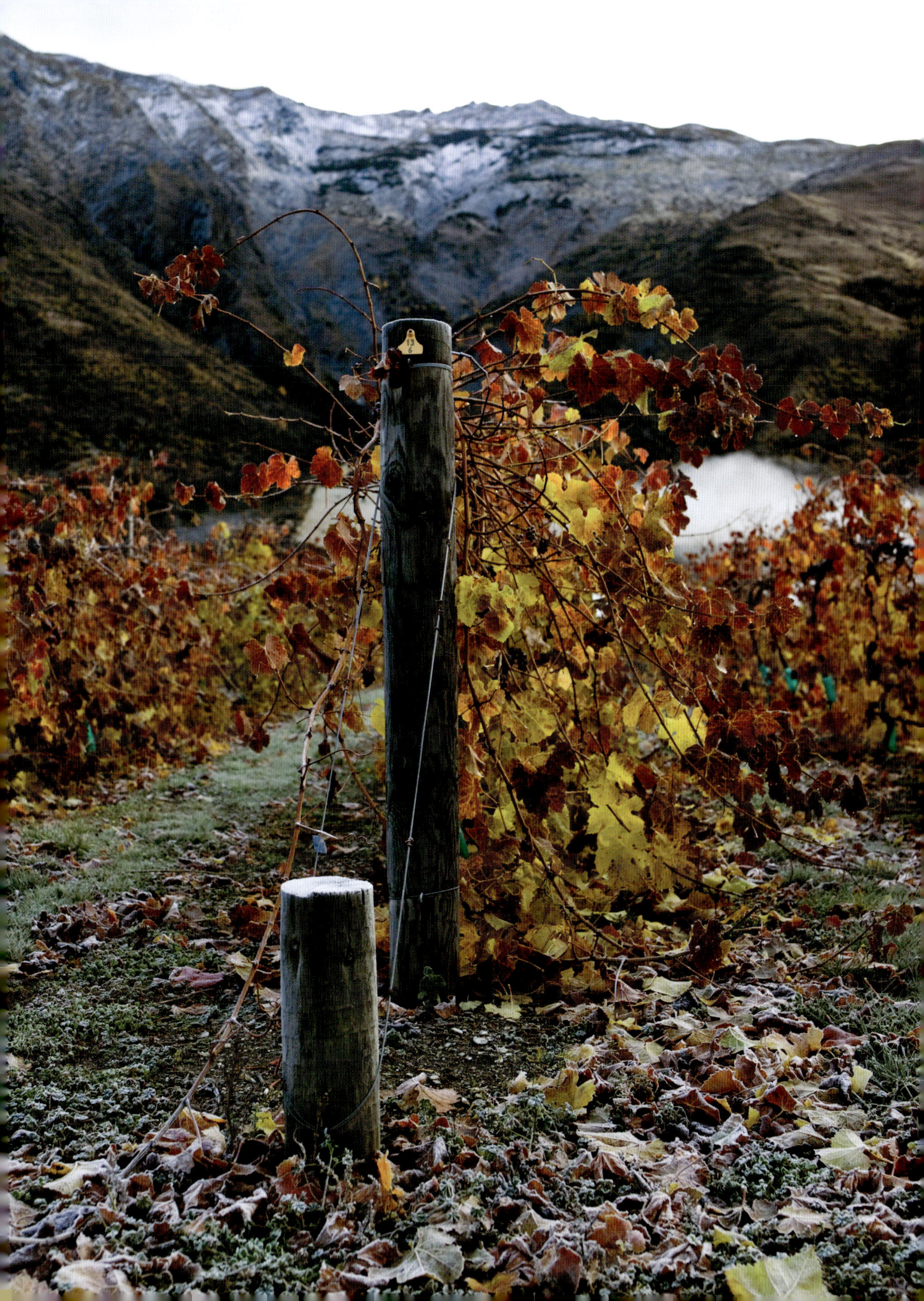

GIBBSTON VALLEY WINES

Few names are as closely entwined with the short, exciting ride of Central Otago Pinot as that of Gibbston Valley Wines. What began as Alan Brady indulging an idle fantasy in the early 1980s has become a brand that is a byword for fine Pinot, and a place that has become the hub of wine tourism in the region. The punters flock to the Gibbston Valley winery. A convenient distance from Queenstown, it offers a fine restaurant, great scenery, tours of its underground cellar and, of course, excellent wine.

Neither Brady nor Grant Taylor, the winemaker who did so much to unlock the potential of Gibbston Pinot, is still with Gibbston Valley Wines. (They haven't strayed far; both are involved with nearby boutique labels in the Gibbston.) Majority ownership of the winery has passed to Americans Mike Stone and Phil Griffith, while the winemaking duties now rest with Hawke's Bay-raised Christopher Keys, who started in 2006.

Gibbston Valley Wines currently produces around 18,000 cases of wine annually, 70% of which is Pinot Noir. Like most other wineries in the Gibbston, it draws fruit from other subregions, though in its case there is a marked preference for Bendigo.

Perhaps the winery's best-known Pinot — certainly its most awarded — is the Gibbston Valley Reserve Pinot Noir. Only made in good years, it was first produced in 1995 and during the early years it was very much an essay in power and extract. Grant Taylor's goal was to create a complex Pinot with longevity and in that he was successful. The wine evolved during the Taylor years and that is continuing with Keys, who has ideas of his own.

> My inclination is to bring the Reserve away from power towards intensity, grace and perfume — simply as a personal preference. Bendigo has become an increasingly important part of the wine. The 2008 Reserve is 100% Bendigo from the School House Vineyard. It marks a turn towards a closer relationship with the vineyard as we identify the lots, clones and soil types that provide the character we're after.

The Gibbston Valley Central Otago Pinot Noir is the winery's standard Pinot, now made exclusively from company-owned blocks in Bendigo, Gibbston and Alexandra. It is generally a warm, full-structured, ripe style, more reflective of Bendigo than anywhere else. 'We do not add enzyme or tannin, and treat the fruit gently,' says Keys. 'We are quite comfortable allowing the vintage to express itself rather than the opposite.'

The Gibbston Valley Gold River Pinot Noir is the early-release label, aimed to be lush and approachable in style. Again, the provenance of most of the fruit is Bendigo.

A fascinating single-vineyard Pinot was made in 2007, on the twentieth anniversary of the winery's first commercial vintage. One of the wines in the Expressionist series, it is called Le Maître Pinot Noir and features a sketched portrait of Alan Brady on the label. Made using fruit from Brady's original home block vines in the Gibbston, it exhibits the depth and suave texture only old vines can provide. Another Expressionist Series single-vineyard Pinot from Bendigo was made in 2008, Le Mineur d'Orient. Declares Keys:

> I aspire to move ourselves outside the traditional Central Otago mould, i.e. big black-cherry-fruited Pinot. I want to do something more sophisticated, based on impeccably ripe fruit, beauty and delicacy. Gibbston Valley Wines was a creator of the former style — now owning all our vineyards, we want to listen to them and craft something beautiful.

Following pages: pp260-1 Hand-harvesting in the Gibbston; pp262-3 Pinot berries in Central Otago are typically larger than those in other regions.

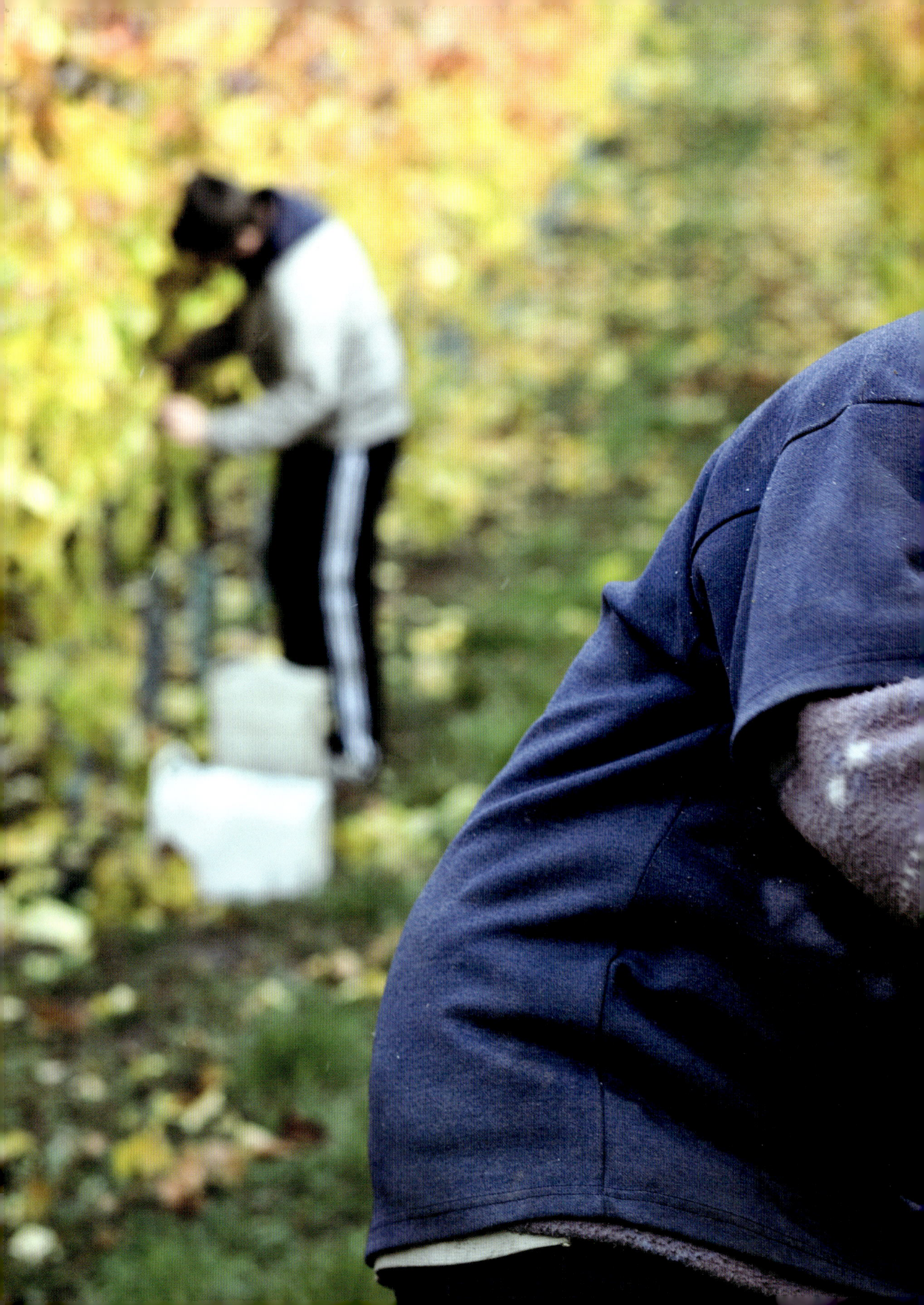

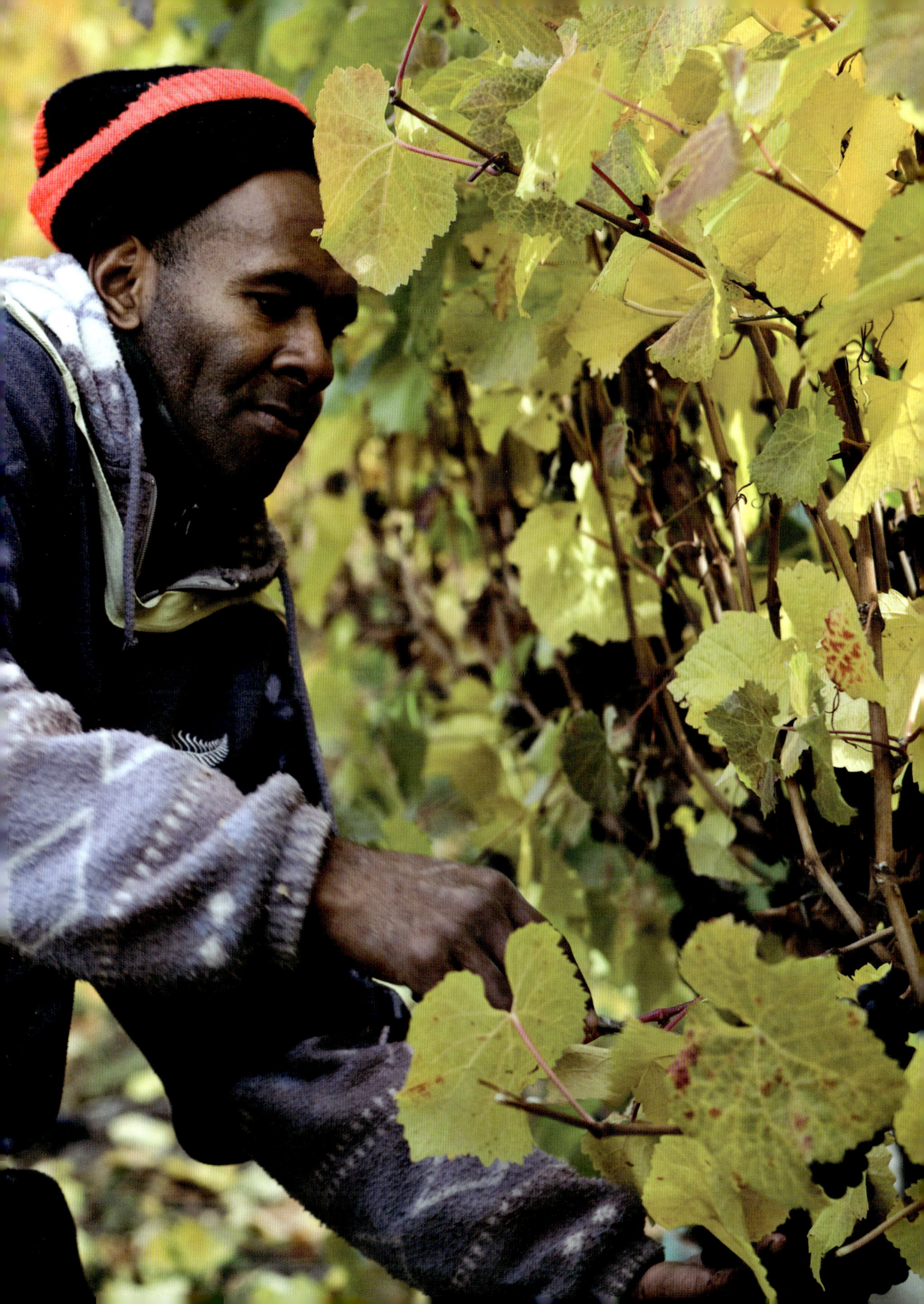

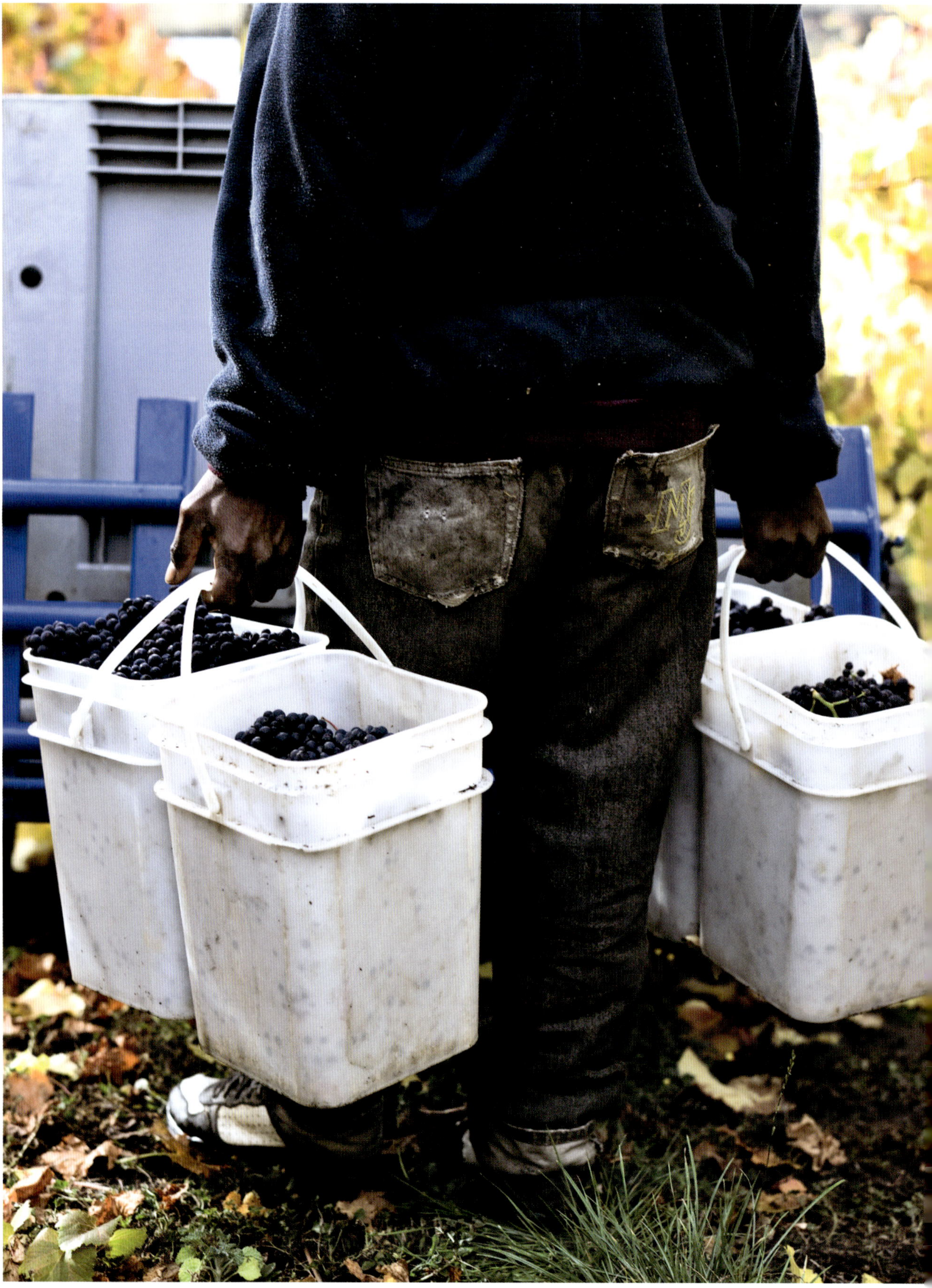

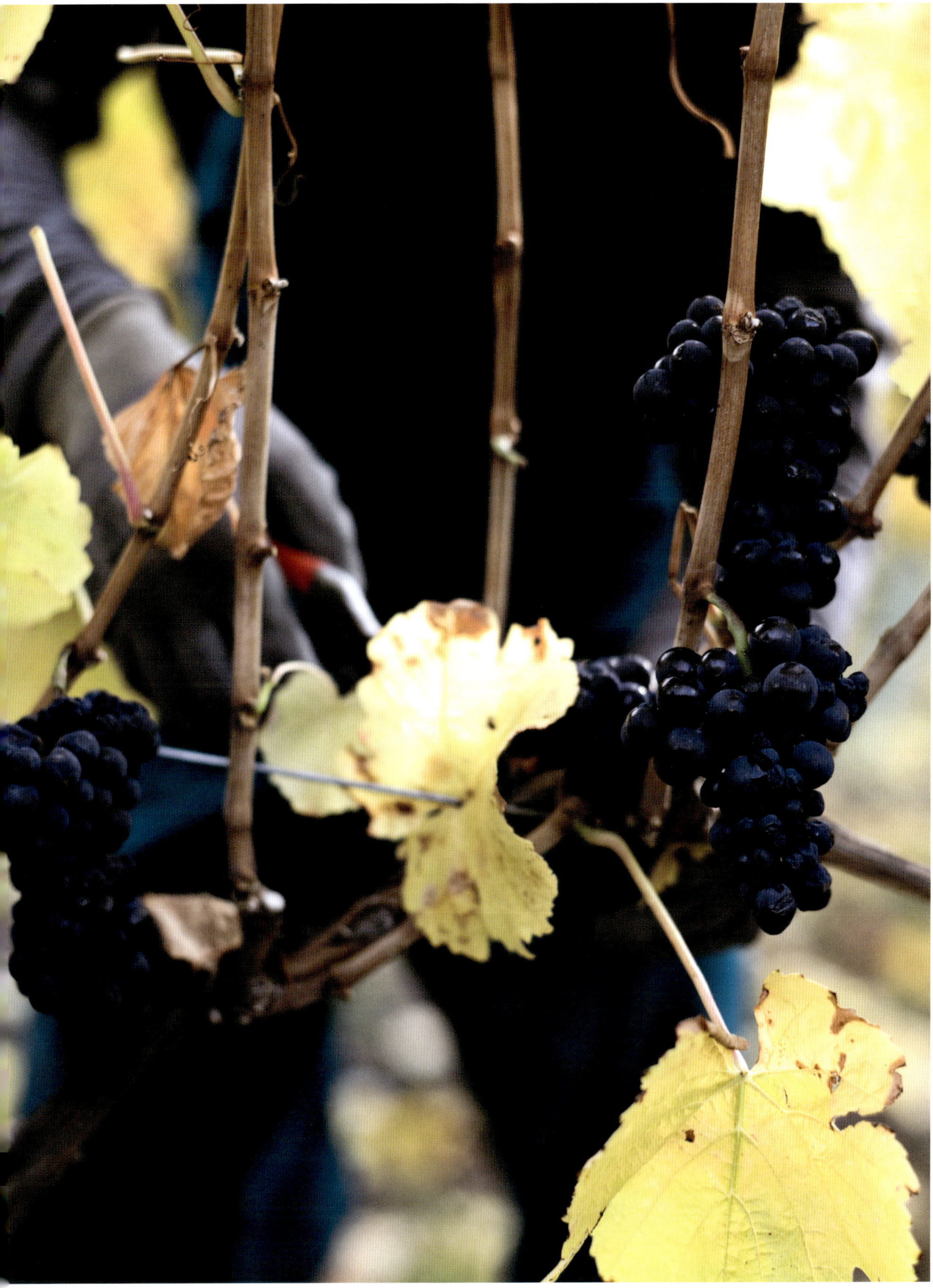

THE PLAN... IS TO KEEP THINGS SMALL, CULTISH AND OFF THE BEATEN TRACK.

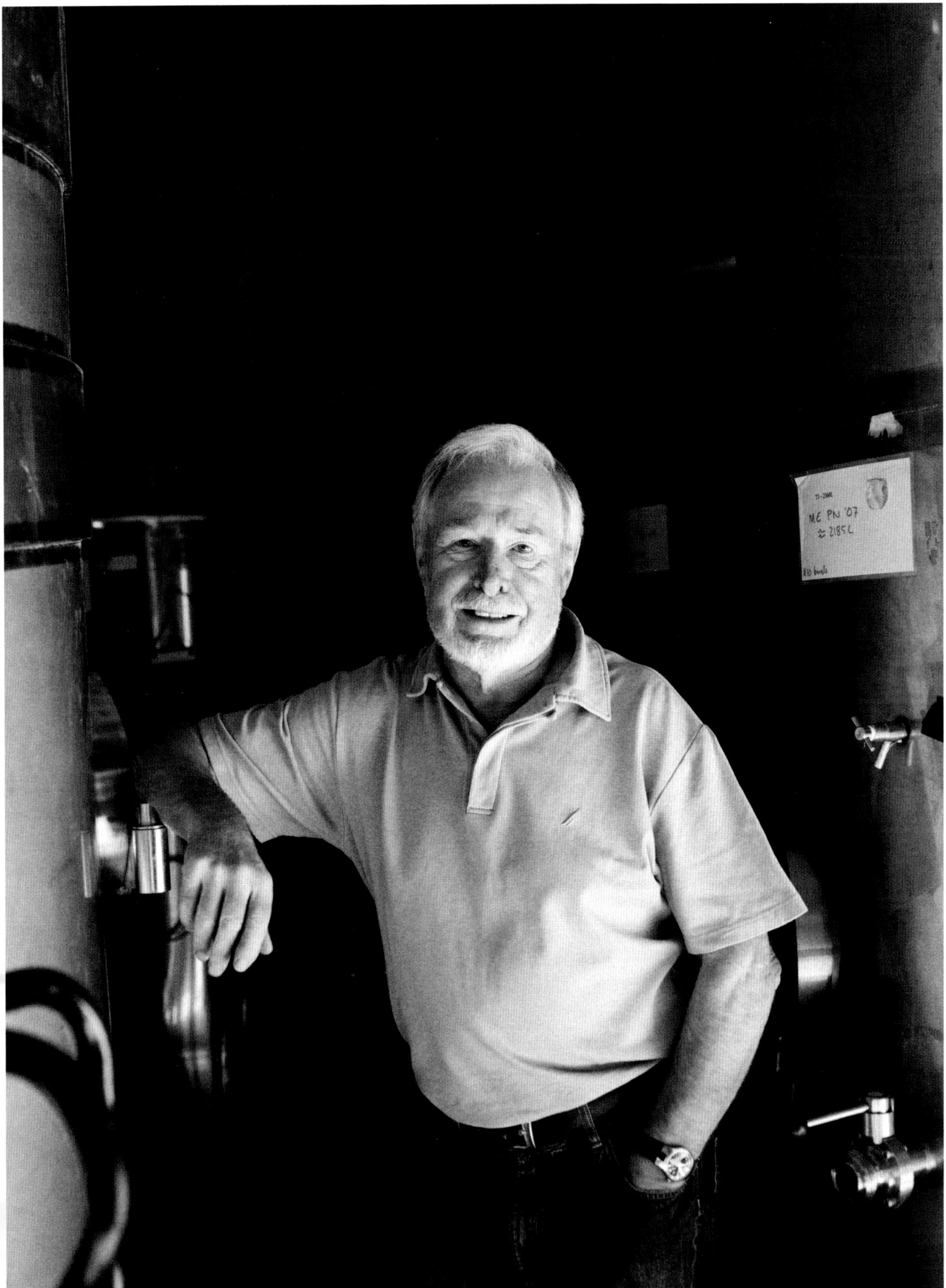
ME PN '07
≈ 2185L

MOUNT EDWARD

Alan Brady's embrace with the Gibbston began when he planted the first vines in the valley in 1981. Today the relationship is as firm and fond as ever, and has earned him the unofficial title 'Godfather of the Gibbston'.

In 1997, soon after selling his stake in pioneering Gibbston Valley Wines, Brady established the boutique Mount Edward beside his Gibbston home. He joined forces with experienced winemaker Duncan Forsyth (formerly with Chard Farm) and another partner, John Buchanan, in 2004. This increased the winery's vineyard holdings and led to a recent expansion of the winery itself. The plan, though, says Forsyth, 'is to keep things small (peaking at around 10,000 cases annually), cultish and off the beaten track. This is an intensely personal project dedicated to Pinot Noir and Riesling.'

'Pinot is what drives me,' says Brady. 'I wasn't a winemaker before [at Gibbston Valley Wines]; I was a manager. I learned everything about winemaking from Grant Taylor [an early winemaker at Gibbston Valley]. Pinot Noir drove me to be a hands-on winemaker … it is a variety that expresses itself more subtly and artistically than any other.'

Mount Edward wines are not easy to come by; 80% of production is exported and domestically only a few fine-wine stores and restaurants stock the brand. The winery's uncompromising attitude to quality, and to making more savoury, less fruit-driven styles of Pinot, makes searching for them worthwhile.

The winery currently produces the Mount Edward Central Otago Pinot Noir and three single-vineyard wines when the fruit merits it: Susan's Vineyard (Gibbston Valley), Muirkirk Vineyard (Bannockburn) and Morrison Vineyard (Lowburn). A second-tier label, Earth's End, is the recipient of fruit not deemed of sufficient quality for the Mount Edward label.

The Mount Edward Central Otago Pinot Noir is produced with fruit drawn from a range of vineyards in the Gibbston Valley and Cromwell, some owned by the winery, some by contract growers. The composition changes annually, depending on whether single-vineyard wines are made and which ones.

The very small (0.6 ha) Susan's Vineyard is owned by contract growers in the Gibbston Valley and provided the fruit for the winery's first-ever single-vineyard

Pinot. Planted in UCD5, UCD6 and 115, it offers the ebullient, red-fruit qualities typical of the Gibbston.

The Morrison Vineyard is estate-owned and situated close to the Wanaka Road in Lowburn. Planted in 1997, it is the winery's oldest Pinot Noir vineyard. There are 6 hectares planted in Pinot, primarily Abel and UCD5, plus a selection of Dijon clones.

The Muirkirk Vineyard is located on Felton Road in Bannockburn and is Mount Edward's newest acquisition. Seven hectares have recently been planted in 10/5, Abel, UCD6 and UCD5. Forsyth is excited at the prospect of what the site will deliver.

> The area is of proven quality. There is some soil variance within the site but predominantly we've a dense clay loam over gravels. We'd expect the clay to have excellent water-holding capacity and reduce the need overall for irrigation soon after planting.

Viticulturist Tim Austin Moorhouse has been charged with converting all of Mount Edward's vineyards to organics. The Morrison Vineyard has already received BioGro certification; the Muirkirk is soon to follow. The winery, too, is seeking BioGro certification.

'We're about layers of mystery, not muscle,' says Forsyth. 'We try to avoid any heavy-handed intervention. Our goal is to produce single wines from each of our vineyards, as we learn more about our sites. We see ourselves as winegrowers rather than producers.'

Previous page: p265 Alan Brady, Mount Edward.

MT DIFFICULTY

Among Central Otago's many riches sits a verbal one. The place names left behind by the early goldminers and settlers are unmatched elsewhere in New Zealand in their quirky, evocative use of language. Many of the region's new wine producers have made good use of this store of poetic signposts. Mt Difficulty, with its brand name and labels such as Roaring Meg (a creek feeding the Kawarau River), is one of them.

Mt Difficulty itself is a muscular, sheltering peak that rises over the southwest corner of Bannockburn. The winery of the same name was begun in 1998 as a joint venture between the owners of several vineyards, who each took a share in the brand. A key player in this collective was Robin Dicey, the South African viticulturist who became interested enough in Central Otago to begin planting a vineyard along Felton Road in 1992. His son Matt became Mt Difficulty's first winemaker, a position he still holds.

Mt Difficulty evolved into a privately owned company and has grown into one of Central Otago's largest wineries, producing 30,000 cases of Pinot Noir annually.

The earth of Bannockburn was ravaged for gold over 100 years ago, the miners' brutal sluicings leaving a weirdly sculpted landscape. Today there is immense pride at Mt Difficulty in the attributes of those soils and the Bannockburn district as a whole. Says Matt Dicey:

> We believe Bannockburn does have special qualities for growing grapes. The soils are diverse … some vines are on heavy clays, others on gravels. In hot summers the clay soils do better; a cooler summer favours gravels. What they all share are high pH levels, tailor-made for wine growing — grapes produce their best wines on sweet soils.

Mt Difficulty owns 40 hectares of vineyards in Bannockburn, and obtains fruit from another 20 hectares in the district which they manage. There are two tiers of Pinot Noir produced under the Mt Difficulty label, for which there are two criteria: (1) the fruit must be 100% Bannockburn, and (2) the vineyards must be managed by the Mt Difficulty team.

The Mt Difficulty Central Otago Pinot Noir was first produced in 1998 and remains the label's hallmark wine. It is a blend of fruit from the winery's more aged Bannockburn vineyards where many of the vines are now 12 years and older. There is much in this consistently excellent wine that typifies fine Central Otago Pinot and its development. Recent vintages have been concentrated, structured wines, with fresh dark fruit and spice notes prominent.

The Mt Difficulty single-vineyard series was started in 2000. Its aim, in the words of Matt Dicey, is to showcase 'the pinnacle of what we produce by highlighting the unique expression of each vineyard'. These wines are only made when a vineyard performs up to expectations, and have so far featured three Bannockburn vineyards: Pipeclay Terrace, Long Gully and Target Gully. The different nuances are fascinating — the Long Gully is dense and robust, the Pipeclay more refined. There is clonal diversity in each vineyard and cropping levels are kept to 4 tonnes per hectare.

The Mt Difficulty Roaring Meg Pinot Noir is the lower-tier wine, made from fruit grown in other parts of the Cromwell Basin. It is a fresh, bright, approachable style, not particularly refined, which is to be expected given cropping levels of 7–8 tonnes per hectare.

Matt Dicey again:

We're only starting to discover some of what Pinot can do down here. We live in a place where Pinot Noir in its youth expresses itself with great clarity and vibrancy. Now we have to see what happens when our vines get some age and start exploring the soils they inhabit. We have exciting soils with high minerality and individuality, with great potential for interest in the resultant wines.

PEREGRINE

When it opened in 2003, the new Peregrine winery building in the Gibbston Valley seemed to sum up the confidence that was accompanying Central Otago Pinot into the new millennium. The building's sweeping, curving canopy roof was, and still is, an original, 'out there' architectural statement. Importantly, it is also one that sits in harmony with its environment.

The roof's aerodynamic shape was inspired by the wing of a falcon, the bird that in turn inspired Greg Hay, himself a flyer (albeit inside small planes and helicopters), when it came time to name his winery. Hay had spent 11 pioneering years helping his brother Rob establish Chard Farm when he decided to strike out alone. Peregrine's first vintage was 1998.

The experience at Chard Farm taught Greg Hay that Pinot Noir was the variety to pursue, and that having vines in the warmer Cromwell Basin was a sound hedging policy if you also grew fruit in the cooler, more marginal Gibbston.

Peregrine started small with the intention of staying that way. Burgeoning demand saw that plan quickly jettisoned and the winery now produces around 10,000 cases of Pinot Noir annually. Peregrine quickly established a reputation for providing intense, true expressions of Central Otago Pinot at a reasonable price. Hay has a marketing degree and is focused on producing wine that meets consumer expectations. 'Over-delivering on quality, price and consistency is a concept that drives us along,' he said.

Between 2004 and 2010, Peter Bartle was the winemaker at Peregrine. He was preceded by the former star turn at Villa Maria, Michelle Richardson, and was recently succeeded by Nadine Cross, formerly of Wither Hills.

There are currently four Pinots in the Peregrine stable, though not all are produced every year. All are blends of grapes from the 12 vineyards from which the winery sources fruit, some of which are owned by contract growers. Most are located in the Cromwell Basin and include a large planting in Bendigo recently aquired by Peregrine. The average vine age is 10 years and the clonal representation is broad.

The Peregrine Pinot Noir is the perennial standard-bearer. For its first few vintages it was made entirely of Gibbston Valley fruit, but it has since evolved into a blend, with Cromwell Basin fruit adding flesh and weight to the more feminine Gibbston component. It is typically dark and seductive and has built up a wide following.

Saddleback is Peregrine's second-tier label and it, too, is a popular wine in its price bracket. It tends towards the red-fruit end of the spectrum. 'It is made with the barrels that remain once the Peregrine-style barrels have been found,' explains Hay. 'Basically all the fruit is treated the same in the vineyard and winery. We then wait and see which parcels of fruit from the 60-odd ferments we do go in which direction.'

Peregrine made its Pinnacle Pinot Noir for the first time in 2005 as a limited-quantity, super-premium wine. With a price tag of NZ$150 a bottle at the cellar door, it was, for a while, New Zealand's most expensive Pinot Noir. Greg Hay: 'It's made from a selection of roughly four barrels that not only stand out on their own, but also make 1 + 1 +1 +1 = 5 when blended.' It is a rich, dark wine, black cherry balanced with savoury depth. The 2007 Pinnacle was released in late 2009.

The Karearea Pinot Noir, named after the native falcon, is also made only in exceptional years. It is a blend of small parcels of different clones from different Central Otago subregions.

....

Following pages: p272 *Colonial cottage made from schist in the Gibbston Valley;* p273 *Greg Hay, Peregrine;* pp274-5. *Peregrine's new winery (2003) is a bold, award-winning architectural statement.*

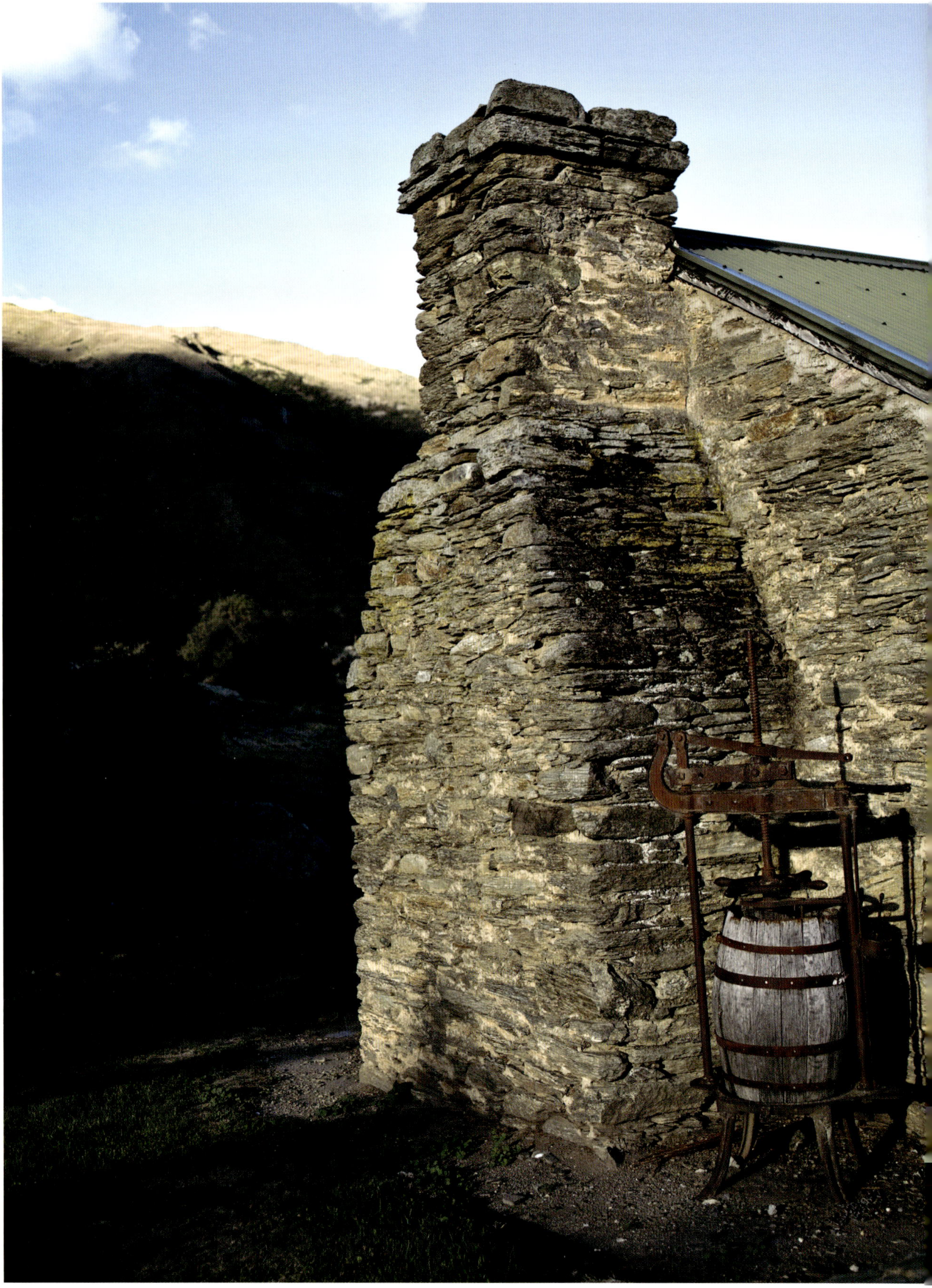

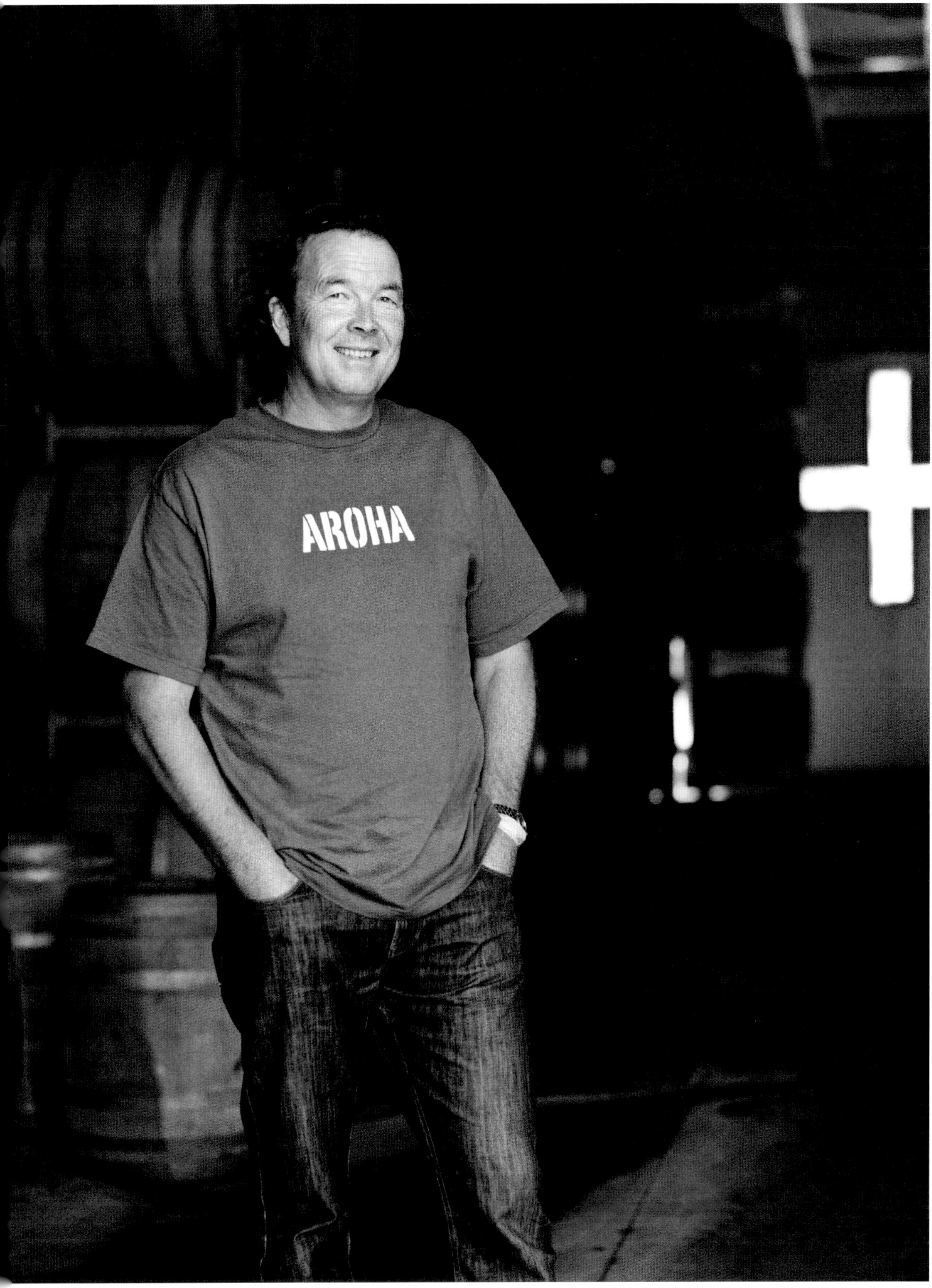
AROHA

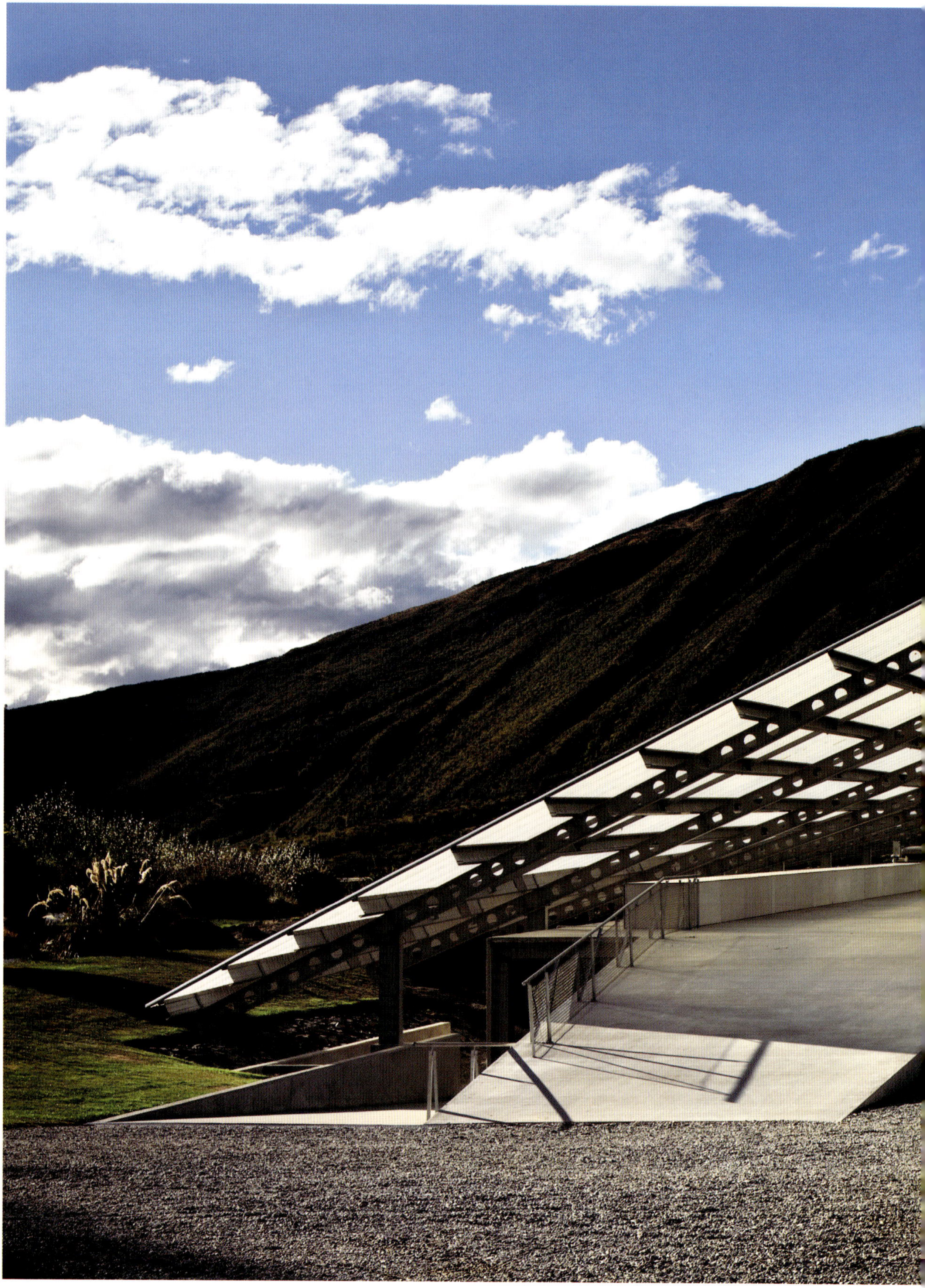

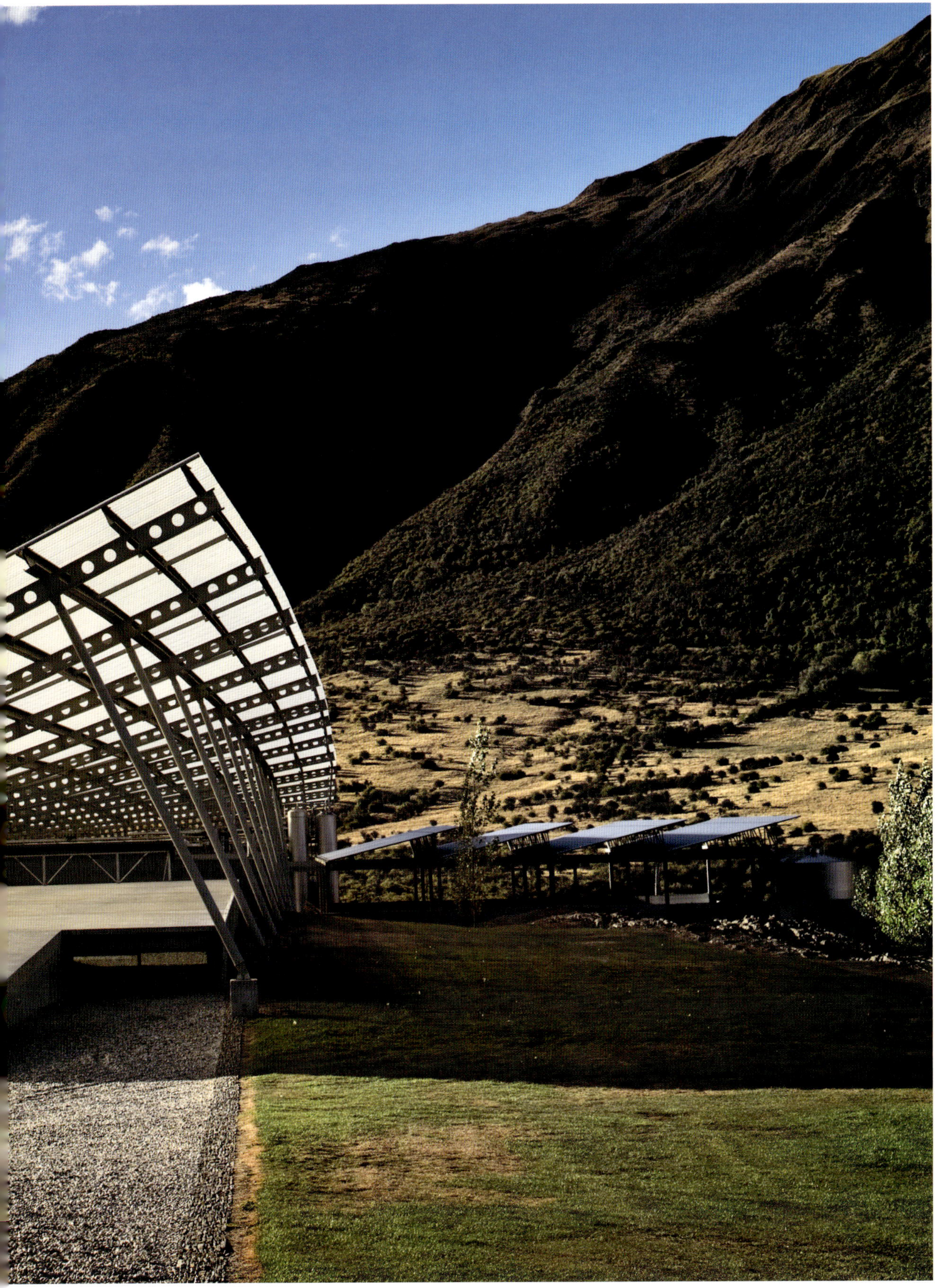

QUARTZ REEF

In the early 1990s, Rudi Bauer became convinced that Central Otago and Pinot Noir were destined for special things, and that he wanted to be a part of it. Bauer, who was born and raised in Salzburg, Austria, assumed the winemaking role at Rippon Vineyard in Wanaka in 1989. Already acquainted with Pinot after having worked with it in Oregon, it was after making the Rippon Pinot Noir 1992 that the rich potential of the 'Pinot–Central' partnership became clear to him.

'I remember tasting that wine in the winery with winemakers Grant Taylor [now with Valli] and Steve Davies [formerly with Carrick] and I saw something … that wine was a glimpse into the future,' he says. 'I knew then that this was Pinot Noir country.'

Today, as well as being a pillar of the Central Otago — and national — Pinot scene, Bauer is co-owner and winemaker of Quartz Reef. Bauer had his eyes on the rabbit-infested, heat-drenched, north-facing slopes of Bendigo years before he first began planting vines there in 1998. His partners in the venture are Bendigo Station owner John Perriam and businessman Trevor Scott. An original partner, Clotilde Chauvet of the Champagne house Marc Chauvet, was bought out by Bauer and Scott in 2008. The winery has always dedicated a portion of its Pinot Noir fruit to the making of sparkling wine, over which Chauvet exerted influence during the early years.

Quartz Reef began producing the Quartz Reef Pinot Noir in 1998, using fruit grown at Pisa Range Estate for the first three vintages. Since 2001, when fruit from the label's own vineyard came on-stream, Quartz Reef has been all about Bendigo, and from 2007 all the fruit used has been 100% estate grown.

Quartz Reef's Bendigo Estate vineyard was the first to be planted in Bendigo. Lying beneath it is what is believed to be New Zealand's largest quartz deposit, hence the name. The original vineyard, planted in 1998, is a 15-hectare north-facing slope comprising arid clay, fine gravel and quartz soils. Pinot Noir clones 10/5, UCD5, 115, 667, 777 and Abel were planted to a density of between 5000 and 8000 vines per hectare. Recently, Bauer made the decision to commit to biodynamics to 'achieve an even better reflection of the site and its variety'.

Right: Rudi Bauer, Quartz Reef.

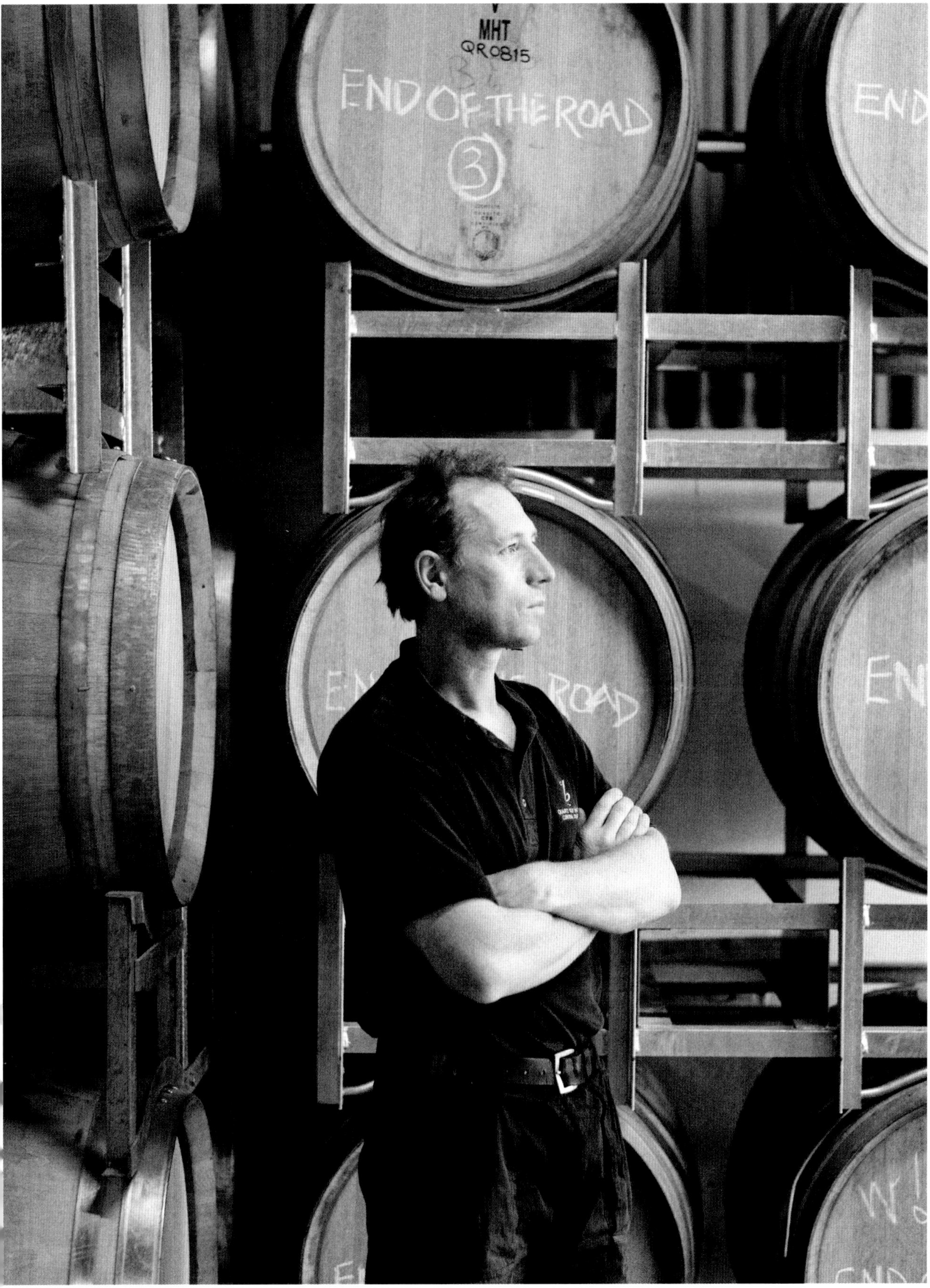
MHT
QR0815
END OF THE ROAD
3

The Quartz Reef Bendigo Central Otago Pinot Noir is made from the lion's share of the vineyard's annual issue. Another Pinot, the Quartz Reef Bendigo Estate Pinot Noir, is a limited-release wine.

'For the Bendigo Estate we choose a ferment that we really like, one that highlights the growing season and the definition of the vineyard,' explains Bauer. 'It's not always the best [ferment]; it's the most exciting and intriguing. It varies from year to year. In 2005, the Abel [clone] ferment was the best. The clone make-up of that wine changes from year to year.'

The Quartz Reef Pinots are sturdy, quite rich wines, which reflect the heat and aspect of the site. 'I'm personally fond of more feminine styles, like Chambertin,' says Bauer. 'What I didn't know when I began was that Bendigo had a different idea … the wines are quite firm and masculine. I have to follow what is given to me, so my tack has changed. The style I'm aiming for I describe as being like a well-dressed Italian or Argentinian man walking down a boulevard … a masculine wine with a wonderful ability to carry itself in a gentle, stylish way.'

The evolution of Central Otago Pinot Noir has been as he foresaw, and is something to which he has made a significant contribution.

The decision to concentrate purely on Pinot Noir has now, after 10 years, given Central Otago the edge as being known as a serious, committed, high quality-producing region. Our success has been fast-tracked through the work ethic and willingness of viticulturists and winemakers to share their experiences. I think in another five to seven years we will achieve profound international status.

RIPPON VINEYARD

If the land could speak, the first thing it would tell you is that it doesn't really want 5000 vines per hectare on its back. That's about the last thing it wants — it happens nowhere in nature. If that is your goal, you must offer something in return. When creating a relationship on these terms, one starts to see the problems on the property as not just an intrusion or a weakness that needs to be combated or corrected, but the land asking you something.

Thus spoke Rippon Vineyard's winemaker Nick Mills at the Soil Dynamics session at Pinot Noir 2007 in Wellington, revealing much about his core winegrowing philosophy. Few on the New Zealand Pinot scene are as passionate as Mills about the importance of conversing Cistercian-like with the land, and few make such singular and honest expressions of Pinot Noir.

When his father Rolfe Mills began planting experimental vines in 1974 on land beside Lake Wanaka that had been in the family for generations, Nick Mills was one year old. He went on to attend the University of Otago and become a fearless champion freestyle skier before throwing himself into wine. After four years studying in Burgundy and working at some of its great domaines, Mills has a more intimate knowledge of Pinot Noir's ancestral seat than most other New Zealand producers. It was in Burgundy that he was drawn to biodynamics. He took over the winemaking role at Rippon in 2002, two years after his father's death, and immediately became a presence both as a vigneron and as a strong advocate for biodynamics, his region and New Zealand Pinot in general. Helping him at Rippon — Mills is insistent on the importance of the team element — are other family members, including his wife, Jo (administration), sister Charlie (vineyard manager) and brother David (property manager).

Until very recently, the winery produced two Pinot Noirs. The flagship has been the Rippon Pinot Noir, typically a beautifully perfumed, delicate yet incisive, memorable wine. Its stablemate is the Rippon Jeunesse Young Vines Pinot Noir, a lighter, fresher style.

'Rippon is a place, a single vineyard … no fruit is bought in,' says Mills. That place is one of affecting beauty — a 15-hectare north-facing sweep down to the deep blue

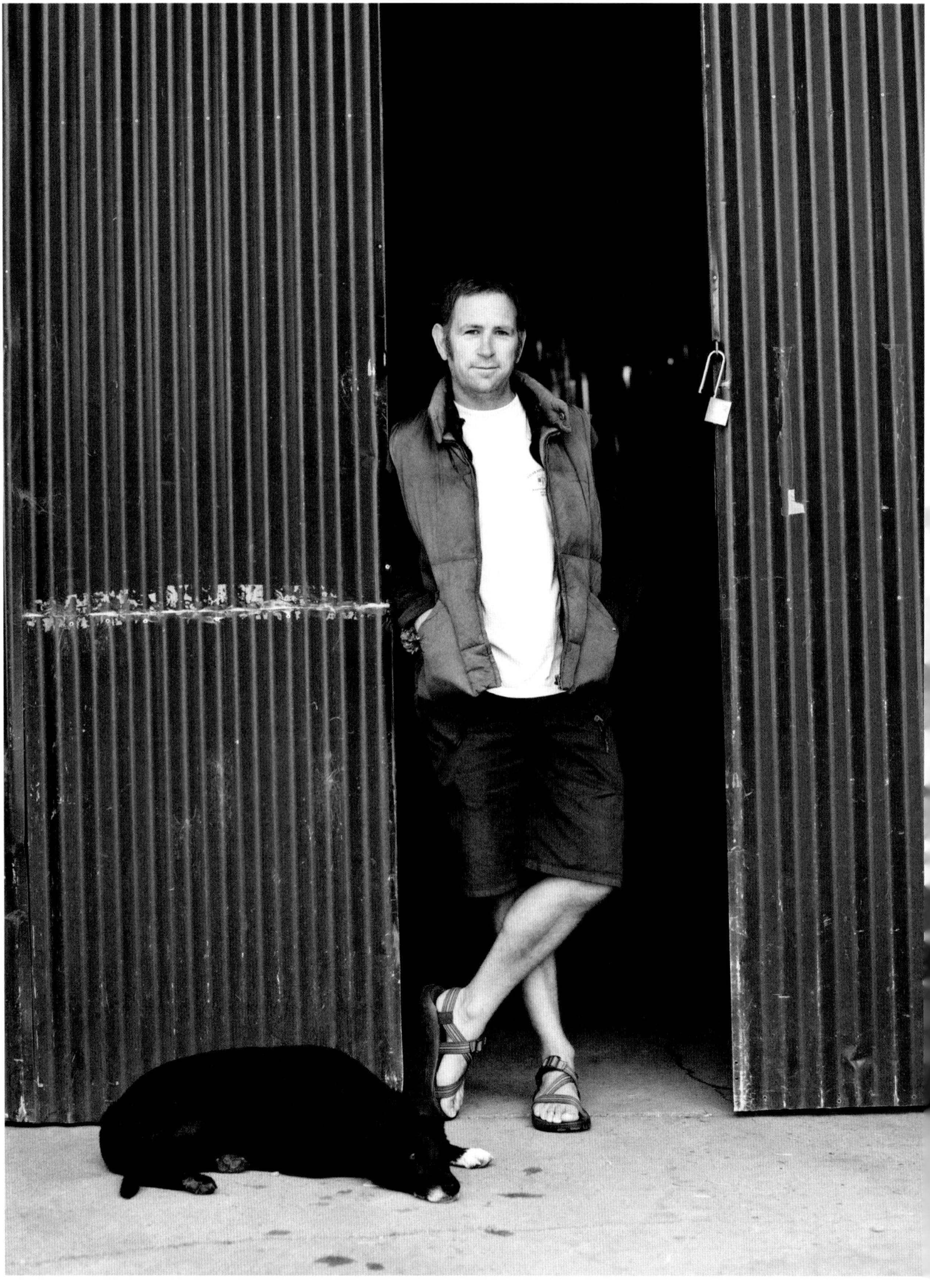

waters of Lake Wanaka. The Rippon vineyard is one of the region's highest (up to 330 metres above sea level) and the closest to the continental divide. The hillside is made up of free-draining schist gravel.

The vineyard is also home to Central Otago's oldest vines. The fruit destined for the Rippon Pinot Noir is grown on vines planted between 1986 and 1991, all on their own roots and representing clones UCD5, UCD6, 2/10, 10/5, 13 and Lincoln (drawn from Danny Schuster's mass selection). The vine density ranges between 3500 and 5000 vines per hectare.

The fruit is hand-picked into small cases, allowing it to arrive at the sorting table undamaged. Small parcels representing different clones, vine ages and micro-sites are fermented (using naturally occurring yeasts) and matured apart. The proportion of fruit left as whole bunches varies between 15% and 40%. The wine is then racked to barrel (30% new) where it spends 15 to 18 months, passing through malolactic fermentation unprompted in the spring. Bottling occurs typically 18 months after vintage and the wine is only released six months after that.

The Jeunesse Pinot is made in similar fashion, although without any new oak component. The oldest vines in the Jeunesse were planted in 1994; the more recent plantings include Dijon clones 114, 115, 667 and 777.

In 2010, Rippon added two new wines to its Pinot releases. A pair of special parcels were deemed mature and distinctive enough to be bottled separately as Rippon Emma's Block Mature Vine Pinot Noir 2008 and Rippon Tinker's Field Mature Vine Pinot Noir 2008. Only 100 cases were made of each. 'Over several years these parcels captured our attention for emerging characters specific to their subsoil,' says Mills.

••

Left: Nick Mills, Rippon. **Following pages:** *The Rippon Vineyard, a Central Otago jewel.*

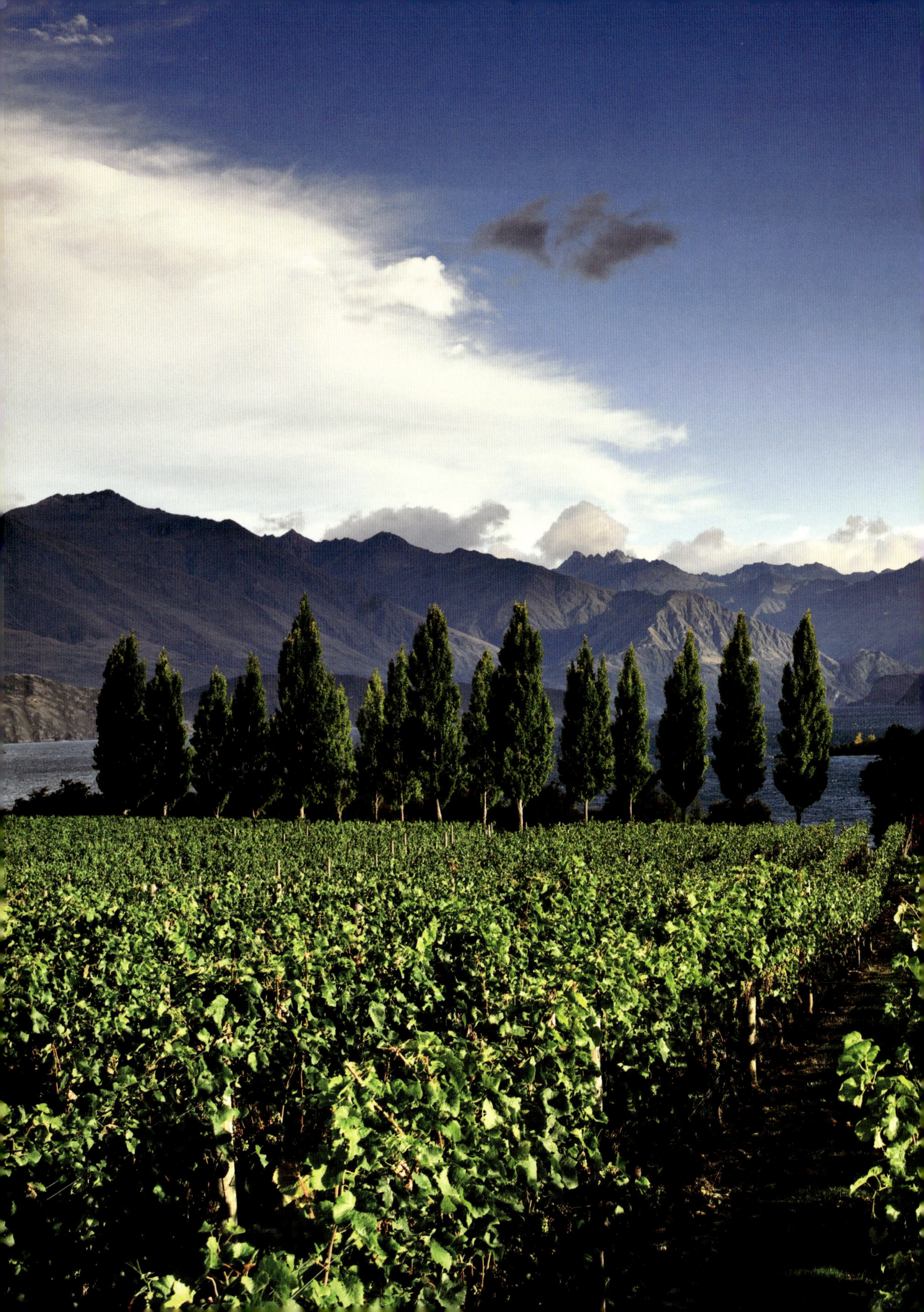

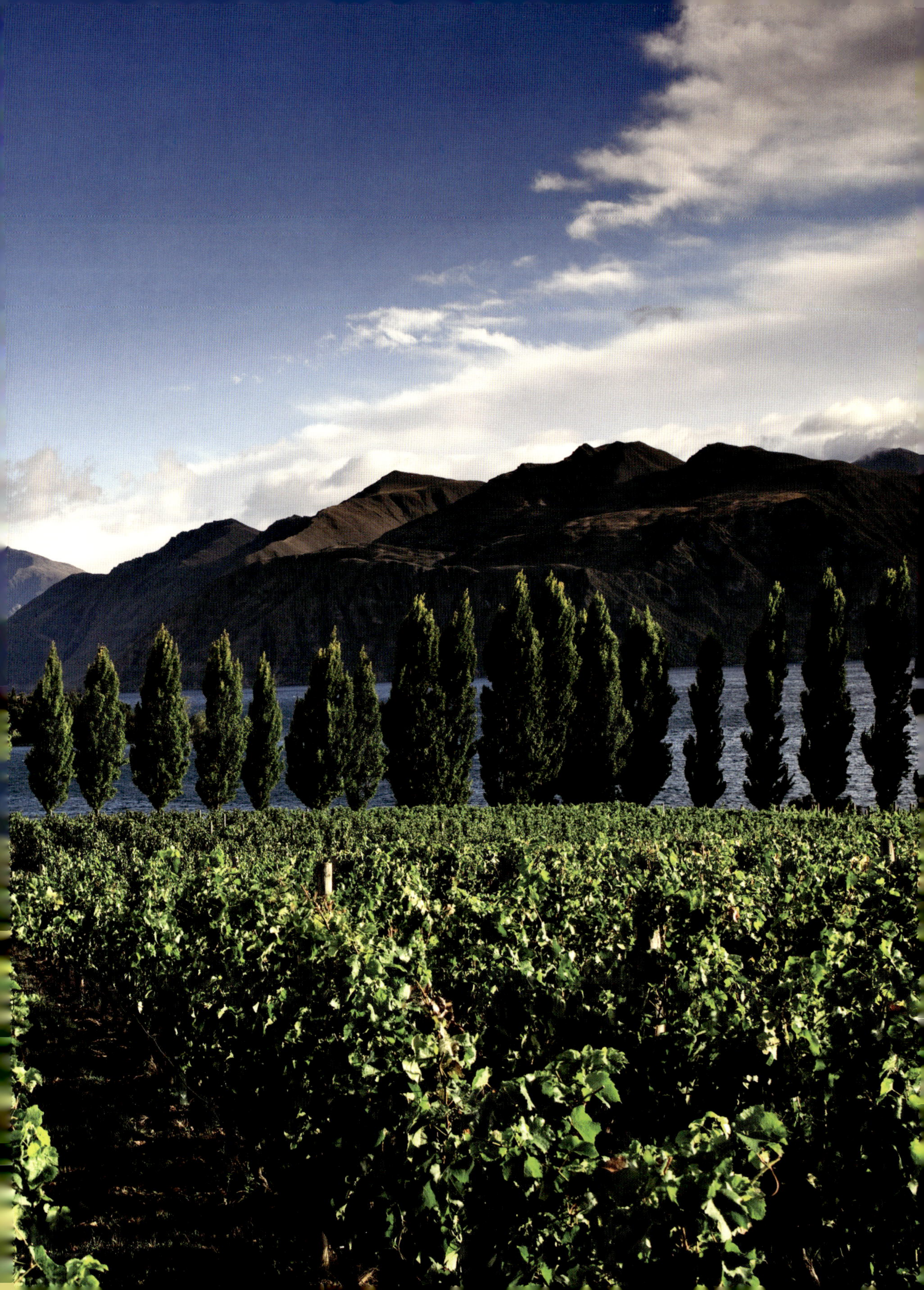

ROCKBURN

'I don't believe in over-extracted, huge wines that want to be Shiraz and I am always chasing greater purity, minerality and transparency, but fear the market doesn't care for such wines and is alienated by them.'

With these words, Rockburn winemaker Malcolm Rees-Francis expresses a concern shared by many New Zealand Pinot producers, particularly in small wineries. In his case, however, the market — and certainly the wine-show system — cares very much for his wines. Since his arrival at Rockburn in 2005 after four years as Blair Walter's understudy at Felton Road, Rees-Francis has produced a string of critically acclaimed Pinots that have bolstered the standing of Rockburn considerably.

The Rockburn brand was established in 2001. Before that it was Hay's Lake, a label that grew out of Dunedin surgeon Dick Bunton's small vineyard beside photogenic Lake Hayes. Expansion brought two other shareholders on board, businessmen Chris James and Paul Halford. The winery operates from a new building, which incorporates an attractive, sun-filled cellar door, in Cromwell's winery-studded industrial sector.

Rockburn makes two Pinots annually: the Rockburn Pinot Noir and the second-tier Devil's Staircase Pinot Noir. In 2006 Rees-Francis produced Rockburn Eight Barrels Pinot Noir, a wine of which he is particularly proud and which he hopes to make again when conditions allow.

The winery draws all its fruit from two very different sites — a 9-hectare vineyard in the Gibbston Valley and a much larger (33 hectares) vineyard at Parkburn, looking out across Lake Dunstan. In total, there are 21 hectares planted in Pinot Noir. The Gibbston site is host to just two clones (UCD5 and UCD6), and was planted in 1999. At Parkburn, the clone mosaic is far more complex, with a broad range from 10/5 through to 777. Vine age varies, with the oldest planted in 1999. Typically, 85% of the fruit for both the Rockburn and Devil's Staircase Pinots is sourced from Parkburn, the remainder from the Gibbston site.

Rockburn's cropping levels are mostly dictated by the seasons. At least, they were until the extraordinary 2008 vintage, which blindsided Rees-Francis with a yield of 10 tonnes per hectare. 'Our estimates were way out and we've revised our methods since.'

Rees-Francis picks on flavour and no longer even takes a refractometer (used to gauge sugar levels) into the vineyard. He ferments whole bunches when conditions and quality allow. The Rockburn Pinot Noir lives in French oak (30–40% new) for 10 – 11 months, going through malolactic fermentation naturally in the spring. It is racked, filtered and bottled in March.

Rockburn's cooler Gibbston vineyard produced the fruit in 2006 that inspired Rees-Francis to make the small-quantity, premium Eight Barrels Pinot Noir. It was a year when the fruit ripened earlier than usual in the Gibbston and provided Rees-Francis with phenolic ripeness at remarkably low sugar levels. The finished wine (which included 25% Parkburn fruit to stiffen the sinews) saw 25% new French oak. It was dark, lithe and gently seductive, with an alcohol level of only 12.8%. Declares Rees-Francis:

> I believe that wine to be a potential window into the future of Central Otago Pinot Noir. But most people didn't get it at the time. A bottle opened in December 2008 was just beginning to unfurl and it pains me to think what it will become while nobody's watching. But still … I'm damned lucky to be a winemaker working in one of the best places in the world for one of the greatest varieties. Every day I am thankful I even found wine, let alone got to where I am now.

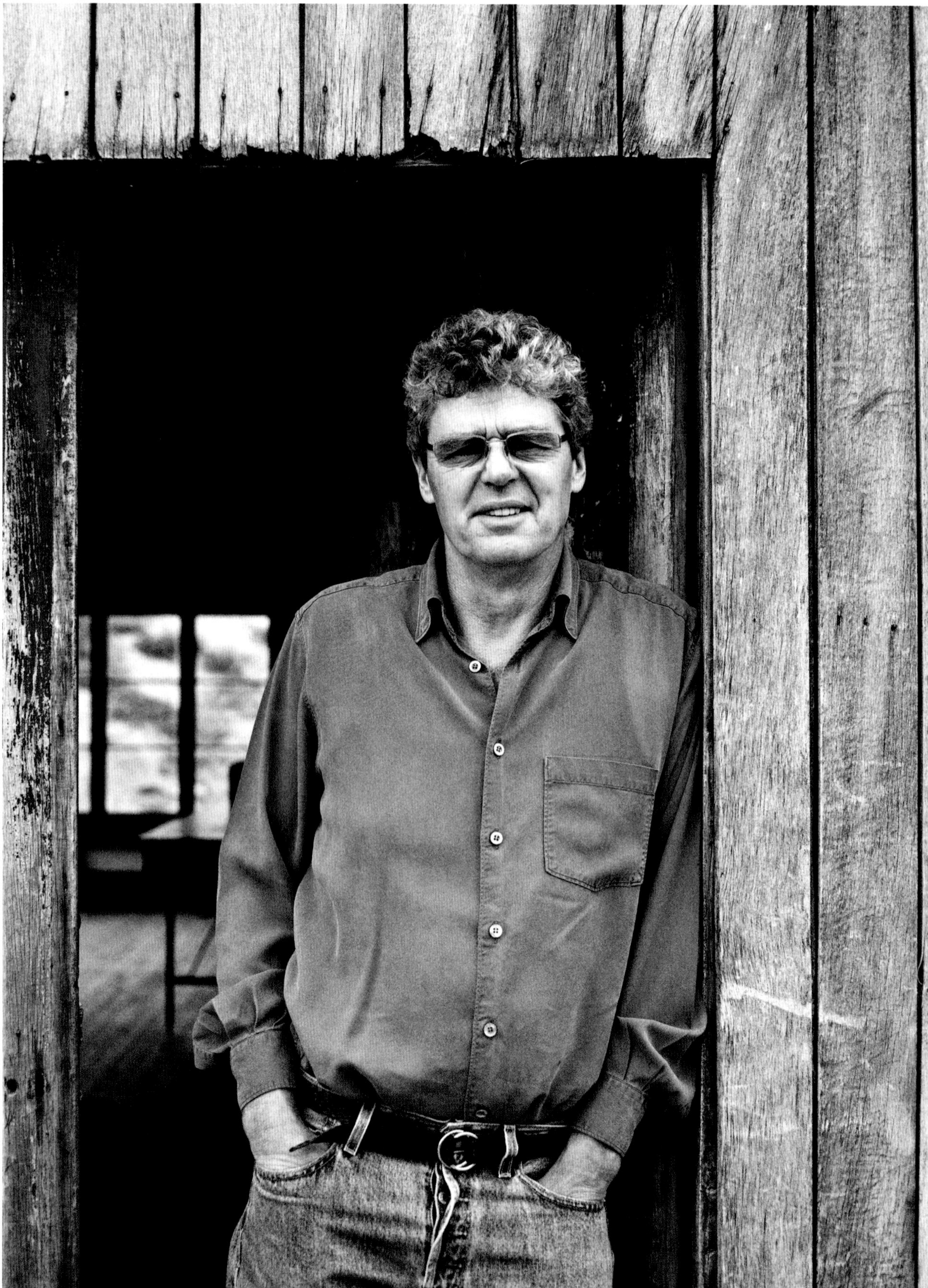

VALLI

Grant Taylor was born in Kurow in the Waitaki Valley. He spent many years making wine in California before returning to the South Island in 1993 to take the winemaking role at Gibbston Valley Wines. There, he was at the forefront of Central Otago's extraordinary journey through the 1990s, when the stature of the region's Pinot Noir grew with every vintage. Some of the Pinots he produced at Gibbston Valley are still rightly regarded as Central Otago classics.

His own label, Valli, was launched in 1998. Not destined to stay a 'moonlit venture', it grew to become Taylor's central focus after he made his last vintage at Gibbston Valley Wines in 2006.

Valli was his mother's maiden name. An ancestor, Giuseppe Valli, was a winegrower in Italy who made the trip out to New Zealand in the 1870s. Valli's home-base is *the* Valley — the Gibbston, to which Taylor remains attached.

Taylor claims to be 'an Otagophile more than a Pinotphile. I just want to put Otago in a bottle.' In fact, it goes deeper than that. He wants to put several small corners of Otago in bottles. He is a great proponent of not blending out the differences between the Central Otago subregions, but letting each have its say.

To this end, Valli makes three single-vineyard Pinots: one from the home vineyard in the Gibbston and two others using contract-grown fruit from sites in Bannockburn and the Waitaki Valley, in North Otago. These wines are typically rugged individuals. In strong vintages the Gibbston is violet-scented with a well-honed acidic blade; the Bannockburn has deep florals and is broader, more savoury and more tannic; the Waitaki Valley wine displays earthier, herbal notes. They are all concentrated, well-structured wines, built for the long haul.

'Understanding these subregions with their different characters is where I'm at … learning that what is suited to one is not necessarily the best for another,' explains Taylor. ' For instance, in the Gibbston we started out with [clone] 10/5, because there was little else at the time. The newer clones are now doing much better there, but I like what I get from 10/5 in Bannockburn.'

Left: Grant Taylor, Valli.

Valli's 3.5-hectare Gibbston block stretches upwards from the road near the eastern end of the valley. A covering of alluvial loess between a half-metre and a metre deep sits over firm, free-draining river gravels. The vines were planted in 2000 to a density of 4040 vines to the hectare; Pommard and the Dijon clone predominate. Taylor aims at crop levels between 3 and 5 tonnes per hectare.

Of the three vineyards from which Taylor obtains fruit, this is the most significant, not least because it accounts for most of Valli's volume, which overall is still tiny. In 2007, 750 cases issued from the Gibbston vineyard, 350 from Bannockburn and just 25 from the Waitaki Valley, which was affected by frost.

The Gibbston is also the vineyard that was planted personally by Taylor and which he manages day-to-day. When it comes to Pinot, he believes he vineyard is all:

> The changes leading to the quality of the wines we have today are primarily due to improvements in the vineyard. Pay attention to every detail in the vineyard and the better the fruit will be and the clearer it will reflect the site. That's one of the great things about Pinot … when you treat it well in the vineyard it says 'thank you'.

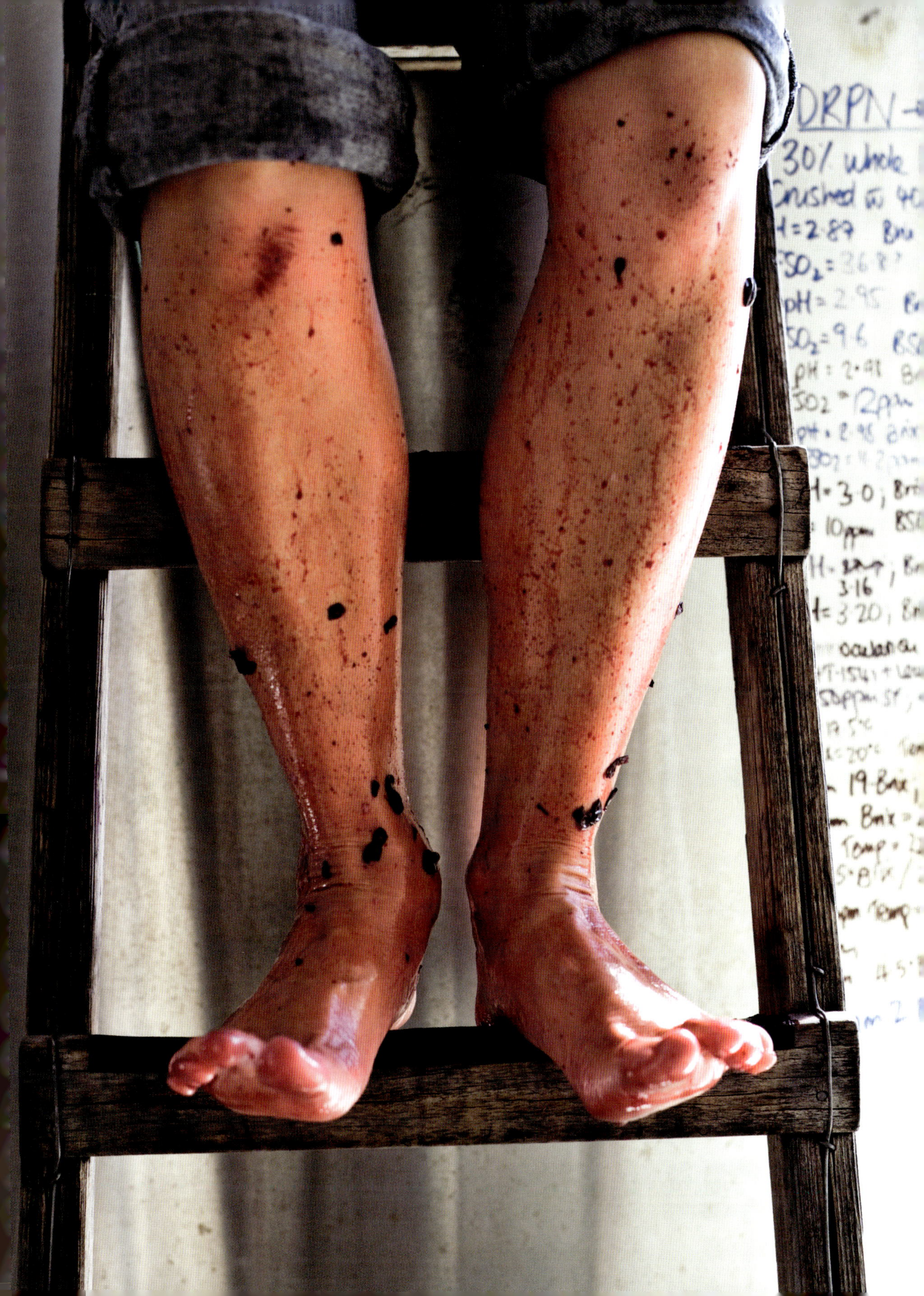

INDEX

Page numbers in **bold** refer to the main sections on each region and winery.

A M 10/5 55
Abel clone 61, 110, 113, 118, 127, 130, 138, 144, 197, 216, 240, 244, 267, 276, 278
Abel, Malcolm 61
Abyss Pinot Noir (Churton) 165
Air New Zealand Wine Awards 59–60, 62, 161
Akaroa 204
Albany Surprise 20, 51, 55
Alexandra subregion 64, 93, 233, 234, 235, 259
Aloxe-Corton area 56–7
Alphonse La Valée 63
Angel Flower Pinot Noir (Pyramid Valley) 227
Arnst, Bart 188
Aroha Pinot Noir (Craggy Range) 118
Ata Rangi winery 18, 60, 61, 66, 70, 73, 105, **108–15**
   Ata Rangi Pinot Noir 63, 94, 110, 113, 121
   Célèbre Pinot Noir 63
   Crimson Pinot Noir 113
   McCrone Vineyard Pinot Noir 2007 113
Atkins, Annette 138, 139
Auntsfield Heritage Pinot Noir 158
Auntsfield Winery 158
Austin Moorhouse, Tim 267
Australia 20, 59, 69, 241; *see also* Tasmania; Victoria
Australian National Wine Show 198
Auxerrois grape 32
Avery, John 59, 60
Awatere Valley 163, 182, 198, 199, 214
Awatere Valley Pinot Noir (Kaituna Valley) 214–15

Babich, Joe 50, 57–8, 59
Babich Pinot Noir 1981 57–8
Bachtobel 55, 57, 58, 65, 158, 159
Baco 22A 51, 55
Bakano 54
Baker, Adrian 119
Banks Peninsula 204, 214
Bannockburn 227, 234, 240–1, 251, 267, 268–9, 287, 288
Bannockburn Sluicings Vineyard Pinot Noir (Craggy Range) 119
Barracks Vineyard, Cloudy Bay 168–9

barrels 79, 121, 146, 177, 184, 197
Bartle, Peter 270
Bauer, Rudi 19, 58, 64, 161, 251, 276, 278
Beaune 60
Beaven, Don 58, 59–60
Beetham, Marie Zélie Hermanze 46, 104
Beetham, William 45–8, 49, 104, 130, 134
Bell Hill 204, **206–13**
   Bell Hill Pinot Noir 206
Ben Morven Valley 158, 198
Bendigo 119, 234–5, 258, 259, 270, 276
Bendigo Central Otago Pinot Noir (Quartz Reef) 278
Bendigo Estate Pinot Noir (Quartz Reef) 278
Bendigo Estate Vineyard, Quartz Reef 276
Berrysmith, Frank 55, 65
Bicknell, Brian 180, 181, 188
Bicknell, Nicola 180
biodynamic production 22, 119, 165, 188, 206, 226, 227, 241, 251, 276, 279
Bish, Tony 64
Blackridge 64, 229
Bladier brothers 45
Bone Hill Pinot Noir (Kaituna Valley) 215
Bordeaux 18, 19, 31, 48, 61, 62, 164
Bothwell, Phil 182
botrytis 106
Bouchard Finlayson Trophy for Pinot Noir 94, 110
Boundary Vineyards 184
Brady, Alan 64, 229, 232, 258, 259, 266
Bragato, Romeo 49, 50, 63
Brancepath 46, 47–8
Brancott Estate Vineyard 182
Brancott Valley 159, 163, 168, 171, 177, 181, 196
Brancott Valley Pinot Noir (Fromm) 177
Brightwater 152
Buchanan, John 266
Bunn, Carol 232
Bunton, Dick 284
Burgundy (region) 17–20, 22, 30–1, 33, 36, 43, 50, 56–7, 60, 61, 62, 68, 69, 70, 71, 79, 90, 97, 98, 196, 199, 226, 232, 279
Burgundy (wine) 19–20, 21, 31, 48, 55, 60, 130, 161, 164
'Burgundy Wine', produced by Feraud 45, 49
Burnham 206
Burnt Spur Pinot Noir (Martinborough) 133
Busby, James 42, 43

Byrne Pinot Noir (Mahi) 180

Cabernet Sauvignon 18, 20, 54, 71, 138, 168
Cairnmuir Vineyard, Carrick 240–1
California 20, 21, 56, 69
Calvert Pinot Noir (Craggy Range) 118–19, 252
Calvert Pinot Noir (Felton Road) 251
Calvert Vineyard, Felton Road 252
Calvert Vineyard Pinot Noir (Pyramid Valley) 227, 252
Campbell, Claire 130
Camshorn brand 184
Camshorn Vineyard, Pernod Ricard 182, 184, 203
canopy management 67, 71, 73, 215
Canterbury 58–9, 88, **202–7**; *see also* individual placenames and vineyards/wineries
Canterbury Pinot Noir (Kaituna Valley) 215
Canterbury Plains 204
Carrick **240–1**
   Carrick Central Otago Pinot Noir 241
   Excelsior Pinot Noir 241
   Unravelled Pinot Noir 241
   Célèbre Pinot Noir (Ata Rangi) 63
Central Otago 19, 45, 49, 63–4, 69, 70, 80, 88, 90, 97, 118, 182, 184, 203, **228–88**; *see also* individual placenames and vineyards/wineries
Central Otago Pinot Noir Celebration 67, 232
Central Otago Pinot Noir Limited 232–3
Central Otago Wine Company 78, 232
Central Otago Winegrowers Association (COWA) 64
Chambers, Bernard 48
Champagne (region) 31, 50
Chapuis, Claude 56–7
Chard Farm 229, **242–9**, 270
   Finla Mor Pinot Noir 244
   River Run Pinot Noir 242, 244
Chard, Richard 242
Chardonnay 32, 36, 43, 55, 138, 170, 216, 227
Chasselas 44, 45
Chauvet, Clotilde 276
Chifney, Stan 61
chlorosis 207
Churton **164–7**
   Churton Abyss Pinot Noir 165
   Churton Pinot Noir 165
Cistercian monks, Burgundy 17–19, 22

Clarke, Oz 18, 21, 99
Clayvin Vineyard, Fromm 159, 196
Clayvin Vineyard Pinot Noir (Fromm) 177
climate 21, 36, 69, 93–4, 106–7, 141, 162, 163, 203, 204, 214, 233, 234
clones 33, 55, 57, 58, 59, 60, 61, 65–6, 69–70; see also Abel clone; Bachtobel; Dijon clones; 10/5 clone; 2/10 clone; UCD clones
Clos de Vougeot vineyard 43
Clos de Ste. Anne Pinot Noir (Millton) 89
Cloudy Bay **168–9**
 Cloudy Bay Pelorus 161, 168
 Cloudy Bay Pinot Noir 168, 169
 Cloudy Bay Sauvignon Blanc 168, 169
Coates, Clive 19
Columella, *De Re Rustica* 31
Comerford, John 56
Cook Block, Chard Farm 244
Corbans 57, 164
Cornish Point Pinot Noir (Felton Road) 251–2
Cornish Point Vineyard, Felton Road 252
Côte d'Or 17, 19, 50
Côte Rotie 44
Coutts Island 59
Craggy Range 89, 105, **118–19**, 127, 227, 252
 Aroha Pinot Noir 118
 Bannockburn Sluicings Vineyard Pinot Noir 119
 Calvert Pinot Noir 118–19, 252
 Te Muna Road Pinot Noir 118
 Zebra Vineyard Pinot Noir 119
Craighall Vineyard, Dry River 121
Creech, Wyatt 134
Crimson Pinot Noir (Ata Rangi) 113
Cromwell Basin 234, 266, 269, 270–1
Cross, Nadine 270
Cuttance, Brother John 51

Dalefield 63
Davies, Steve 240, 276
Deaker, Blair 240, 241
Delegat's 61, 62
Department of Agriculture 63
Devil's Staircase Pinot Noir (Rockburn) 284
Dicey, Matt 268, 269
Dicey, Robin 229, 268
Dijon clones (113, 114, 115, 667 and 777) 65–6, 69–70, 113, 118, 121, 130, 134, 146, 156, 165, 180, 197, 214, 216, 221, 241, 244, 267, 276, 281, 288

Docherty, Jane 240, 241
Dog Point **170–3**
 Dog Point Pinot Noir 170–1
Domaine de la Romanée-Conti 19, 61
Domaine Leflaive 90
Donaldson family 161, 220
Donaldson, Ivan 58, 220
Donaldson, Matt 71, 204, 220, 221
Dougall, Clive 188, 190
Dry River 61, 66, 80, 98, 105, **120–1**
Dry River Pinot Noir 1989 61
Dumont d'Urville, Jules Sébastien César 43

Earnscleugh 63
Earth Smoke Pinot Noir (Pyramid Valley) 227
Earth's End label, Mount Edward 266
East, Jane and Michael 218
Eaton, Buffy 216
Eaton Family Vineyard Pinot Noir (Pyramid Valley) 227
Eaton, Jo 196
Eaton, Michael 216
Eaton, Mike 159, 196, 197
'The Edge' Pinot Noir (Escarpment) 128
Edwards, Verdun and Sue 64
Eight Barrels Pinot Noir (Rockburn) 284, 285
Elie-Regis, Brother 44
Elms, Stewart 251
The Elms Vineyard, Felton Road 252
Emma's Block Mature Vine Pinot Noir (Rippon) 281
Escarpment Vineyard **127–9**
 'The Edge' Pinot Noir 128
 Escarpment Pinot Noir 127
 'Insight Series' 127–8
 Kiwa (Moana) Pinot Noir 128
 Kupe Pinot Noir 127
 Pahi (Voyager) Pinot Noir 128
 Te Rehua Pinot Noir 128
Excelsior Pinot Noir (Carrick) 241

Fairhall Estate Vineyard 158
Fairhall Valley 184
Felton Road 97, 113, 118–19, 227, 229, 234, 240, **250–5**, 284
 Calvert Pinot Noir 251
 Central Otago Pinot Noir 252
 Cornish Point Pinot Noir 251–2
 Felton Road Pinot Noir 251
 Felton Road Pinot Noir Block 3 251, 252

Felton Road Pinot Noir Block 5 251, 252
Feraud, John Désiré 45, 49, 63
Finla Mor Pinot Noir (Chard Farm) 244
Finn, Judy 152
Finn, Tim 58, 66, 97, 152, 156
Fistonich, Sir George 198
Forrest Estate 89
Forrest, John 89
Forsyth, Duncan 266, 267
France 31, 32, 43, 46, 65; see also Bordeaux; Burgundy; Champagne; Clos de Vougeot vineyard
Francis, Tom 156
Frère, Marie Zélie Hermanze 46, 104
Fromm, Georg 159, 174, 196
Fromm, Ruth 174
Fromm Winery 65, 161, **174–9**
 Brancott Valley Pinot Noir 177
 Clayvin Vineyard Pinot Noir 177
 Fromm Vineyard Pinot Noir 177
 La Strada Reserve Pinot 1994 161, 174, 177

Gamay (Beaujolais) grape 30, 31, 32, 146
Geris, George 199
Germany 31, 32, 64, 218, 242
Giants Winery 118
Gibbston subregion 234, 235, 242, 258–9, 266–7, 270, 284, 287–8
Gibbston Valley Wines 64, 229, 242, **256–63**, 287
 Gibbston Valley Central Otago Pinot Noir 258–9
 Gibbston Valley Pinot 64
 Gibbston Valley Reserve Pinot Noir 258
 Gold River Pinot Noir 259
 Le Maître Expressionist Series Pinot Noir 259
Giesen 161
Giesen, Marcel 206–7
Gisborne 65, 66, 89
Gladstone subregion 105
Glovers 141
Gold River Pinot Noir (Gibbston Valley) 259
Gould, Belinda 218–19
The Gradient Pinot Noir (Mountford Estate) 217
Grant, Bill 64
Grape Vision 229
The Great Bear Pinot Noir 2002 (Palliser Estate) 137
The Great George Pinot Noir 2005

(Palliser Estate) 137
The Great Harry Pinot Noir 2006 (Palliser Estate) 137
Green, Steve and Barbara 240, 241
Greenhough, Andrew 66, 98, 144, 146–7
Greenhough Vineyard 66, 98, 141, **144–51**
   Hope Vineyard Pinot Noir 144, 146
   Nelson Pinot Noir 146
Greening, Nigel 251
Greenmeadows 48, 51
Griffith, Phil 258
Growers Collection Pinot Noir (Pyramid Valley) 227

Haeger, John Winthrop 32
Halford, Paul 284
Hammond, Katy (Poppy) 120
Hammond, Shayne 120
Hancock, John 62, 89
Hanson, Anthony 19
Hare's Breath Pinot Noir (Muddy Water) 218, 219
Hawke's Bay 44, 64, 65, 66, 89, 99, 118, 119; see also individual placenames and vineyards/wineries
Hay, Greg 242, 270, 271
Hay, Rob 64, 242, 270
Hay's Lake label 284
Healy, James 168, 170, 171
Healy, Wendy 170
Heathcote, William 48
Henderson 50, 57
Herd, David 158
Hermitage 48
Hillside Reserve Pinot Noir (TerraVin) 197
Hoare, William 174, 177
Hohnen, David 168
Home Vineyard, Neudorf 152, 156
Home Vineyard, Seresin 188, 190
Hope Vineyard, Greenhough 144, 146
Hope Vineyard Pinot Noir (Greenhough) 144, 146
Huapai 54, 55, 57, 158
Huchet, Brother Cyprian 44, 47
Hudson, Lynnette 220
Hunters 164
hybrid grapes, American 51, 55

Ihumatao Peninsula 198
'Insight Series' (Escarpment) 127–8
International Wine and Spirit Competition, London 94, 110
Irwin, Bill 65

Isabel Estate 161
Italy 31

Jackson, David 58, 63
James, Chris 284
Jeunesse Young Vines Pinot Noir (Rippon) 279, 281
Johnson, Allan 134, 137
Johnson, Hugh 90
Joseph Soler block, Villa Maria 241
Judd, Kevin 168, 169

Kaipara harbour 48
Kaituna Valley 58–9, 204, **214–15**
   Awatere Valley Pinot Noir 214–15
   Bone Hill Pinot Noir 215
   Canterbury Pinot Noir 215
   Kaituna Valley Pinot Noir 59, 214
Kaituna Vineyard, Pernod Ricard 182
Kalberer, Hätsch 65, 161, 174, 177
Karearea Pinot Noir (Peregrine) 271
Kasza, Denis 55
Kavanagh, John 142
Kelly, Geoff 47–8
Keys, Christopher 258, 259
Kirby, Robert and Mem 127
Kiwa (Moana) Pinot Noir (Escarpment) 128
Kotinga Vineyard, Voss Estate 138
Kramer, Matt 229
Kumeu 61
Kumeu River (San Marino) 54
Kupe Pinot Noir (Escarpment) 127

La Strada Reserve Pinot 1994 (Fromm) 161, 174, 177
La Tâche vineyard, Burgundy 61
Lamb, Brother Joe 51
Lamb, William 48, 50
Lane, Nick 169
Lansdowne Vineyard 47, 48
Le Maître Expressionist Series Pinot Noir 259
Leah Pinot Noir (Seresin) 190
Leaning Rock 229
Leflaive, Anne-Claude 90
Lett, David 20
Lin, CP (Chung Pin) 216–17
Lincoln College (University) 58, 59, 63, 130, 204, 214
Lindauer (Montana) 66, 159
Lloyd Vineyard Pinot Noir (Villa Maria) 199
Long Gully Single Vineyard Pinot Noir (Mt Difficulty) 269

Louis Vuitton Moët Hennessy 168
Lovat Vineyard, Dry River 121
Lowburn 234, 242
Lyon 44

McCallum, Dawn 120
McCallum, Neil 58, 61, 66, 98, 120
McCrone, Don and Carol 113
McCrone Vineyard Pinot Noir 2007 (Ata Rangi) 113
McDonald, Tom 20
McKean family 203
McKenna, Larry 61–2, 66–7, 78, 79, 127–8, 130
McKenna, Sue 127
McWilliams 20, 54
Madeleine Royale 63
Mahi **180–1**
   Byrne Pinot Noir 180
   Marlborough Pinot Noir 181
   Rive Pinot Noir 180–1
Main Divide Pinot Noir (Pegasus Bay) 221
Main Divide TEHAU (Pegasus Bay) 221
Main Divide TIPINUI (Pegasus Bay) 221
Malbec 37
Maling, Alistair 199
Marie Zélie Pinot Noir (Martinborough) 104
Marist Brothers (Society of Mary) 44, 47, 51
Marlborough 49, 57, 65, 66, 88, 93, 94, 106, **158–201**
   southern valley flesh 163
   see also individual placenames and vineyards/wineries
Marsden, Samuel 42
Martin, Neal 97
Martinborough 18, 21, 60–3, 65–6, 80, 93, 104–5, 106, 118, 128, 133, 198, 203
   appellation 105
   see also Ata Rangi winery; Dry River; Escarpment Vineyard; Martinborough Vineyard
Martinborough Vineyard 60–2, 104, 105, 127, **130**, **133**, 134
   Burnt Spur Pinot Noir 133
   Marie Zélie Pinot Noir 104
   Martinborough Vineyard Pinot Noir 62, 130
   Martinborough Vineyard Reserve Pinot Noir 133
   Russian Jack Pinot Noir 133
   Te Tera Pinot Noir 133

Mason, Paul 130, 133
Masters, Helen 70, 73, 113
Masterton subregion 105
Matawhero Wines 65
   Matawhero Pinot Noir 1987 65
Materman, Patrick 184
Meeanee 44
Merlot 18, 36, 37, 168
Mills family 279
Mills, Lois 64, 66
Mills, Nick 279, 281
Mills, Rolfe 58, 64, 281
Millton Vineyard 89
   Clos de Ste. Anne Pinot Noir 89
Milne, Derek 60, 61, 104, 105, 120, 130
Milne, Duncan 130
Ministry of Agriculture and Fisheries, Hastings 58
Mission Estate 44, 51
   Mission Reserve Pinot 51
Momo label, Seresin 190
   Montana 57, 65, 96, 158, 162, 182, 196
   Lindauer 66, 159
   Marlborough Pinot Noir 96, 159
   Montana Showcase series (formerly Terroir series) 184
   Montana South Island Pinot Noir 182, 184
   Montana 'T' Terraces Marlborough Pinot Noir 184
   see also Pernod Ricard
Monte Christo 45
Montpellier 59
Moran, Warren 56–7, 93, 94
Morillon 31
Morris, Jasper 96–7, 161
Mt Difficulty 229, 240, **268–9**
   Long Gully Single Vineyard Pinot Noir 269
   Mt Difficulty Central Otago Pinot Noir 268–9
   Pipeclay Terrace Pinot Noir 269
   Roaring Meg Pinot Noir 269
   Target Gully Pinot Noir 269
Mount Edward **265–7**
   Morrison Vineyard Pinot Noir 266, 267
   Mount Edward Central Otago Pinot Noir 266
   Muirkirk Vineyard Pinot Noir 266
   Susan's Vineyard Pinot Noir 266–7, 267
Mountford Estate **216–17**
   The Gradient Pinot Noir 217
   Mountford Estate Pinot Noir 217
   Mountford Liaison Pinot Noir 217
   Mountford Village Pinot Noir 217

Moutere Hills subregion 93, 140–1, 141–2, 146, 152
Moutere Pinot Noir (Neudorf) 152
Muddy Water **218–19**
   Hare's Breath Pinot Noir 218, 219
   Muddy Water Waipara Pinot Noir 218, 219
   Slow Hand Pinot Noir 218
Muirkirk Vineyard Pinot Noir (Mount Edward) 266
Mulholland, Claire 130
Müller-Thurgau 20, 144, 152, 159, 170
Mundy, Trevor 59
Mustang Vineyard 168, 169

Naboth's Vineyard, Millton 89
Nelson Pinot Noir (Greenhough) 146
Nelson region 88, 98, **140–56**
Neudorf 66, 97, 141, 142, **152–7**
   Moutere Pinot Noir 152
   Tom's Block Pinot Noir 152, 156
Nga Waka Vineyard 63
Nobilo Claret 1976 56–7
Nobilo, Mark 55
Nobilo, Nick 20–1, 54–6, 57, 59, 158
Nobilo, Nikola 54, 55
Nobilo Pinot Noir 1973 55
Nobilo Pinot Noir 1976 55–6
Nobilo's Dry Red 54
Noirien 31
Norman, Remington 161
North Otago 88–9

Oberlin 55
Old Weka Pass Road label 206, 207
Oliver, Reg 120
Olssen's 229
Omaka Valley 161, 163, 168, 171, 188, 196
Omihi hills 202
Omihi Vineyard, Pernod Ricard 203
Oregon 20, 21, 22, 56, 66, 69, 113, 127, 141, 234, 240, 241, 251, 276
organic production 22, 146, 165, 171, 180, 188, 218, 241, 267
Otago, *see* Central Otago; North Otago
Otago Station Vineyard 119

Pahi (Voyager) Pinot Noir (Escarpment) 128
Palliser Estate 63, 105, **134–7**
   The Great Bear Pinot Noir 2002 137
   The Great George Pinot Noir 2005 137
   The Great Harry Pinot Noir 2006 137
   Palliser Estate Pinot Noir 134, 137
Parkburn 242, 284, 285

Parker, Robert 94, 97
Paterson, Howard 89
Paton, Alison 110
Paton, Clive 18, 60–1, 63, 66, 110, 113
Pattie, Phyll 110
Pegasus Bay 71, 161, 188, 202, **220–5**
   Calvert Vineyard Pinot Noir 227
   Main Divide Pinot Noir 221
   Main Divide TEHAU 221
   Main Divide TIPINUI 221
   Pegasus Bay Pinot Noir 220–1
   Prima Donna 221
Pelorus (Cloudy Bay) 161, 168
Pencarrow Pinot Noir 137
Peregrine **270–5**
   Karearea Pinot Noir 271
   Peregrine Pinot Noir 270–1
   Pinnacle Pinot Noir 271
   Saddleback Pinot Noir 271
Pernod Ricard **182–7**, 203; *see also* Montana
Perriam, John 276
Philip Duke of Burgundy (Philip the Bold) 30
phylloxera 49, 50, 70, 240
pigeage 19
Pinckney, Anne 63–4
Pinnacle Pinot Noir (Peregrine) 271
Pinot Blanc 18, 33
Pinot Gris 18, 33, 163, 214
Pinot Meunier 33, 48, 49, 50–1
Pinot Noir grape
   challenges of growing 36
   genetic make-up 33
   introduction to New Zealand and early plantings 42–3, 44–51
   names 31
   origins 31–2
   picking 71, 73
   planting and growing 67, 68–71, 73, 98, 171, 196–7, 206, 215
   plantings, statistics 66, 88, 141, 161–2, 202–3, 228
   spelling 'pineau' 30, 31
   varieties parented by 32
   vine age 89, 97, 99, 121, 128, 134, 139, 170, 171, 214, 240, 281
   yield 37, 71, 105, 138, 156, 168, 241, 284
   see also clones
Pinot Noir wine
   biodynamic production 22, 119, 165, 188, 206, 226, 227, 241, 251, 276, 279
   early New Zealand winemakers 45–51
   exports 96
   organic production 22, 146, 165, 171,

180, 188, 218, 241, 267
production 55–6, 62, 73, 78–80, 98, 121, 133, 138–9, 156, 165, 177, 181, 207, 219, 220, 221, 253, 281, 285
sales 96
scents, appearance and flavours 37
terroir 19, 37, 56, 90, 93–4, 96, 97–9, 121, 128, 130, 162, 177, 180, 188, 203, 226
Pinotage 56
Pipeclay Terrace Pinot Noir (Mt Difficulty) 269
Pisa Flats 234
Pisa Range Estate 276
Pommard (UCD5) clone 65, 113, 118, 121, 130, 138, 146, 156, 161, 197, 216, 240, 241, 244, 267, 276, 281, 284, 288
Pompallier, Bishop 44
Ponoma Vineyard, Neudorf 152
Prima Donna (Pegasus Bay) 221
prohibition movement 49–50
Prophet's Rock 80
Pyramid Valley 93, 119, 204, **226–7**
　Angel Flower Pinot Noir 227
　Calvert Vineyard Pinot Noir 227, 252
　Earth Smoke Pinot Noir 227
　Eaton Family Vineyard Pinot Noir 227

Quartz Reef 19, **276–8**
　Bendigo Central Otago Pinot Noir 278
　Bendigo Estate Pinot Noir 278
　Quartz Reef Pinot Noir 276
Queen of the Vineyard 63

Rabbit Ranch label 242
Rachel Pinot Noir (Seresin) 190
Ranfurly, Lord 47
Ranzau winery and vineyard 144
Rattray, Mark 218
Raupo Creek Pinot Noir (Seresin) 190
Raupo Creek Vineyard, Seresin 188, 190
Rees-Francis, Malcolm 284–5
resveratrol 36
Rhône region 50
Richardson, Michelle 89, 198–9, 270
Riddiford, Richard 134, 137
Riedel, Georg 97–8
Riesling 170, 203, 266
Rippon 64, 66, 80, 97, 229, 276, **279–83**
　Emma's Block Mature Vine Pinot Noir 281
　Jeunesse Young Vines Pinot Noir 279, 281
　Rippon Pinot Noir 64, 235, 276, 279, 281

Tinker's Field Mature Vine Pinot Noir 281
Rive Pinot Noir (Mahi) 180–1
River Run Pinot Noir (Chard Farm) 242, 244
Roaring Meg label, Mt Difficulty 268
Roaring Meg Pinot Noir (Mt Difficulty) 269
Robertson, Julian 120
Robinson, Jancis 20, 32, 37, 64, 121
Rockburn **284–5**
　Devil's Staircase Pinot Noir 284
　Eight Barrels Pinot Noir 284, 285
　Rockburn Pinot Noir 284
rootstock 50, 70, 207
Ross, Colin 188, 190
Ruamahanga River 105, 106
Ruby Bay vineyard 141, 161
Russian Jack Pinot Noir (Martinborough) 133
Rutherford Vineyard Pinot Noir (Villa Maria) 199
Ryan, Kathryn 216–17

Sacred Hill 64
Saddleback Pinot Noir (Peregrine) 271
St Helena 59–60, 130, 202, 204, 214, 218
　St Helena Pinot Noir 1982 59–60, 202
San Marino (Kumeu River) 54
Sauvignon Blanc 22, 37, 49, 71, 88, 94, 96, 99, 158, 159, 162, 163, 164, 165, 170, 214
Schultz, Russell and Sue 130
Schuster, Danny 58, 59, 60, 63, 130, 161, 202, 214, 281
Scott, Trevor 276
Seddon, 'King' Dick 48
Seddon Vineyard Pinot Noir (Villa Maria) 199
Seifried vineyard 141
Seresin Estate 180, **188–95**
　Home Pinot Noir 190
　Leah Pinot Noir 190
　Momo label 190
　Rachel Pinot Noir 190
　Raupo Creek Pinot Noir 190
　Sun & Moon Pinot Noir 190
　Tatou Pinot Noir 190
Seresin, Michael 188
Shaw, Dean 232
Shiraz (Syrah) 20, 36, 37, 44, 61, 99
Showcase series (Montana; formerly Terroir series) 184
Sideways effect 17, 21
Sinclair Vineyard, Chard Farm 244

Slow Hand Pinot Noir (Muddy Water) 218
Smart, Richard 68
Smith, Steve 89, 118. 198
Society of Mary (Marist Brothers) 44, 47, 51
soils 69, 93, 133, 141–2, 161, 162, 163, 203–4, 206, 214, 226, 233, 234, 240, 268, 288
Southern Pinot Workshop 66–7, 98, 144, 152
Spain 43
Speargrass Flat 64
Steamboat conference, Oregon 66
'Steamboat New Zealand' 66–7
Steans, Graeme 58, 214
Stone, Mike 258
Stoneleigh Winery 164, 184
Sun & Moon Pinot Noir (Seresin) 190
Susan's Vineyard Pinot Noir (Mount Edward) 266–7, 267
Sutherland, Ivan 170–1
Sutherland, Margaret 170
Sweetwater 44
Switzerland 31
Syrah (Shiraz) 20, 36, 37, 44, 61, 99

'T' Terraces Marlborough Pinot Noir (Montana) 184
Taramea winery 64, 229
Tararua Vineyard 48, 50
Target Gully Pinot Noir (Mt Difficulty) 269
Tasmania 21
Tatou Pinot Noir (Seresin) 190
Tatou Vineyard, Seresin 188, 190
Taylor, Grant 89, 240, 258, 266, 276, 287–8
Taylor's Pass Vineyard Pinot Noir (Villa Maria) 199
Te Kauwhata Viticultural Station 54–5, 58, 63
Te Mata Estate 48
Te Muna Road Pinot Noir (Craggy Range) 118
Te Muna Road Vineyard, Craggy Range 118
Te Muna Valley 105, 127
Te Rehua Pinot Noir (Escarpment) 128
Te Tera Pinot Noir (Martinborough) 133
10/5 clone 58, 59, 60, 65, 69, 113, 121, 128, 130, 135, 138, 146, 152, 161, 163, 170, 180, 197, 204, 216, 218, 221, 240, 241, 244, 252, 267, 276, 281, 287
Terraces Vineyard, Pernod Ricard 182

TerraVin **196–7**
    Hillside Reserve Pinot Noir 197
    TerraVin Pinot Noir 197
terroir 19, 37, 56, 90, 93–4, 96, 97–9, 121, 128, 130, 162, 177, 180, 188, 203, 226
Terroir restaurant, Craggy Range 118
Teviotdale hills 203, 204
Thomas, Wayne 158
Tiffen, Henry 48, 49, 51
Tiger Vineyard, Chard Farm 244
Tiller, Mike 159, 161, 174
Tiller, Robyn 159, 161
Tinker's Field Mature Vine Pinot Noir (Rippon) 281
Tipuranga Teitei o Aotearoa award 113, 253
Tirohanga 138
Tom's Block Pinot Noir (Neudorf) 152, 156
Transcaucasia 31–2
Trinity Hill High Country Pinot Noir 89
Triplebank brand 184
Turner, AC 48
2/10 clone 58, 59, 65, 204, 221, 244, 281

UCD5 (Pommard) clone 65, 113, 118, 121, 130, 138, 146, 156, 161, 197, 216, 240, 241, 244, 267, 276, 281, 284, 288
UDC6 clone 65, 216, 241, 267, 281, 284
UCD13 clone 65, 216
UCD22 clone 65, 141, 146, 152, 216
University of California, Davis 32, 65
Unravelled Pinot Noir (Carrick) 241
Upper Moutere 140–1, 142

Valli 89, **286–8**
Valli, Giuseppe 287
Veldhuizen, Sherwyn 206–7
Victoria 21
Villa Maria 162, **198–201**, 241
    Lloyd Vineyard Pinot Noir 199
    Marlborough Pinot Noir 198
    Rutherford Vineyard Pinot Noir 199
    Seddon Vineyard Pinot Noir 199
    Taylor's Pass Vineyard Pinot Noir 199
    Villa Maria Cellar Selection Pinot Noir 199
    Villa Maria Private Bin Pinot Noir 198, 199
    Villa Maria Reserve Pinot Noir 199
Villaine, Aubert de 19
Vinpro 232
Viognier 44
Viper Vineyard, Chard Farm 244
Vitis labrusca 51
Vitis vinifera 31–2, 42, 43–4, 55
Voss Estate 63, **138–9**
    Voss Estate Pinot Noir 139
Voss, Gary 138–9

Waedenswil viticultural research centre 55
Wai-ata Vineyard 203
Waihopai Valley 163, 164, 198
Waimea Plain subregion 141, 142, 144, 152
Waipara subregion 71, 88, 182, 202, 203–4, 218–19
Waipara Winegrowers 203
Wairarapa region 45–8, 65–6, 88, **104–34**, 203; *see also* Martinborough; and names of individual vineyards/wineries
Wairarapa Vine Improvement Group 65
Wairau Valley 93, 158, 162–3, 177, 180, 182, 188, 190
Waitaki Valley 88–9, 118, 119, 287, 288
Wallace, John 244
Walter, Blair 251, 284
Wanaka subregion 64, 234, 235, 279, 281
Weaver, Sam 164–5
Weersing, Claudia 226
Weersing, Mike 226–7
Wellington 63
Wellington Pinot Noir Conference 21, 67, 68, 93, 96–7, 99, 113, 134, 253, 279
West Auckland 54, 56, 62; *see also* Henderson; Huapai; Kumeu
Whangaroa 44
Wheeler, Jenny 144
Whelan, Grant and Helen 59, 214
Willamette Valley, Oregon 20–1, 234
William Hill 64, 229
William Thomas label 174
Wisor, Doug 118
Wither Hills Pinot Noir 1998 161
Wolter, Mike 78
Woollaston Estate winery 141
Wright, Ken 113

Yukich, Frank 57, 158

Zebra Vineyard Pinot Noir (Craggy Range) 119
Zeestraten, Kees 216–17

BIBLIOGRAPHY

Coates, Clive *Côte d'Or A Celebration of the Great Wines of Burgundy* Weidenfield & Nicolson Ltd; London, 1997

Cooper, Michael *Wine Atlas of New Zealand* (2nd Edition) Hodder Moa; Auckland, 2008

Cooper, Michael *The Wines and Vineyards of New Zealand* Hodder and Stoughton; Auckland, 1984

Cull, Dave *Vineyards on the Edge: the Story of Central Otago Wine* Longacre Press; Dunedin, 2001

Haeger, John *Winthrop North American Pinot Noir* University of California Press; Berkley and Los Angeles, 2004

Hanson, Anthony *Burgundy* Faber and Faber; London, 1982

Johnson, Hugh *The Story of Wine* Mitchell Beazley; London, 1989

Judd, Kevin with Bob Campbell *The Landscape of New Zealand Wine* Craig Potton Publishing; Nelson, 2009

Keogh, Doreen *Fruit of the Vine: The Story of St Mary's Parish* Hawke's Bay, 2004

Oram, Ric *Pinot Pioneers* New Holland; Auckland, 2004

Pitiot, Sylvain and Servant, Jean Charles *The Wines of Burgundy* (10th Edition) Presses Universitaires de France; Paris, 1999

Robinson, Jancis *Tasting Pleasure – Confessions of a Wine Lover* Penguin Books; New York, 1997

Robinson, Jancis *Vines, Grapes and Wines* Mitchell Beazley; London, 1986

Seward, Desmond *Monks and Wine* Mitchell Beazley; London, 1979

Schuster, Danny; Jackson, David & Tipples, Rupert *Canterbury Grapes & Wine 1840-2002* Shoal Bay Press; Christchurch, 2002

Scott, Dick *Pioneers of New Zealand Wine* Reed Books/Southern Cross Books; Auckland, 2002

Stewart, Keith *Chancers and Visionaries: A History of Wine in New Zealand;* Random House; Auckland, 2010

Stewart, Keith *The Great Wines of New Zealand* Penguin Viking; Auckland, 2005

'NO MEAN-SPIRITED BASTARD EVER MADE A DECENT PINOT NOIR'